THE DISTRIBUTION AND
DIVERSITY OF SOIL FAUNA

The Distribution and Diversity of Soil Fauna

JOHN A. WALLWORK

*Department of Zoology, Queens Building,
Westfield College, London, England*

WITHDRAWN

1976

Academic Press

LONDON   NEW YORK   SAN FRANCISCO

A Subsidiary of Harcourt Brace Jovanovich, Publishers

ACADEMIC PRESS INC. (LONDON) LTD.
24/28 Oval Road,
London NW1

United States Edition published by
ACADEMIC PRESS INC.
111 Fifth Avenue
New York, New York 10003

Library of Congress Catalog Card Number: 75 19684
ISBN: 0 12 733350 9

Printed in Great Britain by
Ebenezer Baylis and Son, Ltd.
The Trinity Press, Worcester, and London

Preface

This book is something of an Adam's rib, a companion volume to my earlier *Ecology of Soil Animals* which was published in 1970. Indeed, many of the ideas on distribution patterns and diversity which appear in the following pages had their inception while the earlier work was in preparation. During the last five years, these ideas have been refined and expanded to include the more recent developments in soil ecology.

In this present essay, the framework within which the ecological information is assembled is provided by the gross vegetational characteristics of the various terrestrial formation types. The faunal, "group-by-group" approach of the first volume is replaced, here, by the "habitat" approach. After a brief introduction to the main components of the soil fauna (Chap. 1) and methods for studying patterns of distribution and diversity (Chap. 2), these patterns are traced through a procession of habitats, ranging from grasslands, through arable soils, moorlands, heaths and forests, to hot and cold deserts (Chaps. 3–11). Here are examined the effects of environmental variables, biotic and abiotic, direct and indirect, natural and Man-made. These effects are illustrated by a list of examples drawn from a range of faunal groups, and selected to include representatives of all the major elements of the soil fauna.

To undertake a survey of this kind, much of the material has had to be presented in a condensed form. But a mere condensation of information is not enough. To be of any real value, such an exercise must permit a synthesis —an identification of major trends in the patterns of distribution and diversity. While it would be presumptuous to claim that this exercise has been completely successful, it is reasonable to suggest that certain major themes have emerged during this analytical process, and these are highlighted at the end of each chapter. Finally, an attempt to explain, in general terms, the diversity of the soil fauna is the subject of Chapter 12.

The objectives of this book are, then, two-fold. Firstly, to draw together a considerable amount of basic information on distribution and diversity which is widely scattered in scientific journals throughout the world. Secondly, to present this information in an ordered fashion which will allow some kind of general synthesis of the factors governing distribution and diversity patterns. This second objective is certainly the more difficult of the two, for the soil/litter subsystem is immensely complex. Ecological "pegs" on which to hang information from a variety of situations are notoriously difficult to define. However, perhaps it is not a vain hope that the efforts made here may provide a starting point and a stimulus for further enquiry into this fascinating field of ecological study.

As indicated earlier, this book has been several years in the making. During this time, a great deal of new information has accumulated and has necessitated frequent revisions of the original manuscript. Even so, it has not been possible to keep pace with every new development. Several important works have appeared too late to be included in the reference lists. Among these may be mentioned the proceedings of the Second International Congress of Myriapodology (entitled: *Myriapoda*), edited by J. G. Blower for the Zoological Society of London, and the proceedings of the Fifth International Colloquium on Soil Zoology (*Progress in Soil Zoology*), edited by J. Vanek. However, an effort has been made to make the reference lists as up-to-date as possible in the hope that the information they contain will be accessible to undergraduates and research workers alike.

I am particularly grateful to those authors whose work has guided and shaped the ideas developed here. Acknowledgement of their contributions is given in the text and in the illustrations, and my only regret is that the limitations of space have prevented me from doing full justice to their efforts. Permission to quote from published works has been given generously by a number of individuals and institutions. Singled out for special mention are the editors of *Acarologia, American Zoologist, Annales Entomologicae Fennicae, Journal of Animal Ecology, Journal of Applied Ecology, Oikos, Pacific Insects, Pedobiologia, Science, Societas Zoologica Botanica Fennica "Vanamo"* and *Vie et Milieu,* and the following publishers: Blackwells Scientific Publications, Cambridge University Press and VEB Gustav Fischer Verlag (Jena). I am also grateful for the permissions provided by the American Association for the Advancement of Science, the Ministry of External Relations of the Republic of Chile, the Royal Society, the Systematics Association and the Trustees of the British Museum. Drs. Ludwig Beck (Bochum), Eero Karppinen and Veikko Huhta (Helsinki), M. D. Mountford (U.K.) and Prof. Nelson G. Hairston (University of Michigan) have kindly allowed me to quote material from original publications, and my special thanks go to Professor L. C. Bliss and Mr. James K. Ryan (University of Alberta) and to Dr. Veikko Huhta for access to unpublished information.

A number of people have provided invaluable technical assistance during the preparation of this work and, here, I owe a special debt of gratitude to Mr. Gerhard Ott (Pomona College, California), Mrs. Bernardette Parry (British Museum of Natural History) and to Prof. J. P. Harding, Miss Celia

Earle, Mrs. Diana Herrett and Mrs. Haidee Newman-Coburn, all of Westfield College. Their efforts are greatly appreciated.

Interest in the ecology of soil animals has gathered momentum over the last two decades and it shows no signs of slackening. As our knowledge of the events occurring in the soil/litter subsystem continues to improve, there is a real need to take an inventory of what we have learned. This book is no final statement, but if it meets this need, even in only a small way, it will have achieved its purpose.

October, 1975   JOHN A. WALLWORK

Contents

1*

DEDICATION

To the memory of a good friend

Ian Healey
1940–1972

this book is dedicated

For

JANE, MICHAEL AND SUSAN

for all their help
and patience

And for

CE, DI, AND HAIDEE

who also helped

Chapter 1

An Introduction to the Soil Fauna

The soil is teeming with life. It is a world of darkness, of caverns, tunnels and crevices, inhabited by a bizarre assortment of living creatures which any science fiction writer would be proud to include in his repertoire. Yet nobody has ever come close to equalling Nature's own design in his description of the unusual. This world is open to anyone who takes the trouble to shake a handful of leaf litter, garden soil or compost through a sieve, and to examine the separate that is produced, under a binocular microscope. Here can be found heavily-armoured eight-legged mites clambering, like miniature tanks, over soil particles, vibrant soft-bodied springtails moving with agility through soil spaces, false scorpions stalking their prey, with massive pincers held aloft, and small snails moving serenely, confident in the knowledge, or so it would appear, that they can withdraw to the safety of their shell should danger threaten. These forms and many others co-exist in the community of the soil. They co-operate and compete, and they interact with each other to form an integrated system which functions in a major way to effect the breakdown of organic material. In this way, the re-cycling of plant nutrients is promoted.

The diversity of the soil fauna is quite remarkable. Every major animal phylum, except the Coelenterata and Echinodermata, is represented somewhere or other in the soil. Even within a small area, species diversity can equal that of a coral reef, although it is less evident. The majority of soil animals are microscopic in size and have to be separated from the solid substratum in which they live before they can be observed.

The ecological stage on which these animals play out their roles is the soil. A stage which has a constantly changing backdrop as we move from grassland to forest, from moorland to heathland, from desert to mountain, and from natural to cultivated sites. Throughout this book, we will be tracing the patterns of diversity and distribution of the soil fauna in these, and other, habitats. This is the plot; but first, the players.

The most important groups of the soil fauna in most sites are: the Protozoa, Nematoda, Annelida, Mollusca and Arthropoda, and each of these is now introduced, briefly, in turn.

Protozoa

The four main groups of soil protozoans are the flagellates, amoebae, rhizopods and the ciliates. They live, in the main, in the water films surrounding soil particles and plant roots, where they mostly feed on bacteria. Some of the amoeboid forms are also known to ingest particles of organic material.

Not all four groups are equally represented in all soil types. Ciliates show a preference for calcareous soils, while rhizopods flourish best in highly organic profiles (Stout, 1963, 1968). The rhizopods are notable among the Protozoa for their sensitivity to pH conditions (see Fig. 4.8, p. 106), and the distribution patterns of individual species are influenced by the extent to which they can tolerate acid conditions (Bonnet, 1961). The species composition of the ciliate fauna, on the other hand, is more closely related to the soil moisture conditions. Ciliate species vary in the efficiency with which they can encyst during unfavourable conditions, and this factor will determine the ecological range of a species.

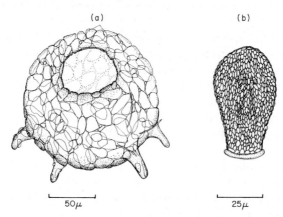

(a)   (b)

50μ   25μ

Fig. **1.1** Two soil-inhabiting genera of rhizopod Protozoa. (a) *Centropyxis*, (b) *Nebela*.

From the ecological point of view, less is known about protozoans than about many other groups of the soil fauna. This is partly due to their microscopic size, and also to the fact that conventional methods for extracting the soil fauna, using funnel or flotation techniques, are usually unsuitable for Protozoa. Rhizopods (Fig. 1.1) can be counted directly in soil washes, but the other groups are best studied using microbiological plating techniques. The results obtained using these different methods are not easily compared, and it is often difficult to get an overall picture of the protozoan fauna of a particular site, and to assess the contribution made by this group to the general functioning of the soil community. Certainly, there are indications that the Protozoa and bacteria form an important predator/prey system in the soil, particularly in the vicinity of plant roots where they may contribute significantly to the turn-over of plant nutrients.

Nematoda

It has often been said that the nematodes are so numerous and so widely distributed that if everything else were to disappear, the outline of the earth would still be visible as a continuous sheet of these animals. Like the Protozoa, the soil-inhabiting members of this group are microscopic in size and live in water films, particularly on the surface of plant roots. Because of a thick cuticle covering the body, these worms are not able to move by peristaltic action. Instead, they have an undulatory method of propulsion which utilizes the surface tension of the water film in which they live.

Although rather uniform in appearance and internal structure, soil nematodes have a variety of modes of life. Some are entirely free-living forms, some free-living for part of their life cycle and parasitic on plants for the remainder, and others, such as the eel-worms, are predominantly plant parasites (see Chapter 5). The free-living forms show diversity in their feeding habits, and various attempts have been made to classify soil nematodes into nutritional types (Nielsen, 1949; Banage, 1963; Yeates, 1971). Distinctions have been made between predators, microbial feeders, omnivores and plant feeders, the latter category being sub-divided according to the degree of parasitism. However, these categories are rarely exclusive, for such is the ecological plasticity of soil nematodes that few have specialized feeding habits.

Nematodes can be extracted from soil or litter samples by a wet funnel method (see O'Connor, 1962; Wallwork, 1970) or by an elutriation process (see Seinhorst, 1962) which uses water currents to separate the animals from the substratum. Using these methods, population densities of soil nematodes have been estimated in the order of millions per m² (see Banage, 1963), and they are often equally abundant in grassland and forest soils. However, the uniformity of appearance of the nematodes, as a group, creates taxonomic problems for the ecologist. It is difficult for a non-specialist to make identifications at the species level, and this is a serious handicap in studies which are designed to investigate distribution patterns and species diversity.

Annelida

In most north temperate localities, the true earthworms belong to two families, the Lumbricidae and Enchytraeidae, with contrasting ecologies.

The Lumbricidae are the familiar earthworms of grassland and garden soils. Because of their relatively large size, these worms often make an overwhelming contribution to the total biomass of the soil fauna in base-rich soils. They are much less frequent in acid, highly organic soils, for relatively few species can tolerate pH conditions of 4·0 or less. Lumbricids move through the soil by means of peristaltic waves of contraction of the body wall musculature. This type of locomotory pattern allows some regions of the body to increase in diameter while others are constricted. This "bulging" has the effect of pushing aside soil particles and enlarging the burrow through which the worm moves. The activities of deep-burrowing species, such as *Lumbricus terrestris* and *Allolobophora nocturna*, may create channels in the soil which extend for several feet below the surface. Such channels increase the possibilities for aeration and drainage in the soil profile, features which improve the texture of the soil. However, not all lumbricids are deep-burrowing species, and some do not produce any well-defined burrows at all. Two of the commonest species of British lumbricids, *Allolobophora caliginosa* and *A. chlorotica*, produce shallow burrow systems in the top 20 cm of the profile, but other surface dwellers, *Lumbricus rubellus* and *L. castaneus* for example, are essentially non-burrowers.

The feeding activities of lumbricid earthworms have a considerable effect on the fertility of the soil (see Satchell, 1967). These animals feed on decomposing leaf litter and generally show a preference for nitrogen-rich materials. Burrowing species transport quantities of organic material from the soil surface through their burrows to deeper parts of the soil profile. They also transport mineral soil, ingested in the deeper parts of their burrows, to the surface where it is deposited as the familiar earthworm cast. As a result of these activities, the mineral and organic components of the soil become thoroughly intermixed, and the formation of organo-mineral complexes, which are such a characteristic feature of fertile soils, is promoted.

Some of the more acid-tolerant species of lumbricids, such as *Lumbricus rubellus*, *Dendrobaena octaedra*, *D. rubida* and *Bimastos eiseni*, are often found associated with cattle dung pats in grassland sites, and evidently play an important part in the dispersion and decomposition of these deposits. In areas of the world where lumbricids are rare or absent, for example Australia, dung pats persist and accumulate on the soil surface. These represent a store of energy and nutrients which is "locked up" and unavailable to other soil organisms. Strenuous efforts are being made, in these areas, to introduce animals, such as dung beetles (Scarabaeidae), which will serve the same function as lumbricid earthworms.

Representative samples of earthworm populations can be obtained from soil by a variety of methods. The most direct, but perhaps the most laborious, is to dry-sieve the soil and sort by hand. Baiting techniques, using dung, may also be employed, but these are selective and may not provide reliable estimates for species, such as *Allolobophora caliginosa*, which are not attracted to dung. Dry funnel extractions of the Tullgren type have also been used, but these are also selective, depending on the mesh size used. Perhaps the most common method used at present is the application of a solution of formalin sprinkled over the surface of the soil. As this chemical diffuses into the soil, lumbricids react vigorously and crawl on to the surface where they can be collected easily.

Enchytraeids are much less conspicuous than their lumbricid relatives. They are generally white in colour, although the dark contents of the gut can usually be seen through the body wall, and they measure no more than a few centimetres in length. Enchytraeids can evidently tolerate acid conditions much better than lumbricids, and establish large populations, of the order of thousands per m², in the highly organic soils of woodlands and moorlands. They are one of the few groups of soil animals (the nematodes are another) which are successful in colonizing the rather sterile peaty layers of bog soils (see Chapter 6). Although locomotion is by peristaltic action, these small worms do not produce well-defined burrows, but rather make use of existing soil spaces.

Studies on the feeding habits of enchytraeids (see O'Connor, 1967) have shown that decomposing plant fragments and mineral soil are ingested in amounts which are roughly proportional to their availability. Some species, such as *Achaeta eiseni* and *Cognettia cognetti*, apparently feed preferentially on fungi, however.

Enchytraeids can be collected from soil and litter samples very efficiently using a wet funnel extractor of the Baermann or Nielsen type (O'Connor, 1962). This is not a very diverse group of soil animals; only eight genera can be regarded as common in the soils of western Europe. However, species identifications are not easy, and this presents an obstacle to detailed ecological studies.

It might be supposed that the enchytraeid contribution to the total soil metabolism is lower than that of the lumbricids, if only because the former are smaller in size and have a much lower biomass than the latter. However, enchytraeids have a much higher metabolic rate (as much as ten times higher, in some cases) than lumbricids, and this may compensate for the relatively low biomass. In rather acidic woodland and moorland soils, enchytraeids can account for about 11% of the total metabolic activity in the soil, and this compares very favourably with the contribution made by *Lumbricus terrestris* in woodlands (Satchell, 1967; O'Connor, 1967; Wallwork, 1970).

Mollusca

The two groups of terrestrial pulmonate molluscs, the snails and the slugs, are readily distinguishable by the fact that the former have an external shell whereas the latter do not (see Fig. 4.6, p. 103 and Fig. 5.3, p. 116). Both groups are widespread in a variety of soil types. However, the snail fauna usually achieves its maximum diversity in sites where calcium carbonate is most readily available, although quite a number of species, such as *Cepaea hortensis*, *Helix pomatia* and members of the genus *Laciniaria*, can tolerate low calcium levels in their environment, and are often found in forest soils. Slugs are most common in loose arable soils, particularly heavy soils which are not susceptible to drying out during the summer, but species such as *Milax budapestensis* and *Arion hortensis* can avoid surface drought and frost by retreating below the soil surface (Runham and Hunter, 1970). *Agriolimax reticulatus* is more of a surface dweller (Hunter, 1966), and apparently is more resistant to drought and desiccation than *M. budapestensis* and *A. hortensis*. Slugs also occur in woodland, and here members of the genus *Limax* are not uncommon, particularly in the vicinity of decaying wood.

Snails and slugs are common pests of gardens and cultivated crops, and their ability to feed on living plant material is a reflection of the fact that some species, at least, possess cellulase enzymes in the gut. These molluscs are among the few members of the soil fauna which can digest the structural components of plant tissue and, thereby, make a significant contribution to its chemical breakdown. Not all species feed in this way, however. Indeed, a range of feeding habit is shown with some species being saprophages (for example, *Helix aspersa*, *Cepaea nemoralis*, *Arion ater* and *Limax maximus*), others fungivores (*Caecilioides* spp.), while species belonging to the genera *Testacella*, *Glandina* and *Vitrina* are predaceous on earthworms.

The activities of the plant feeders, particularly those which retreat into soil burrows, promote the dispersion of organic material through the profile in much the same way as do the activities of the lumbricid earthworms. Molluscs also produce copious amounts of slimy mucus which binds the soil particles together and improves the crumb structure of the soil.

Molluscs can be separated from leaf litter by a wet sieving method (see Southwood, 1966; Wallwork, 1970), but this is laborious and selective. An alternative method, attributed to Vågvölgyi (1952) by Newell (1971), consists of the following procedure:

1. The soil or litter sample is flooded with water. The air-filled shells of dead snails will float to the surface, and can be decanted off for identification and counting.

2. The sediment is drained carefully and a concentrated solution of mag-

nesium sulphate is added. This will separate off the slugs and smaller snails which will float to the surface.

3. After draining off the magnesium sulphate, a molluscicide solution is added to the remaining sample. This will kill the larger snails.

4. The molluscicide can be left for a day before it is drained off. The sample is then dried so that air bubbles form in the shells of the snails.

5. The sample is again flooded with water and the snails can be collected as a "float".

This method can be used for extracting molluscs both from leaf litter and mineral soils.

Arthropoda

Members of this phylum occur in virtually every soil type throughout the world. Arthropods have been recovered from the sandy soils of hot deserts, the embryonic soils of the high alpine zone and the inhospitable gravels of the cold deserts of Antarctica. The phylum is ubiquitous in its distribution and the diversity of its soil-dwelling component is immense.

In temperate soils, five groups of arthropods often make an appreciable contribution to the total biomass and metabolism of the entire community, namely the Crustacea, Myriapoda, Apterygote Insecta, Pterygote Insecta and the Arachnida.

Crustacea

The terrestrial members of this class belong, in the main, to the Isopoda, the familiar "woodlice". Moisture is an important factor in the lives of these animals for their body surface is incompletely waterproofed, and they flourish best in damp organic deposits, such as deciduous woodland leaf litter and compost heaps. In Britain, three of the commonest woodlice in such habitats are *Oniscus asellus*, *Trichoniscus pusillus* and *Porcellio scaber* (see Fig. 8.8, p. 221). They are frequently joined by *Philoscia muscorum* (Fig. 1.2), a species which also inhabits the damper regions of grass tussocks. Some woodlice, such as *Armadillidium vulgare* (see Fig. 3.3, p. 63) which is commonly encountered in chalk and limestone grassland, and the desert-dwelling *Armadillo* spp. and *Venezillo arizonicus*, have developed mechanisms which allow them to resist desiccation and to colonize dry habitats. These mechanisms include: (1) a low basic transpiration rate, (2) the ability to roll up into a ball and thus minimize the area of body surface exposed to the environment, and (3) the restriction of respiratory surfaces to special invaginations of the cuticle (pseudotracheae) located on the ventral part of the body.

Fig. 1.2 The woodlouse *Philoscia muscorum* (Scopoli). This specimen was collected from chalk grassland and measures 8·5 mm in length. (Photo. by A. Newman-Coburn.)

Woodlice generally show a nocturnal activity pattern, hiding in leaf litter, bark crevices or under stones during the day, and emerging to feed at night. Evidently, these animals have a well-developed olfactory sense and respond to the odour of other individuals in the population. It may well be that this response guides them back to the safety of their refuges after their nocturnal wanderings. It may also explain why woodlice often occur in large clusters. This clustering behaviour is another way of controlling water loss in species which cannot roll into a ball. When these animals press together, their ventral surfaces, through which most of the water loss occurs, are effectively shielded from the desiccating effects of the environment.

Vertical population movements of a seasonal nature have been demonstrated in this group. *Armadillidium vulgare* regularly descends to the deeper layers of chalk grassland soil profiles during the British winter, and this same species has also been shown to perform a similar migration during the dry season in Californian grassland (Paris, 1963). Woodland species frequently move out of the soil and on to the aerial vegetation during the summer months, but reverse this trend as winter approaches.

Woodlice are undoubtedly important saprophages in communities where they occur in any numbers. They will feed on a variety of decomposing organic material, from dead wood to carrion. Woodland isopods can be maintained successfully in cultures on a diet of leaf litter, green algae or fungi.

Most isopods are large enough to be seen with the naked eye and usually can be collected by hand. Pitfall trapping can be used to estimate nocturnal activity patterns. The smaller species and juvenile stages of the larger ones can be collected in funnel or bowl extractors. Identification to the species level is relatively easy (see keys by Edney, 1954; Sutton, 1972).

Myriapoda

All of the four groups of myriapodous arthropods (Chilopoda, Diplopoda, Symphyla and Pauropoda) are associated with the soil and all except the pauropods enjoy a relatively wide range of distribution.

The Chilopoda are the centipedes, and in Britain, belong mainly to two

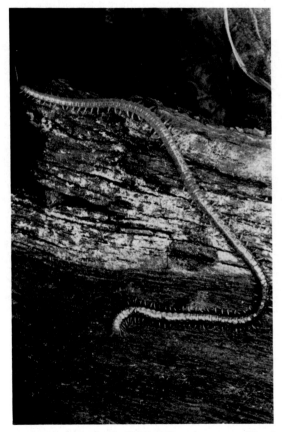

Fig. 1.3 The geophilomorph centipede *Haplophilus subterraneus* (Shaw). This specimen was taken from garden compost in Buckinghamshire, and measures 60 mm in length. (Photo. by A. Newman-Coburn.)

groups, the Lithobiomorpha and the Geophilomorpha. The lithobiomorphs (see Fig. 9.3, p. 247) are the centipedes commonly found under the bark of decaying wood, in leaf litter and under stones. They move with alacrity and show a remarkable agility in wriggling into small crevices when disturbed. Like all centipedes, their mouth parts include a pair of poison claws, but they are quite harmless and can be handled without fear. The geophilomorphs (Fig. 1.3) live a subterranean existence and are rarely seen on the soil surface, in contrast to their lithobiomorph relatives. These are slender, thread-like centipedes, with many more pairs of legs distributed along the ribbon-like body than other chilopods, and they are sometimes mistakenly identified as "wireworms" by the gardener. The large tropical scolopendromorph centipedes, often measuring 15 cm or more in length and with a girth equal to that of a man's middle finger, have a formidable bite and can inject a poison which causes considerable discomfort. The Scolopendromorpha

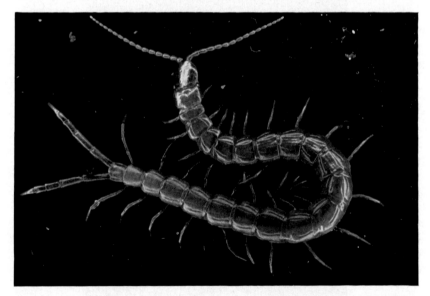

Fig. 1.4 The scolopendromorph centipede *Cryptops anomalans* Newport. This specimen was collected in the garden of a London suburb and measures 45 mm in length. (Photo. by A. Newman-Coburn.)

are not native to Britain but the genus *Cryptops* (Fig. 1.4) has become established here, and is common in gardens around the London area.

Centipedes probably rank among the more important predators in sheltered sites such as woodland leaf litter, decaying wood, hedgerows and compost heaps. However, not all species are exclusively carnivorous, and Lewis (1965) has shown that *Lithobius variegatus* can switch from a carnivorous diet to one comprising mainly leaf litter during winter months in a Yorkshire woodland.

Lithobiomorph centipedes are large enough to be collected by hand or by pitfall trapping. The Geophilomorpha are more difficult to collect because of their smaller size and subterranean mode of life. Bowl or funnel extractors will separate these animals from leaf litter, and they can also be expelled from mineral soil by the application of formalin solution to the soil surface.

The Symphyla, sometimes known as 'glasshouse centipedes', are small, white, delicate myriapods often abundant in cultivated soils particularly where these are moist and rich in organic material. In greenhouses, symphylids can build up enormous densities, and here *Scutigerella immaculata* becomes a serious horticultural pest due to its habit of feeding on living roots of cultivated plants. Other symphylids are saprophages and rarely occur in large enough densities to make any appreciable contribution to the overall community metabolism. A modified flotation apparatus has been designed by Edwards and Dennis (1962) for extracting Symphyla from soil.

The Diplopoda are the millipedes, an important group of saprophages in woodland systems. Familiar examples of this group are the "round-backs" (Iuliformia), the "flat-backs" (Polydesmoidea) and the pill millipedes (Pentazonia), three fairly distinct morphological groups with differing ecologies. The iuloid millipedes are cylindrical forms capable of burrowing in the soil and, in some cases, into decaying wood; resistance to desiccation is well developed in this group, as it is in the other group of burrowers, the pill millipedes. The distribution of the latter extends into grasslands, particularly in limestone and chalk areas. Here, it is often found alongside the woodlouse *Armadillidium vulgare* which it resembles in being able to roll up into a ball. Flat-back millipedes have a cuticle that is incompletely waterproofed. They are much more strongly restricted to the damp habitats of woodland leaf litter and the bark crevices of fallen trees than the other groups just mentioned. Nor are they efficient burrowers, and consequently have to utilize existing spaces between fallen leaves and beneath decaying wood.

Although some millipedes, such as the iuloid *Schizophyllum sabulosum* and *Tachypodoiulus niger* (Fig. 1.5), can feed on living plant material, most prefer to feed on leaf litter and dead wood which is in the process of microbial decomposition. Some chemical degradation of this material occurs as it passes through the millipede gut, but the main effect of this feeding activity is probably to promote the physical breakdown of organic material in the soil. In certain sites, such as chalk grassland (see Chap. 3) and tropical forests (see Chap. 8), this effect can be quite considerable.

Millipedes can be collected in much the same way as woodlice. Hand-sorting and dry sieving can provide good results, and pitfall traps will give estimates of activity patterns. Seasonal shifts in population centres occur,

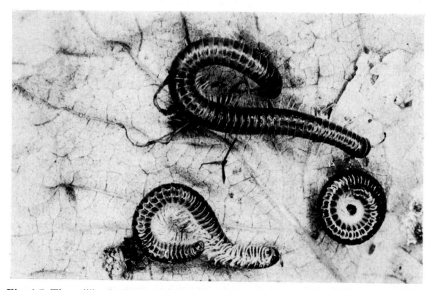

Fig. 1.5 The millipede *Tachypodoiulus niger* (Leach). (Photo. by A. Newman-Coburn.)

however, and these must be taken into account when designing a sampling programme. The pill millipede, *Glomeris marginata*, moves into deeper layers of the soil profile during winter. The iuloid *Cylindroiulus punctatus* moves between the mineral soil, where it spends the winter, and the crevices beneath the bark of decaying logs, where mating and reproduction occur during the summer.

Apterygota

All four groups of apterygote insects, namely the Thysanura, Diplura, Protura and Collembola, are associated with the soil, but only the last-named of these is of major importance.

The Collembola, or "springtails" have a cosmopolitan distribution which ranges from the high alpine snowline of the Himalayas, through equatorial tropical forests to the frozen wastes of the Antarctic continent. This group is one of the most abundant among the fauna of soils supporting temperate and tropical grassland, moorland, heathland and forests throughout the world. In most temperate localities, collembolan densities are measured in thousands per m², and if any generalization can be made about distribution patterns, it is that maximum diversity and densities usually occur in mull-type soils, where drainage is adequate, and where the soil is not base-deficient.

Moisture plays an important part in determining the distribution patterns of soil Collembola, for species vary in their response to this factor. *Podura aquatica*, *Tetracanthella brachyura*, *Isotoma antennalis* and *Sminthurides malmgreni*, for example, habitually show a preference for very wet conditions (hydrophile); others are mesophile and will prefer conditions that are neither too wet nor too dry (*Folsomia brevicauda*, *Friesea mirabilis* and *Isotoma sensibilis* are common members of this group in Britain), while a xerophile element characteristic of dry sites is represented by *Tetracanthella wahlgreni*. In hostile environments, such as the cold Antarctic desert, collembolan populations have a remarkable ability to tolerate extremely low environmental temperatures. The Antarctic collembolan *Isotoma klovstadi*, for example, can survive in temperatures approaching $-30°$ C (Pryor, 1962.)

The collembolan fauna of a given site, particularly in woodland, often shows a vertical stratification of species. Forms with a large body size, such as members of the genera *Entomobrya* and *Tomocerus* (see Fig. 8.8, p. 221), occur mainly in the upper layers of leaf litter, whereas small-sized forms (*Onychiurus*, *Friesea*, *Tullbergia* species) are concentrated at greater depths in the profile. Generally speaking, species which live on or near the surface are strongly pigmented, and have well-developed eyes, antennae and springing organs. The deeper dwellers are often weakly pigmented, with reduced sense and springing organs. Here, we have an example of different "life forms" – morphological types evidently adapted to different microhabitats. These morphological differences are paralleled by physiological differences also. Zinkler (1966) has shown, for example, that the rate of O_2 consumption by small soil-dwellers, such as *Onychiurus armatus*, is significantly higher, under experimental conditions, than that of the much larger litter-dwelling *Tetrodontophora bielanensis*. Further, the rate in soil-dwelling species appears to be unaffected by increased CO_2 concentrations, in contrast to that in litter-dwellers. This is an obvious adaptation to a life spent in confined spaces. Members of the Sminthuridae (see Fig. 8.8, p. 221), on the other hand, have managed to achieve a greater degree of independence from the soil than most other Collembola. Unlike the latter, sminthurids have tracheae and do not need to use the general body surface for respiration; a greater degree of water conservation can then be achieved. This group of collembolans is often encountered in the aerial parts of vegetation and on tree trunks. They are phytophagous and some, the "lucerne flea" *Sminthurus viridis* for example, are agricultural pests.

Most of the Collembola, however, can be regarded as saprophages. Food items include decomposing plant material, carrion and faeces, although it may be the fungi and bacteria associated with these substrates which are being selected. Fungal hyphae and spores frequently form an appreciable part of the diet of these saprophages, while unicellular green algae are taken by entomobryids which live in the upper litter layers. A few species of Collembola are carnivorous (*Friesea* spp., *Isotoma* spp.), feeding on rotifers, tardigrades and eggs of eel-worm nematodes.

Bowl and funnel extractors are very satisfactory for collecting Collembola from leaf litter samples, but flotation methods are to be preferred when there is an appreciable amount of mineral soil present (see Wallwork, 1970).

Pterygote Insecta

The part played by the soil in the lives of the "higher" insects is a very varied one. For some it is their permanent home, from egg to adult; in others it is merely a temporary refuge, a shelter from the cold of winter or the drought of summer. Between these two extremes, there are many insects which depend on the soil for only a part of their life cycle. To this latter category belong a number of Diptera, Lepidoptera and Coleoptera, economically important because their soil-dwelling larvae are pests of cultivated crops (see Chapter 5).

The Coleoptera also contribute to the permanent soil fauna in no small measure. Ground beetles (Carabidae, see Fig. 8.8, p. 220) and rove beetles (Staphylinidae) are important predators on the soil surface and in leaf litter in moist temperate regions. Not all members of these two families are carnivorous, however; some are phytophagous or saprophagous, habits which are shared by the Tenebrionidae (see Fig. 10.10, p. 285), one of the more conspicuous groups of beetles in hot desert soils (see Chap. 10), and also by the chafers and dung beetles (Scarabaeidae, see Fig. 5.4, p. 117), carrion beetles (Silphidae), the tiny feather-winged beetles (Ptiliidae), and click beetles (Elateridae). Decaying wood often contains a varied insect fauna among which the beetles are again prominent (see Chapter 9), although in dry temperate and tropical regions they are often overshadowed by the termites. These two groups, the beetles and the termites, feed on the decaying wood and make an important contribution to its decomposition. Not all species of termite feed on wood, however; phytophagy and saprophagy occur in the genera *Trinervitermes*, *Hodotermes* and *Odontotermes*, while the Macrotermitidae cultivate subterranean fungus gardens.

Decaying wood also provides a home for certain ant species, such as *Formica rufa* and *Camponotus* spp. Ants, as a group, are widely distributed in many parts of the world, in forest and grassland alike (Gaspar, 1971), although they achieve their maximum diversity in tropical forests (see Chapter 8) and arid grasslands (see Chapter 10). Some species produce conspicuous mounds (*Lasius flavus* in grassland and *Formica rufa* in woodland) while others, such as the predatory ponerine ants of Australia, have subterranean nests. "Harvester" ants, so-called because they feed on seeds which they collect and store in underground nests, are common in hot desert soils where the nest is marked by a low, saucer-shaped mound.

In woodland leaf litter, particularly of deciduous species such as ash, oak and beech, insect larvae of various kinds often occur in large numbers. Here can be found the saprophagous larvae of the dipteran Tipulidae, Chironomidae, Muscidae, Mycetophilidae and Dolichopodidae, together with the predatory immatures of the dipteran Tabanidae and coleopteran Carabidae and Staphylinidae. The feeding activities of these larvae undoubtedly contribute to the decomposition of organic material and the flow of energy through the soil/litter system, but surprisingly little attention has been paid to these effects so far.

This insect fauna is so diverse that it is difficult to comprehend in its entirety. A variety of sampling and extraction methods have been devised to investigate it. Pitfall trapping, combined with a mark/recapture technique, can be used to make estimates of activity rhythms and population densities of surface-active carabid, staphylinid and tenebrionid beetles. Quadrat sampling and traffic census methods often provide a guide to population densities of ants and termites, while emergence traps may be used to estimate densities of the temporary members of the soil fauna, such as larval Tabanidae, Tipulidae and Scarabaeidae. Flotation methods have been found useful for separating larval Elateridae ("wireworms") from grassland soil, and the eggs of economically important Diptera, such as the frit fly *(Oscinella frit)* and the wheat bulb fly *(Leptohylemyia coarctata)*, from arable soil. Some of the smaller soil insects, such as the ptiliid and pselaphid beetles, can be collected from dry funnel extractors. A detailed account of these techniques is given by Southwood (1966).

Arachnida

Of the ten or eleven sub-classes comprising this Class of arthropods, only four are sufficiently diverse and widely distributed to merit our attention here. The remainder are either rare or unusual forms, such as the whip scorpions and ricinuleids, or strongly localized in their distribution to tropical and sub-tropical regions, for example the solifugids and scorpions. The four with which we are concerned are the spiders (Araneida), the harvestmen (Opiliones), the false scorpions (Chelonethi, or Pseudoscorpiones) and the mites (Acari). The first three of these groups are exclusively carnivorous in habit; the Acari is a very diverse group which includes carnivores, phytophages, saprophages and parasites.

Araneida. Many spiders live in the aerial zone above the ground and cannot really be considered as members of the soil fauna. Among the groups which live close to the ground or in leaf litter can be numbered the wolf spiders (Lycosidae), the crab spiders (Thomisidae, see Fig. 3.5, p. 64), the Agelenidae (see Fig. 8.8, p. 222), the Gnaphosidae, Clubionidae, and the "money" spiders of the family Linyphiidae. These predators take their toll of smaller soil animals, particularly when the latter venture on to the soil surface, but little is known about the ecological significance of these predator/prey systems. Certainly, they would seem to be of importance in sand dune ecosystems, for example, where spiders form a conspicuous component of the ground fauna (see Chapter 7).

Microclimate seems to be an important factor in determining distribution patterns. Some species, at least, show narrow tolerances of environmental temperature and relative humidity. However, these preferences may change with the seasons and with the completion of mating (Vlijm and Kessler-Geschiere, 1967). The diversity of the spider fauna in a given site, on the other hand, is often related to the structural diversity of the habitat, and this applies particularly to those groups of spiders which spin webs. Grass stems, low shrubs, rock crevices, tussocks of vegetation all provide surfaces and structural members from which webs can be suspended, and the greater the diversity of these surfaces, the greater is the opportunity for niche diversification.

Opiliones. This is a relatively small group but it enjoys a wide distribution in temperate regions. The degree of association with the soil varies, and is more strongly marked in members of the families Trogulidae and Nemastomatidae, which tend to be associated with calcareous soils, than in the Phalangiidae which are equally common in the vegetation growing on base-rich and acid sites. The degree of dependence on the soil appears to be related to the degree of susceptibility to desiccation. Intolerant species, such as *Nemastoma lugubre*, are nocturnal in habit, but rarely venture far from the ground surface. Resistance to desiccation is more pronounced in *Phalangium opilio* and *Mitopus morio*, and these are active during the day in the vegetation above the ground. They feed mainly on small insects and their larvae and some of the smaller species of centipedes (Phillipson, 1960).

Seasonal vertical migrations occur in the Opiliones which resemble, in some respects, those of the woodlice discussed earlier. The pattern of these movements is generally an upward extension of the species range from the soil into the vegetation during the summer as individuals develop towards maturity. This trend may be reversed but it is not always complete. *Oligolophus agrestis* (Fig. 1.6) moves high into the tree foliage during the summer in southern Britain, but moves back to the soil in winter. This downward movement is much less marked in another arboreal species *Leiobunum blackwalli*.

Surface-active opilionids can be taken in pitfall traps in woodland areas.

In grassland and permanent pastures, members of the Phalangiidae can be collected by sweep nets.

Fig. 1.6 The harvestman *Oligolophus agrestis* (Meade). This female, which was taken from the foliage of an apple tree in Buckinghamshire, during the month of October, measures 5 mm in length. (Photo. by A. Newman-Coburn.)

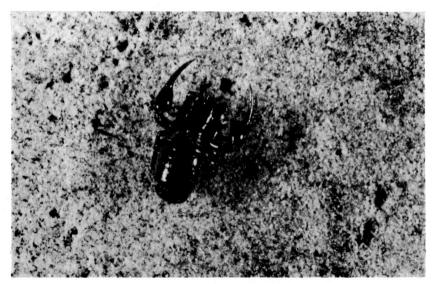

Fig. 1.7 The false scorpion *Neobisium muscorum* (Leach). (Photo. by J. P. Harding.)

Chelonethi. The false scorpions (Fig. 1.7) are one of the more fascinating groups of the soil fauna, although there is still much to be learned about their ecology. Within a given site, the pseudoscorpion fauna is never very diverse although the group, as a whole, has a wide distribution throughout the world with its maximum diversity occurring in the tropics and sub-tropics.

According to Weygoldt (1969), two of the most important features of the microhabitats of pseudoscorpions are high humidity and the availability of small crevices into which these animals can retire. Decomposing leaf litter in temperate and tropical forests provides both of these requirements and pseudoscorpions often occur in high densities. Gabbutt (1967) recorded numbers consistently in excess of 500 per m² on a beech site in Oxfordshire. Commonly occurring species in this litter type are *Neobisium muscorum*, *Chthonius ischnocheles*, *C. orthodactylus* and *Roncus lubricus*, and these are most easily collected in spring, early summer and autumn. They are much less evident in winter when they may withdraw to the soil or into silken hibernation chambers (Gabbutt, 1969).

The seashore also offers a multitude of rock crevices and humid spaces beneath debris on the drift line. Here, in Europe, can be found examples of habitat separation based on moisture preferences, with *Dactylochelifer latreillei* inhabitating dry sand dunes, and *Neobisium maritimum* living lower down the shore in the inter-tidal zone where it is submerged periodically. In the tropics and sub-tropics large pseudoscorpions belonging to the genus *Garypus* are common in crevices along rocky shores. In North America, a marked zonation of pseudoscorpions occurs along the coast of North Carolina (see Chapter 7).

Members of the Neobisiidae and Chthoniidae are adapted for life in caves. These forms are weakly pigmented and often blind, and have long legs and pedipalps.

Distribution patterns often extend beyond the boundaries of the soil and species such as *Dactylochelifer latreillei*, *Chelifer cancroides* and *Chernes cimicoides* are more commonly encountered in bark crevices. The nests of birds, small mammals and insects often house pseudoscorpions. *Lasiochernes pilosus*, for example, habitually occurs in the nests of moles and voles. *Neobisium sylvaticum* and *Chelifer cancroides* occur in bird nests, and the latter species also frequents beehives. Some chernetids and cheliferids live in the nests of ants and bumblebees where they may kill and feed upon their hosts.

Cheiridium museorum is often called the "book scorpion" from its habit of frequenting old libraries. It is one of several species which are often found in houses, cellars and farm buildings; *Allochernes videri* and *Toxochernes panzeri* are other examples. The efficient dispersal of many of these pseudoscorpions may be facilitated by their ability to become phoretic. Certainly, species which live in association with other arthropods or with small mammals have been found attached to their hosts, and probably disperse from site to site in this way.

Pseudoscorpions feed on a variety of prey. Soil-dwellers of the genera *Neobisium* and *Chthonius* feed readily on the collembolan *Lepidocyrtus lanuginosus* and on psocids. Chernetids and cheliferids also prey on psocids and other insects and their larvae. The effects of this predation on soil populations are not really known, but they certainly merit further study.

Soil- and litter-dwelling pseudoscorpions can be collected quite satisfactorily by dry sieving or dry funnel extraction, although due allowance must be made for the fact that at least a part of this fauna may not emerge from the sample if hibernation is occurring. Crevice faunas can be collected by hand, although greater success can probably be achieved by using a pooter.

Acari. The mites, as a group, have a world-wide distribution which surpasses even that of the Collembola. These two groups of microarthropods

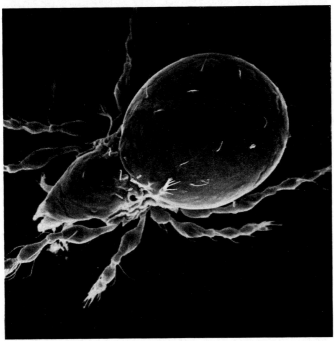

Fig. 1.8 A cryptostigmatid mite of the oppioid type ($\times$ 208). (Photo. by B. Parry and reproduced through the courtesy of the British Museum, N.H.)

2

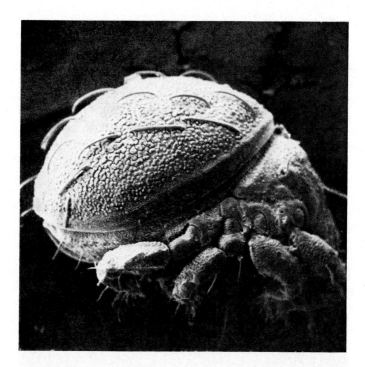

Fig. 1.9 A cryptostigmatid mite belonging to the genus *Hermannia* ($\times$ 56). (Photo. by B. Parry and reproduced through the courtesy of the British Museum, N.H.)

make a major contribution to the total faunal diversity in a wide range of soil types, and there are many similarities in their ecology. Moisture preferences play a large part in determining the distribution patterns of these arachnids, and hydrophile, mesophile and xerophile categories can be defined which parallel those of the Collembola. Vertical stratification of species in the soil profile is related to body size and the ability to resist desiccation, and here a very close parallel can be drawn between the Collembola and one group of mites, the Cryptostigmata or oribatid mites, which are predominantly soil-dwellers. Cryptostigmatids such as *Oppia* species (Fig. 1.8) which live in the soil and deeper layers of the litter have no water-proofing layer on the surface of the cuticle. The larger-sized species occur mainly in the upper litter layers and in the aerial vegetation, and they include such groups as *Hermannia* (Fig. 1.9), with a strongly sclerotized cuticle, and species such as *Steganacarus magnus*, *Damaeus onustus* and *Humerobates rostrolamellatus* which, like the sminthurid Collembola, use a tracheal method of respiration (Madge, 1964). This method permits the general body surface to be water-proofed, since it is no longer required as a respiratory surface.

Cryptostigmatid mites are saprophages in the broad sense, although a distinction can be made between species which feed on decomposing litter fragments (macrophytic feeders, e.g. *Steganacarus magnus*, see Fig. 8.5) and those which feed on fungi and bacteria (microphytic feeders, e.g. *Ceratoppia bipilis* and *Scheloribates* spp. – see Figs. 6.13 and 4.5, respectively). These two major categories can be subdivided further. For example, the macrophytophages include species which feed on woody tissue (xylophages) and species which feed on leaf tissue (phyllophages). Microphytophages can feed on fungi and yeasts (mycophages), bacteria (bacteriophages) or algae (phycophages). The scheme can be extended further to include panphytophages, which show a combination of some or all of the above categories, zoophages which feed on living animal material, necrophages which take carrion, and the faecal-feeding coprophages. A fresh approach to this topic has been developed by Luxton (1972), who studied the enzyme complements of various species of cryptostigmatid mites. His main conclusions may be summarized as follows:

1. Enzymes capable of digesting some of the more complex structural polysaccharides of plants were identified, including a cellulase, a pectinase and a xylanase. This suggests that cryptostigmatid mites can attack the structural components of plant tissue. However, since these components are often coated with lignin which is not attacked, there remains a query as to how much of the structural cellulose is, in fact, digested.

2. The species which feed mainly on fungi (microphytophages) do not possess a cellulase, but they can digest at least one other polysaccharide (amylose), and possibly a second (chitin). Fungi have a high percentage of chitin in their cell walls, and the ability to digest this substance will allow microphytophages to break down cell walls and attack the cell contents.

3. Microphytophages possess another enzyme, an alpha-glucoside capable of digesting trehalose. Trehalose is not found commonly in higher plants, but it does occur in appreciable amounts in fungal material. Here is an example of the way in which the digestive system of a soil animal may be correlated with its feeding habit. It also suggests that these soil mites are more specific in their food preferences than has been believed.

4. A more general conclusion is that the range of carbohydrase enzymes occurring in a particular species is related to its feeding specificity. Species which feed only on a narrow range of food materials have fewer enzymes than those which feed more indiscriminately.

The Cryptostigmata is only one of four Orders of mites which occur in the soil, although it is the one that shows the greatest degree of restriction to this habitat. The remainder are the Mesostigmata, Prostigmata and Astigmata.

The Mesostigmata includes a number of saprophagous forms, such as

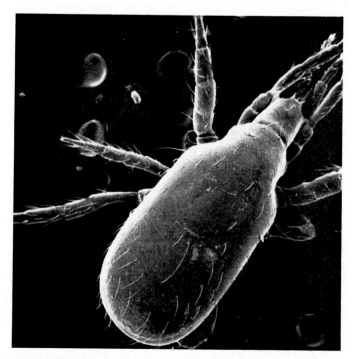

Fig. 1.10 The predatory mesostigmatid mite *Pergamasus crassipes* (L.) ($\times$ 83). (Photo. by B. Parry and reproduced through the courtesy of the British Museum, N.H.)

members of the genus *Cilliba* (see Fig. 8.8, p. 222), and a number of important predators, such as *Pergamasus crassipes* (Fig. 1.10) which inhabits beech litter, feeding on Collembola, other small arthropods and enchytraeid worms. Mites belonging to the family Macrochelidae are frequenters of compost heaps and dung piles where the nematodes and insect eggs on which these predators feed are in plentiful supply.

The Prostigmata is a heterogeneous group which includes, in addition to soil-dwellers, plant-feeding spider mites and water mites. There is little information about the ecology of this group in the soil, largely because the taxonomy has not yet been worked out. There is some evidence (see Loots and Ryke, 1967) that the ratio Cryptostigmata to Prostigmata leans towards the latter in soils with a high mineral content, and to the former in organic soils.

The Astigmata occur rather sporadically in soils, and never achieve great diversity. They usually occur in greatest densities in rather dry sites, such as arable land and hot desert soil (see Wallwork, 1972). However, this group has achieved greater fame outside the soil environment. It includes a number of pests of stored products, such as the flour mite *Acarus siro*, and various parasites, such as the scabies mite, the mange mites, feather and follicle mites.

Soil mites can be collected by the same extraction techniques used for Collembola. Funnel and bowl methods are effective for separating these animals from leaf litter. Separation from mineral soil is usually carried out by a flotation method.

Other groups

The soil fauna also includes a number of groups which are rare, or much more localized in their distribution than those discussed above. We need spend little time on these for their roles in the soil community are often minor, and they are incidental to the theme of this book. However, they deserve a passing mention.

Members of the soil water fauna include, besides the Protozoa and Nematoda, smaller groups or phyla such as the rotifers, gastrotrichs and tardigrades. The rotifers, a group more commonly associated with fresh water than with soil, are suspension feeders, ingesting particles of organic debris which are drawn into the mouth by ciliary action. The gastrotrichs are small, spiny animals of creeping habit which are predaceous on other members of the soil water fauna. Tardigrades have a higher level of organization than the rotifers and gastrotrichs, being considered by some authors as "degenerate" arthropods. Sometimes known as "water bears", these bizarre animals show a variety of feeding habit, some being saprophages, others mycophages, carnivores or necrophages (Hallas and Yeates, 1972).

Two groups of very distantly-related forms which have in common a distribution which is mainly localized in the moist tropical and subtropical forests of South Africa, South America and Australasia, are the terrestrial planarians of the phylum Platyhelminthes, and the curious "arthropod" group, the Onychophora, to which *Peripatus* belongs. These two groups have very similar ecologies and, indeed, individuals of each have been found in close association under natural conditions (Lawrence, 1953). Both feed on living prey or carrion, which is digested extracellularly and then sucked into the mouth or pharynx.

Although the woodlice are the predominant group of terrestrial Crustacea, othere members of this Class occur in soils usually, but not always, in moist sites. Harpacticoid copepods are surprisingly abundant in moist beech litter in southern England, while amphipods are common in soils and debris of seashores in many parts of the world. Some amphipods, such as members of the genera *Talitrus* and *Orchestia* have achieved a sufficient degree of "terrestrialness" to have extended their distribution inland to grassland habitats in New Zealand (Hurley, 1968). Terrestrial ostracods and decapods have been less successful, however, and are mainly confined to the moist leaf litter of tropical and subtropical forests.

Our preoccupation with the invertebrates must not allow us to overlook the fact that several vertebrates can be classified among the true soil fauna. There are many vertebrates which use the soil as a place of retreat when conditions above the ground become unfavourable, of course. These have little or no impact on the soil community. However, true soil-dwellers occur among the Amphibia, Reptilia and Mammalia.

Caecilians are apodous Amphibia which live a subterranean existence in the tropics. These are curious worm-like vertebrates with reduced eyes and limbs and a compact skull which may aid in burrowing through the soil. They feed on insects and worms.

Burrowing reptiles are more common in desert soils than elsewhere, and are represented by tortoises, such as *Testudo polyphemus*, limbless lizards belonging to the family Amphisbaenidae, and snakes of the genus *Typhlops*. Little is known of the way in which the activities of these forms are integrated into the functioning of desert soil communities, as a whole.

Among the mammals, rodents such as the gerbils, kangaroo-rats, ground squirrels, gophers and voles may be counted as members of the soil fauna. Again, these are mainly inhabitants of desert soils and while they present many interesting features to the physiological ecologist (see Schmidt-Nielsen, 1964), there is very little information about their relationships with other members of the soil fauna. These animals are essentially herbivorous, feeding mainly on dry seeds which may be stored underground. They may have a role in dispersing organic material in the soil, but this is unlikely to be more than a local phenomenon.

Moles, on the other hand, produce extensive burrow systems in temperate soils, often depositing mounds of mineral soil on the surface. These activities increase aeration and drainage of the soil, and also promote organo-mineral mixing. Of even greater ecological interest is the feeding habit of the mole, for the diet consists mainly of lumbricid earthworms and, to a lesser extent, insect larvae. Accordingly, the distribution of moles tends to parallel that of the earthworms; both are more abundant in base-rich mull soils than in acid forest soils.

REFERENCES

BANAGE, W. B. (1963). The ecological importance of free-living soil nematodes with special reference to those of moorland soil, *J. Anim. Ecol.*, **32**, 133–40.

BONNET, L. (1961). Caractères généraux des populations thécamoebiennes endogées, *Pedobiologia*, 1, 6–24.

EDNEY, E. B. (1954). British woodlice, *Synopses of the British Fauna*, no. 9, Linnean Society of London.

EDWARDS, C. A. and DENNIS, E. B. (1962). The sampling and extraction of Symphyla from soil. *In:* "Progress in Soil Zoology" (Murphy, P. W., ed.), 300–4. Butterworth, London.

GABBUTT, P. D. (1967). Quantitative sampling of the pseudoscorpion *Chthonius ischnocheles* from beech litter, *J. Zool. Lond.* **151**, 469–78.

GABBUTT, P. D. (1969). Life-histories of some British pseudoscorpions inhabiting leaf litter, *In:* "The Soil Ecosystem", 229–35 (Sheals, J. G., ed), Syst. Assoc. London.

GASPAR, C. (1971). Les fourmis de la Famenne. II. Une etude zoosociologique, *Rev. Écol. Biol. Sol*, **8**, 553–607.

HALLAS, T. E. and YEATES, G. W. (1972). Tardigrada of the soil and litter of a Danish beech forest, *Pedobiologia*, **12**, 287–304.

HUNTER, P. J. (1966). The distribution and abundance of slugs on an arable plot in Northumberland, *J. Anim. Ecol.*, **35**, 543–57.

HURLEY, D. E. (1968). Transition from water to land in amphipod crustaceans, *Am. Zool.*, **8**, 327–53.

LAWRENCE, R. F. (1953). "The Biology of the Cryptic Fauna of Forests", Balkema, Cape Town and Amsterdam.

LEWIS, J. G. E. (1965). The food and reproductive cycles of the centipedes *Lithobius variegatus* and *L. forficatus* in a Yorkshire woodland, *Proc. zool. Soc. Lond.*, **144**, 269–83.

LOOTS, G. C. and RYKE, P. A. J. (1967). The ratio Oribatei:Trombidiformes with reference to organic matter content in soils, *Pedobiologia*, **7**, 121–24.

LUXTON, M. (1972). Studies on the oribatid mites of a Danish beechwood soil. I. Nutritional biology, *Pedobiologia*, **12**, 434–63.

MADGE, D. S. (1964). The water relations of *Belba geniculosa* Oudms. and other species of oribatid mites, *Acarologia*, **6**, 199–223.

NEWELL, P. F. (1971). Molluscs, *In* "Methods of Study in Quantitative Soil Ecology" (Phillipson, J., ed.), IBP Handbook no. 18, chapter 9.

NIELSEN, C. O. (1949). Studies on the soil microfauna II. The soil-inhabiting nematodes, *Natura jutl.*, **2**, 1–131.

O'CONNOR, F. B. (1962). The extraction of Enchytraeidae from soil, *In* "Progress in Soil Zoology", 279–85 (Murphy, P. W., ed.), Butterworth, London.

O'CONNOR, F. B. (1967). The Enchytraeidae, *In* "Soil Biology", 213–57 (Burges N. and Raw, F., eds.). Academic Press, London and New York.

PARIS, O. H. (1963). The ecology of *Armadillidium vulgare* (Isopoda: Oniscoidea) in California grassland: food, enemies and weather, *Ecol. Monogr.*, **33**, 1–22.

PHILLIPSON, J. (1960). A contribution to the feeding biology of *Mitopus morio* (F.) (Phalangida), *J. Anim. Ecol.*, **29**, 35–43.

PRYOR, M. E. (1962). Some environmental features of Hallett Station, Antarctica, with special reference to soil arthropods, *Pacif. Insects*, **4**, 681–728.

RUNHAM, N. W. and HUNTER, P. J. (1970). "Terrestrial Slugs", Hutchinson University Library, London.

SATCHELL, J. E. (1967). Lumbricidae, *In* "Soil Biology", 259–322. (Burges, N. and Raw, F., eds). Academic Press, London and New York.

SCHMIDT-NIELSEN, K. (1964). "Desert Animals", Oxford, New York and Oxford.
SEINHORST, J. W. (1962). Extraction methods for nematodes inhabiting soil, *In* "Progress in Soil Zoology", 243–56 (Murphy, P. W., ed.). Butterworth, London.
SOUTHWOOD, T. R. E. (1966). "Ecological Methods", Methuen, London.
STOUT, J. D. (1963). Some observations on the Protozoa of some beechwood soils on the Chiltern Hills, *J. Anim. Ecol.*, **32**, 281–87.
STOUT, J. D. (1968). The significance of the protozoan fauna in distinguishing mull and mor of beech (*Fagus silvatica* L.), *Pedobiologia*, **8**, 387–400.
SUTTON, S. (1972). "Woodlice", Ginn, London.
VÅGVÖLGYI, J. (1952). A new sorting method for snails, applicable also for quantitative researches, *Annls. hist-nat. Mus. natn. hung.*, **3**, 101–4.
VLIJM, L. and KESSLER-GESCHIERE, A. M. (1967). The phenology and habitat of *Pardosa monticola*, *P. nigriceps* and *P. pullata* (Araneae, Lycosidae), *J. Anim. Ecol.*, **36**, 31–56.
WALLWORK, J. A. (1970). "Ecology of Soil Animals", McGraw-Hill, London.
WALLWORK, J. A. (1972). Distribution patterns and population dynamics of the micro-arthropods of a desert soil in southern California, *J. Anim. Ecol.*, **41**, 291–310.
WEYGOLDT, P. (1969). "The Biology of Pseudoscorpions", Harvard, Cambridge, Mass.
YEATES, G. W. (1971). Feeding types and feeding groups in plant and soil nematodes, *Pedobiologia*, **11**, 173–79.
ZINKLER, D. (1966). Vergleichende Untersuchungen zur Atmungsphysiologie von Collembolen (Apterygota) und anderen Bodenkleinarthropoden, *Z. vergl. Physiol.*, **52**, 99–144.

Chapter 2

Methods for Studying Distribution and Diversity

The study of horizontal distribution patterns and species diversity of the soil fauna presents certain problems. Soil populations are large in size and frequently are not localized in isolated habitat units. Rather, they are dispersed through a continuum which is the soil, and their distribution in this continuum is, in most cases, non-random or 'clumped'. The very large size of various soil populations (Table 2.1) precludes the collection of entire

Table **2.1**
Densities of the major groups of soil animals (compiled from various sources).

Group	Numbers per m²	Habitat	Authority
Testate Protozoa	16×10^6	*Sphagnum*	Heal (1962)
Nematoda	4–20×10^6	Grassland	Nielsen (1949)
Lumbricidae	390	Grassland	Reynoldson (1955)
Enchytraeidae	$134 \cdot 3 \times 10^3$	Conifer soil	O'Connor (1957)
Collembola	35×10^3	*Calluna* moor	Hale (1967)
Acari: Cryptostigmata	176–410×10^3	Hemlock mor	Wallwork (1959)
Diplopoda	100–300	Sycamore/ash wood	Blower (1970)
Pseudoscorpiones	500	Beech soil	Gabbutt (1967)

faunas. Instead, samples have to be taken, in the form of soil cores, from which the animals can be extracted, and from which a representative picture of distribution and diversity can be obtained. In order to avoid personal bias, the selection of sampling sites must be a random operation, and due allowance must be made in the subsequent analyses for the natural clumping of individuals in the population. It is not the purpose of this chapter to dwell in detail on these problems, which are mainly statistical in nature, for they have been treated in detail by, for example, Morisita (1962, 1971), Lloyd (1967), Pielou (1969), Usher (1969, 1971) and Iwao (1972). Similarly, the technical problems of extracting animals from soil cores are adequately documented in the literature (see, for example, Murphy, 1962; Southwood, 1966; Wallwork, 1970). Here, we are concerned more with the methods used in defining natural associations of species, and in establishing where sharp faunal discontinuities occur in nature, so that subsequently these may be interpreted in terms of environmental discontinuities.

Some basic definitions

Let us suppose that we take a sample consisting of ten soil cores, or sampling units, distributed at random over a study area. We count and identify the species of a particular faunal group present in each sampling unit, and also make a note of the number of individuals of each species in each sampling unit. In addition to noting that the number of individuals of any one species varies from one sampling unit to another, we will, in all probability, be able to make other observations. For example, each sampling unit may not contain all of the species present in the total sample, some species being present in some units, absent in others. From this we deduce that individual species show a *frequency of occurrence*. Again, it may be evident that all of the species present in one sampling unit are not equally represented in terms of numbers of individuals. Perhaps two or three species may be represented by 20–30 individuals each, a rather larger number of species by 10–20 individuals, and the majority by less than 10 individuals each. In other words, the species present may have different *relative abundances*.

It may be possible to go even farther than this. Thus, we may find that two, three or more species always, or nearly always, occur together, i.e. when species A is present in a sampling unit, species B and C are also present. Similarly, when species A is absent, so are species B and C. One hypothesis suggested by this observation is that species A, B and C are interdependent in some way. Alternatively, the three species may have similar preferences for, or tolerances of, certain micro-environmental conditions. In other words, these species may constitute an ecological grouping, or *association*.

The definition of species associations suggests the possibility that they may serve as indicators of particular environmental conditions. With this objective in mind, the horizontal distribution of soil animals is often described in relation to some variable feature of the environment, such as vegetation type, soil type, pH, water content or micro-climate. Much of what follows in this book is concerned with these relationships.

Natural groupings of species populations which have any ecological significance are notoriously difficult to define. In theory at least, the boundaries of such natural groupings must be drawn where clear discontinuities occur in the horizontal, or vertical, distribution of the constituent species. Macfadyen (1963) made the distinction between methods which seek to determine the specific composition of the natural grouping ("*intra-community studies*"), by estimating the extent to which the distributions of various species are similar, and those that endeavour to delimit the boundaries of distribution of the grouping as a whole ("*inter-community studies*"). While this division is certainly a logical one, it must be stressed that it is not always a simple matter to determine where to draw the line between one natural grouping and another, for there is a considerable amount of overlap in the distributions of species of soil animals, as we will see. Furthermore, the division into "intra-group" and "inter-group" studies is often arbitrary, for the same method may be applicable to determine the amount of diversity within one natural grouping, and the amount of affinity between two natural groupings. This really reflects a lack of knowledge about where one natural grouping ends and another begins. Nevertheless, we can distinguish, in the methods described below, those which estimate the degree of association between pairs of species, and those which measure the amount of faunal similarity between samples taken from different sites, habitats or natural groupings.

Frequency, Abundance and Fidelity

One simple way to describe the soil fauna of a particular habitat, or to compare the fauna of two or more habitats, is to identify the species present and compile check lists. However, this method does not provide any information about distribution patterns within the habitat, nor does it convey any idea of the relative importance of individual species, i.e. it does not distinguish between rare and common species. We must then turn to other, more critical, methods of description and comparison which, as a general rule, make use of the concepts of *frequency*, *abundance* and *fidelity*.

Frequency. The first observation made in the simple sampling routine mentioned at the start of this chapter was that any given species may be present in some sampling units, and absent from others. Within the context of this sampling programme, the species may then be said to exhibit a frequency of occurrence which can be expressed in numerical terms, usually as a percentage, in the following way:

$$\text{Frequency} = \frac{a}{n} \times 100$$

where a is the number of sampling units in which the species occurs, and n is the number of sampling units comprising the total sample.

This criterion can provide some information about the distribution of a species if the frequency values are obtained from different sites or habitats. The frequency of any one species may then be compared with that of others, and an index expressing the degree of similarity between various species frequencies can be calculated (Kontkanen, 1950; Agrell, 1963). However, the use of frequency data, even in the more sophisticated techniques discussed below, is subject to important qualifications, for the following reasons. It is often the case that as n increases, so does the probability of occurrence of the species, i.e. the more sampling units taken, the greater is the chance of finding the species more often, and thus the frequency value will increase. Certainly, the estimates of frequency become more reliable as the number of sampling units is increased. Obviously, there is a limit to the number of units which can be included in a sample, and, indeed, there is a point beyond which the taking of additional sampling units has little or no effect on the frequency value. This critical point then fixes the most suitable number of units to be included in each sample, and its determination should be one of the most important objectives of any preliminary sampling programme.

This point can be determined very simply by taking a series of samples, each containing a different number of units, and calculating the frequency of occurrence of the particular species concerned, in each sample. These frequency values can then be plotted against the number of units in each sample. At the point where the curve for each species begins to flatten out, indicating that the frequency value becomes relatively constant, the number of units per sample to achieve this can be read off. On the other hand, if the investigation is concerned with comparisons of entire groups of species, as it would be in studies seeking to delimit one natural grouping from another, the

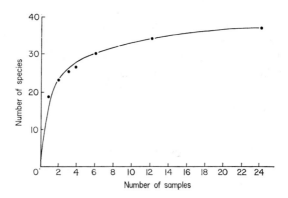

Fig. **2.1** Species-area curve for soil arthropods. (From Hairston and Byers, 1954.)

effective number of sampling units (or samples) will be determined from a curve such as that shown in Fig. 2.1. This type of curve is known as the "species-area" or "minimal-area" curve, and it is used to determine the minimum number of sampling units, or samples, which will yield a high proportion of the total number of species present in the sampling site. Further, the most suitable size of the sampling unit can be obtained in a similar way, for it is obvious that the frequency value for a particular species, and the total number of species recorded, will increase with increasing size of the sampling unit, up to a certain point. This is of particular importance in vegetation analyses involving quadrat sampling (Greig-Smith, 1964) where it is possible, at least in theory, to increase the size of the unit to produce 100% frequency for a given species. Carried to its absurd limits, the unit size could be increased to correspond to the whole study area, thus ensuring 100% frequency for all species. Clearly, such a procedure would fail to distinguish between common and rare species, even if it were possible from a practical point of view. The success of any sampling programme designed to collect frequency data depends, to a large extent, on this distinction being made.

The species-area curve may be used to characterize a particular association, grouping or community, for the angle of its slope may be peculiar to that grouping. Two or more different groupings may be compared on this basis, when this is the case (Smith, 1966). This method of comparison is sometimes used in botanical surveys, but it could be applied equally well to studies of the soil fauna.

Constancy. This term is often used to denote frequency of occurrence, but it has been interpreted in several different ways and has caused some ambiguity. Balogh (1958) suggested that the term should be applied to frequency values calculated from samples taken from two or more similar sites within the same "habitat type", for example the soil under beech forest in two geographically distinct localities. Tischler (1949) applied the concept of "frequency" to comparisons of samples within a single habitat, and "constancy" for comparisons between unit areas of different habitats. This author translated constancy values into ecological terms in a classification which categorized species as "accidental", "accessory", "constant" and "absolutely constant" in the following way:

"Accidental" species: Constancy 0–25%
"Accessory" species: Constancy 25–50%
"Constant" species: Constancy 50–75%
"Absolute" species: Constancy 75–100%

The difficulty with this scheme is that the limits between the various ranks are arbitrarily drawn. There is also a tendency to confuse these rankings, which do not imply any degree of restriction to a particular association, with those of the concept of fidelity, which do.

Fidelity. This concept expresses the degree of restriction of a species to a particular set of environmental conditions. Thus, a species with a high fidelity, expressed in mathematical terms (Goodall, 1953), will be strongly restricted to a well-defined set of environmental conditions, and can be regarded as a good *indicator* species. Braun–Blanquet (1932) in developing this concept for use in botanical studies, defined five "fidelity classes". Plant species which are rare or accidental in the sampling units are termed "strange species"; those without any clear association with any community are "indifferent". "Preferential" species are those which show a clear affinity with one particular community, although they may also occur outside it. "Selective" species show a stronger restriction to one community than preferential species although, occasionally, they may be found in other communities, while "exclusive" species are, to all intents and purposes, restricted to one community. Once again, these rankings are largely subjective.

Abundance and Relative Abundance. Earlier, we noted that all species present in any one sampling unit may not be equally represented in terms of numbers of individuals. This inequality of representation may be described in various ways. For example, the number of individuals of a species in a sampling unit may be related to the surface area or volume of the soil included within that unit. This gives a measure of the *abundance* or *density* of the species in question, and this is particularly useful for estimating population sizes.

Estimates of absolute abundance are very important in productivity studies, but of lesser significance in the delimitation of natural groupings of species where we are more interested in establishing the amount of similarity between the horizontal distribution patterns of various species. For this purpose use may be made of the relationship between the number of individuals of any one species in a sampling unit, and the total number of individuals of all species occurring in this unit. This relationship gives a measure of the *relative abundance*, or relative density, of any one species, and it is usually expressed as a percentage.

It is common practice, particularly among European workers, to use the term "dominance" to denote relative abundance. Unfortunately, this term also carries with it the implication of ecological importance. Perhaps it should be emphasized that a species with the highest relative abundance is not necessarily the one which has an over-riding influence on the functioning of a particular trophic level or on the community as a whole. Numerical "dominance" should not, automatically, be equated with ecological dominance.

Relationship between Frequency and Abundance

Frequency and abundance vary independently as functions of the horizontal distribution pattern, and yet they usually behave in very similar ways: as the frequency of a species increases, so does its abundance. This fact,

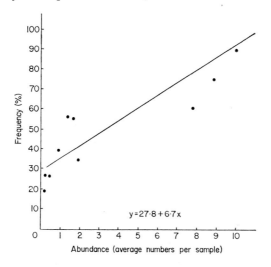

Fig. 2.2 Regression of frequency on abundance for nine species of mites collected from a desert soil in southern California. (From Wallwork, 1972.)

which is illustrated by the data presented in Fig. 2.2, makes the task of the soil ecologist easier. Determinations of abundance and relative abundance involve time-consuming sorting and counting of all individuals present. Determinations of frequency, on the other hand, merely require records of the presence or absence of a species, a much less laborious task. Thus, if it is desired to describe a habitat in terms of its faunal composition, or to compare two or more habitats on this basis, the picture emerging using frequency (or constancy) as the criterion would not differ appreciably from that obtained using relative abundance.

Indices of association between pairs of species

In this section, we are concerned with methods which compare the distribution of pairs of species. The examples chosen reflect two different approaches, namely (1) the use of frequency data, and (2) the use of abundance data.

Correlation

The existence of an association between two species can be detected, statistically, by calculating a correlation coefficient *(r)*. The procedures used in this calculation need not be outlined here, for they can be found in standard statistical texts and, furthermore, this statistic can only be used to estimate the *degree* of association between two species populations if their distributions are random. Random distributions are unusual among soil populations, and it would be necessary to transform raw data before a correlation coefficient could be determined. As an alternative, non-parametric methods which do not rest on the assumption of randomness may be used to test for correlation, when data are expressed as frequencies. An example of one such method is the Contingency Chi-Square, the applications of which have been discussed by Greig-Smith (1964), Debauche (1962) and Kershaw (1964) among others.

The Chi-Square (X^2) test may be used to compare the distribution of pairs of species in one or more series of samples. The comparison is based on the fact that, by observing the frequency of the species in question, and assuming that they are not associated in any way, the number of samples (or sampling units) containing both may be calculated to give an expected value. This may then be compared with the observed frequency of joint occurrences by means of the Chi-Square test. For example, consider two species A and B, distributed in a given number of sampling units (n) such that: (1) a certain number of these units (a) contain both species, (2) a certain number (b) contain species B but not species A, (3) a certain number (c) contain species A but not species B and, (4) a certain number (d) contain neither

Table 2.2

The 2 × 2 contingency table showing frequency distribution of species A and B.

		SPECIES A		
		Present	Absent	Total
S P E C I E S B	Present	a	b	a + b
	Absent	c	d	c + d
	Total	a + c	b + d	n

species. The frequency distribution of the two species may then be compared by means of a 2 × 2 contingency table (Table 2.2). The number of sampling units expected to contain both species, assuming non-association between the two, is calculated from the expression:

$$\frac{(a + b)\,(a + c)}{n}$$

Similarly, the number of sampling units expected to contain neither species can be calculated. Thus, we obtain *expected* values for joint occurrences and non-occurrences. The corresponding *observed* values are (a) and (d). The Chi-Square equation which compares these values is:

$$X^2 = \frac{(ad - bc)^2\, n}{(a + b)\,(c + d)\,(a + c)\,(b + d)}$$

in which the numerator is formed by squaring the difference of cross products *ad* and *bc* and multiplying by the total number of sampling units (n), and the denominator is the product of all the marginal totals shown in Table 2.2. Tables of Chi-Square are available which allow us to test the Null hypothesis that the difference between observed and expected values for the joint occurrence of the two species is not significant. At the 0·05 level of probability, and with one degree of freedom which is appropriate for this particular comparison, the tabulated value for X^2 is 3·841. Thus, if the X^2 calculated from the above equation is equal to, or greater than, 3·841, we may conclude that there is only a 5% chance or less that the difference between observed and expected joint occurrences is due to random variation. This would suggest that the difference is significant; in other words, there is an association between the two species.

If an association exists, it is then necessary to determine its trend, i.e. whether it is positive (associative) or negative (repulsive). This can be done by comparing the observed and expected values for joint occurrences and joint non-occurrences of the two species concerned. If the two species occur together more often than is suggested by the expected values, then there is a positive association. In this case, the values *a* and *d* in Table 2.2 will be large in comparison to *b* and *c*. If the species occur together less frequently than expected, then there may be repulsion, or negative association. The values *b* and *c* will then be large in comparison to *a* and *d*. Again, if correlation is established, the degree of association or repulsion may be estimated from the contingency coefficient *(C)* which, after Debauche (1962), can be defined as:

$$C = \sqrt{\left(\frac{X^2}{n + X^2} \right)}$$

There are, of course, limitations to the use of the X^2. It is based essentially on frequency, and is therefore not independent of sample size. Furthermore, the method depends on some of the sampling units not containing either of the species in question. If all units have either one or both species present it will be necessary to resort to more precise quantitative methods based on regression analyses. In addition, as pointed out by Fager (1957), this method will not detect positive association between two species occurring together in most of the sampling units, i.e. the number of joint non-occurrences will be so low that the X^2 which compares expected and observed joint non-occurrences will not depart appreciably from zero. Conversely, the method could lead to an assumption of positive association between two rare species, provided a large enough number of sampling units in which they did not occur is considered. Finally, the determination of negative affinity (repulsion) by this method implies the absence of a species from samples, and, in view of the probable errors associated with sampling and extraction, is less reliable than the determination of positive association.

On the other hand, the Chi-Square has the important property that it does take into account changes in the relative frequency of occurrence of the two species being compared. In addition, the method is objective in that it fixes a definite level at which differences become significant. Fager (1957) developed an Index which incorporates these desirable qualities, but has fewer limitations than the Chi-Square.

Fager's Index of Affinity

This coefficient is developed from the basic assumption that the probabilities of occurrence of two species are related to the sum of their total number of occurrences, and not to the total number of samples or sampling units taken. As its author states: "The use of this assumption removes the premium on rarity and also makes it possible to find evidence of affinity between two species which occur in most of the samples taken". Proceeding from this, and assuming independent distribution, an expression for the expected number of joint occurrences (J) may be developed:

$$J = \frac{n_a \, n_b}{n_a + n_b}$$

where n_a is the number of samples containing species A, and n_b the number containing species B, and n_a is less than, or equal to, n_b. To determine whether or not the observed joint occurrences are greater than would be allowable on the basis of chance alone, i.e. whether or not they differ significantly from the expected $\mathcal{J}$, a single-ended t-test is used:

$$ t = \left[\frac{(n_a + n_b)\,(2\mathcal{J} - 1)}{2n_a n_b} - 1 \right] \left[\sqrt{(n_a + n_b - 1)} \right] $$

for values of n_a greater than 10, and a ratio n_b/n_a not exceeding 2. The single-ended t-test only investigates positive affinity. The upper limit of 2 for n_b/n_a "ensures that whenever evidence of affinity is obtained the number of joint occurrences will be a considerable proportion of the total number of occurrences of both of the species" (Fager, 1957). This implies that the various species defined as a group form a nearly constant part of each other's environment, and thus this method would not be satisfactory for testing the degree of association between one rare species and another common one. Crossley and Bohnsack (1960) found this upper limit of 2 impracticable, but by extending it to 2·5 and by pooling replicate samples were able to make comparisons between 26 of the most abundant species of cryptostigmatid mite in pine forest litter, and detected a positive association, or recurrent grouping, of 9 species. Fager (1957) tabulated minimum values of $\mathcal{J}$ significant at the 0·05 level of probability, and by this means tested all possible combinations of pairs of species for significant association. The affinities and non-affinities thus determined are made the basis for the definition of recurrent groups. A "recurrent group" is one which satisfies the following requirements:

1. The evidence for affinity is significant at the 0·05 level for all pairs of species within the group.
2. The group includes the greatest possible number of species.
3. If several groups with the same number of members are possible, those are selected which will give the greatest number of groups without members in common.
4. If two or more groups with the same number of species and with members in common are possible, the one which occurs as a unit in the greater number of samples is chosen.

In order to detect recurrent groupings, the affinity information is set out in a trellis diagram and the above criteria are applied in sequence. The use of trellis diagrams is considered later in this chapter.

The Use of Abundance Data

So far, we have been concerned with methods for comparing the degree of association between pairs of species which are based on frequency data. As Southwood (1966) pointed out, such methods may underestimate the amount of association in certain cases. For example, a situation may be visualized in which the number of joint occurrences is small and yet large numbers of individuals of both species are present in these units. The remaining units which comprise the large majority of the sample, may contain very few of one or other of the two species being compared. The large number of these latter units in the sample will tend to mask the importance of the joint occurrences and, accordingly, little affinity will be demonstrated. What is needed, under these circumstances, is an index of affinity which takes into account the abundance of the species concerned. Such an index has been provided by Whittaker and Fairbanks (1958) and modified by Southwood (1966):

$$\text{Index} = 2 \left[\frac{\mathcal{J}_i}{A + B} - 0.5 \right]$$

where $\mathcal{J}_i$ is the total number of individuals of both species present in sampling units containing both, A and B are the total numbers of species A and B in all samples. The index ranges from -1 (no association) to $+1$ (complete association).

This method has been used, in conjunction with a trellis diagram (see Fig. 2.6), to describe natural associations of microarthropod species in a southern Californian desert soil (Wallwork, 1972).

Delimitation of natural associations

It is unwise to assume that the distribution of soil animals is homogeneous, even though the sampling locality may appear to be so at first sight. Microhabitat differences may be considerable, even within a small area, and it is for this reason that a stratified method of sampling, which ensures that all parts of the locality have an opportunity of being sampled, is the most reliable. A way of measuring, quantitatively, the similarities or differences between two or more samples is needed in this case and, more particularly, when comparisons are made between samples taken from localities showing obvious habitat differences. Such comparisons may lead to the definition of discrete natural groupings and some examples of these are given below.

Indices of Diversity

Many of the methods treated in this chapter are based on presence or absence of species, and it is a fact that different indices, applied to the same data, often provide different estimates of association. Of even greater significance is the fact that because they utilize frequency estimates, and as we have seen these are not independent of sample size or number, these indices will vary. As a rule, they increase with increasing sample size and number, even if the samples are taken from the same locality.

As an alternative, the problem of delimiting natural groupings may be approached by considering variations in numerical abundance. Essentially, this approach seeks to define the faunal characteristics of a particular sample series, locality or habitat, through the relationship between the number of species and the number of individuals present. This relationship can be expressed in mathematical terms and may be used as a basis for faunal comparison. It is known as an *Index of Diversity*, and its implications have been reviewed by MacArthur (1965).

One such index, which is useful in soil ecology, is based on the theory, developed by Fisher *et al.* (1943), that if random samples are taken from a natural grouping of populations, the species frequency distribution approximates a logarithmic series. In simpler terms, this means that a large number of the species present are represented by one individual, a relatively lower number of species by two individuals, a lower number still by three individuals, and so on. This kind of distribution may be described by the series:

$$\alpha x, \frac{\alpha x^2}{2}, \frac{\alpha x^3}{3}, \frac{\alpha x^4}{4} \text{------------------} \frac{\alpha x^n}{n}$$

where αx is the number of species with one individual, $\alpha x^2/2$ is the number of species with two individuals, and so on. The terms α and x are constants which can be determined if the total number of species (S) and the total number of individuals (N) of all species in the sampled population are known, since it can be demonstrated that:

$$S = \alpha \left(-\log_e \overline{1-x} \right)$$

and,

$$N = \alpha x/(1-x)$$

Having determined α and x, the logarithmic series appropriate to the data can be calculated. The constant α has the property of being low when there are relatively few species present, and high when the number of species in relation to the number of individuals is high. Hence, it has been designated as an *index of diversity*. The constant x never exceeds unity, but increases with the size of the sample. Providing species frequency conforms to a logarithmic series distribution, different population groups may be compared in terms of their diversity. Williams (1964) provides a detailed and documented discussion of the applications of this index of diversity and also lists a number of useful relationships between S, N, and α. Since, for example:

$$S = \alpha \log_e(1 + N/\alpha)$$

it will be noted that when N is large in comparison with α, the relationship between the number of species and the logarithm of the number of individuals is linear, and in a large sample the index of diversity corresponds to the number of species represented by one individual. This postulate may be applicable to some natural situations where rare species comprise a considerable proportion of the total number present, but it is equally probable that, universally, the numbers of rare and abundant species are approximately equal, and that the majority of species occur in numbers which fall within these two extremes (Preston, 1948; Hairston and Byers, 1954).

In fact, it has been argued that the logarithmic series is not a particularly satisfactory model for the interpretation of species frequency distribution. According to this model, the index α should be constant and independent of sample size. Hairston and Byers (1954) found that this was not the case, and

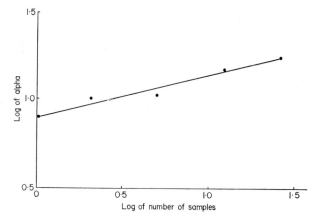

Fig. **2.3** The relation between index of diversity (alpha) and sample size. (After Hairston and Byers, 1954.)

that the value of the index was related to the size (or number) of the samples (Fig. 2.3). These authors rejected the explanation that this trend was due to a greater heterogeneity of the natural species grouping being revealed as more (or larger) samples were taken from it. Similar problems were encountered in attempting to apply log-normal distribution statistics, and it was concluded that such mathematical systems fail to represent accurately the natural situation, because they assume that all species considered in the natural grouping have the same kind of distribution. Hairston and Byers (1954) drew attention to the distinction between the distribution patterns of common and rare species, pointing out that the most numerous species have a more or less random distribution, or if not, at least are present in most of the samples taken, so that "an increase in the number of samples will not have any effect upon a curve describing their distribution". In the case of rare species, clumping, if it occurs, is important for "adding samples will be more likely to add new species rather than individuals of species already recorded". The "broken stick" model proposed by MacArthur (1957) attempts to include all species on a single distribution curve, assuming that all have the same kind of random distribution, and for this reason fails to describe adequately the natural groupings of soil animals in which rare species may be more strongly clumped than common ones (Hairston, 1959).

Several different kinds of distribution curve have been used to determine the limits of natural groupings of species. The simplest of these is the "species-area" or "minimal-area" curve mentioned earlier in this chapter (see Fig. 2.1). This curve may be used to determine the number of species comprising a natural grouping by referring to the point on the ordinate corresponding to the upper asymptote. It will also be recalled that two or more natural groupings may be compared by this method. The use of a scale of logarithms of sample size plotted against the average number of species per sample may result in an S-shaped curve, from which are deduced the spatial limits of the natural grouping (from the point of inflexion of the curve), and the mean number of species per unit area. These characteristics, may be useful in comparing one natural grouping with another (Archebald, 1949; Macfadyen, 1963). Hairston and Byers (1954) obtained a straight line relationship between the logarithm of sample size and the average number of species which did not provide an upper asymptote. These authors suggested that this approach may have more practical application in delimiting natural groupings of a less complex kind than that of the "old field" type studied by them.

Equitability Component of Species Diversity

The preceding discussion has been concerned with the "richness" or "variety" component of species diversity. There is also an "equitability" component (Lloyd and Ghelardi, 1964) which can be understood from the following example. Consider two ecological systems, each of which contains 10 species and 100 individuals. A simple measure of the "richness" component is the index $S/\sqrt{N}$, where $S = $ the total number of species, and N the total number of individuals. Clearly, this index will be the same for both systems. However, it is possible that: (a) in one system, 91 of the individuals belong to one species, while each of the remaining species is represented only by a single individual, and (b) in the second system, each of the 10 species could be represented by 10 individuals. These two systems are then widely different in respect to the way in which the individuals present are apportioned to the species present, i.e. there is a difference in evenness, or *equitability*.

The general diversity of a natural grouping has, then, two components: species variety and equitability. This general diversity is usually denoted by the symbol H, and it will be at a maximum when all species present are equally abundant. In practice, this maximum is never reached, but its value can be calculated if we know the number of species present, for:

$$H_{max} = \log_2 (S)$$

Using information theory (Margalef, 1957), species diversity can be measured by the Shannon-Wiener function if the number of species present and their relative abundances are known. This function has the form:

$$H = - \sum_{r=1}^{s} p_r \log_2 p_r$$

where p_r is the observed relative abundance of the r-th species ($r = 1, 2, 3$ . . . . s).

Working on the mesofauna of beech litter, Lloyd and Ghelardi (1964) recorded 632 individuals distributed among 44 species, mostly micro-arthropods. Now if all the individuals were equally distributed among these 44 species, diversity would be at a maximum and:

$$H_{max} = \log_2 (44) = 5.46$$

However, by applying the Shannon-Wiener function, the observed value for H was 4·16.

As noted above, H provides a measure of the general species diversity and has two components: the number of species and the distribution of the individuals present among these species, i.e. equitability. It is often useful to be able to separate out the equitability component and measure it. To do this it is necessary to provide a theoretical basis for comparison, and MacArthur's (1957) "broken-stick" model serves this purpose. This model distributes individuals among the species as equitably as would be possible in nature, and has the form:

$$\pi_r = \frac{1}{s} \sum_{i=1}^{r} 1/(s - i + 1)$$

where π_r is the theoretical proportion of individuals in the r-th most abundant species ($r = 1, 2, 3 \ldots \ldots s$). Each theoretical proportion is obtained by summing over r terms ($i = 1, 2, 3 \ldots \ldots r$). Applying the Shannon-Wiener function to this model, we can achieve an expression [$M(S)$] which approximates the ecological maximum diversity which can be attained among a given number of species. This expression is defined by the equation:

$$M(S) = - \sum_{r=1}^{s} \pi_r \log_2 \pi_r$$

In the example cited above, $M(S)_{44} = 4.89$; this is higher than the actual diversity obtained in the sample (4.16), but lower than that mathematically possible for 44 species (5.46).

We are now in a position to measure "equitability". Lloyd and Ghelardi (1964) suggest that one possible way to do this is to use the ratio between the observed H and the theoretical $M(S)$ which, in this example, would be $4.16/4.89 = 0.85$. This means that the observed species diversity is only about 85% of that which would be possible if all the individuals were equitably distributed among the 44 species. Another, probably more realistic, approach is "to calculate the number of hypothetical 'equitably distributed' species that would be needed to produce a species diversity equivalent to the observed one". This hypothetical number is denoted by the symbol S' and its information function by $M(S')$. The observed diversity (H) is set equal to $M(S')$ and the value for S' is calculated. In the example we have been pursuing, $H_{44} = 4.16$; the nearest theoretical equivalent to this is $M_{26} = 4.15$. Hence, the value for S' is 26. The index measuring the degree of equitability (ε) is then given by the ratio between S' and the observed number of species present ($S = 44$), and:

$$\varepsilon = \frac{S'}{S} = \frac{26}{44} = 0\cdot59$$

This tells us that the observed distribution of individuals among the species present is only 59% of what it could be if there had been a distribution with maximum equitability, in theoretical terms. In the latter event, no more than 26 species, instead of 44, would have been needed to produce the observed amount of diversity. The time-consuming problem of calculating $M(S')$ for corresponding values of S' has been simplified by Lloyd and Ghelardi (1964) who provide a table of values relating these two quantities. To use this table it is only necessary to calculate the Shannon-Wiener function H which can be re-stated in the following simplified form:

$$H = c \left\{ \log_{10} N - \frac{1}{N} \sum_{r=1}^{s} n_r \log_{10} n_r \right\}$$

where n_r = the numbers of the r-th species, N = the total number of individuals, S = the number of species, and c the conversion factor (3.321928) to change the base of logarithm from 10 to 2. The value for H obtained in this way is set equal to $M(S')$ and the appropriate value for S' is read off from the table. This is then divided by the observed number of species to give a value for ε.

Jaccard's Index

This index is calculated by comparing samples in pairs, noting the number of species common to both samples of a pair, dividing this by the number of species occurring in one or other (but not both) of the pair, and expressing the result as a percentage. If j is the number of species common to both samples, and r the number of species occurring in only one sample of the two, then:

$$\text{Jaccard's Index} = \frac{j}{r} \times 100$$

For example, samples of soil microarthropods were taken from two different habitats, limestone grassland and fenland at Malham, Yorkshire (see Chap. 6, Figs. 6.4 and 6.5). The number of species common to both habitats, based on the sample analyses, was found to be 3; the number of species occurring in only one habitat was 23. Thus, the faunal similarity between these two habitats, expressed as Jaccard's Index is: $3/23 \times 100 = 13\%$.

Clearly, the higher the index, the greater the faunal similarity between the habitats. Total similarity (Index $= 100\%$) is achieved when the number of species common to both is equal to the number restricted to one habitat. If several samples or habitats are compared in this way, the results may be expressed in a coincidence table (see later), which will show at a glance the percentage similarity between any paired combinations (see Balogh, 1958, for more examples of the application of this method).

An alternative, and probably more widely used index, again proposed by Jaccard (1902) for use in botanical surveys, compares the number of species common to two samples with the number of distinct species in the two samples combined, i.e. the number of species that would be recorded in a simple check-list if the two samples were considered as one. This index is expressed by the term:

$$\frac{j}{a + b - j}$$

where $j =$ the number of species common to both samples, $a =$ the number recorded in sample A, and $b =$ the number recorded in sample B. Here again, the index increases with increasing j, but in this case 100% similarity between the two samples is attained when all species are common to both. A variant of this method, proposed by Sørensen (1948), is used commonly in soil zoology.

Sørensen's Quotient of Similarity

This index (Q/S) is defined by the expression:

$$Q/S = \frac{2j}{(a+b)} \times 100$$

where j, a and b express the quantities already noted in the calculation of Jaccard's Index. Sørensen's quotient has the same properties as this index, namely that it increases with increasing j, and it expresses 100% similarity when all species are common to both samples. Haarløv (1960) has demonstrated the application of this quotient to the analysis of microarthropod groupings in various soil habitats in Denmark. This author calculated the similarities between sampling units taken from the same habitat ("internal" Q/S) as well as between different habitats, thus illustrating how this method may be used for intra-group, as well as inter-group, investigations. As a general procedure, internal Q/S values should be calculated whenever possible, since comparisons between different samples or habitats are valid only if the internal Q/S of a given sample or habitat is greater than that occurring when this habitat is compared with another.

Comparisons between natural groupings of species populations involving the quantities j, a and b (as defined above) should have two essential properties, namely that the index of similarity should increase with increasing j and decrease with increasing a and b, and should also be independent of sample size. The indices of Jaccard and Sørensen have the former characteristics but not the latter.

Mountford's Index

To combine these two desirable features, Mountford (1962) proposed an index of similarity (I) defined by the equation:

$$e^{aI} + e^{bI} = 1 + e^{(a + b -)}$$

In comparing the fauna of two sites, the values of j, a and b are inserted in the above equation, and I is obtained by interpolating within the table of exponentials, using the following expression as an approximation to I:

$$\frac{2j}{2ab - (a + b)j}$$

The author also provides a graphical method for determining I.

Mountford's Index is, in fact, the reciprocal of alpha (α), the Index of Diversity considered earlier, and its defining equation is developed mathematically by considering the two samples being compared as one joint sample. It therefore follows that an index of diversity can be proposed for this joint sample, and by substituting (I) for ($1/\alpha$), the defining relationship given above may be obtained. Assuming a logarithmic series species frequency distribution, (α), and consequently (I), should be independent of sample size when samples are taken within the same natural grouping. We have noted earlier that this may not be strictly true in the case of (α). Similarly, Mountford (1962) points out that (I) is not wholly independent of sample size when the samples are drawn from different natural groupings. However, it does provide a better index of similarity than do the methods of Jaccard and Sørensen (Fig. 2.4). In developing this approach further, Mountford (1962) devised an hierarchical method for classifying sites into groups,

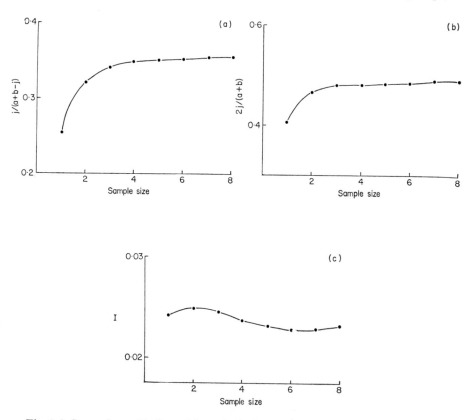

Fig. 2.4 Comparison of indices of Jaccard (a), Sørensen (b) and Mountford (c) to show the effect of increasing sample size. (After Mountford, 1962.)

based on their indices of similarity. Worked examples illustrating the use of this method have been provided by its author and also by Wallwork (1970). Essentially, the procedure involves selecting, from the coincidence table, that pair of sites with the highest index of similarity. These two sites are then grouped together and indices of similarity are calculated between this group and each of the remaining sites. A reduced coincidence table can now be written and, again, the highest index is selected, and the procedure of combining sites and re-evaluating indices is repeated. Eventually, the original sites can be classified into two groupings, within which the extent of similarity between individual sites is shown by the cluster arrangement. This scheme can be applied to botanical, as well as zoological, data.

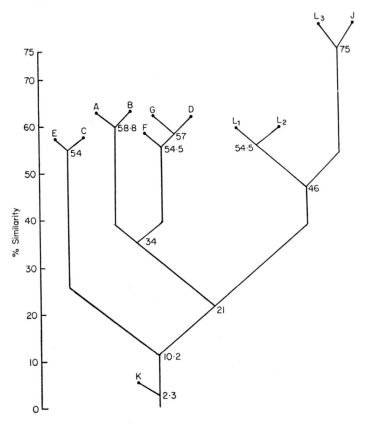

Fig. 2.5 Hierarchical classification of 12 Pennine moorland sites, according to the method of Mountford (1962), using Sørensen's Quotient of Similarity. Sites A and B: *Vaccinium* moor; C: wet fen; D: *Molinia* fen; E: tussock fen; F: fen carr; G: raised bog; J, L_1, L_2, L_3: alluvial grassland; K: grykes in limestone pavement.

The illustration provided in Fig. 2.5 is based on frequencies of plant species in quadrates taken from various sites in the Malham area, Yorkshire.

Mountford's scheme of classification provides an alternative to the "trellis diagram" approach which has been used commonly by soil ecologists for several years.

Trellis Diagrams

By presenting indices of similarity in the form of a coincidence table, we can see at a glance the extent of the similarities when all possible combinations of samples, habitats or species, are compared in pairs. It has been suggested that similarities become even more apparent when this tabulation

Table 2.3

Coincidence table giving the coefficients of floral similarity, calculated by the method of Sørensen, between 12 sites in the vicinity of Malham Tarn, Yorkshire. The sites, which are those identified in Fig. 2.5, are listed in the order in which they were sampled.

	A	B	C	D	E	F	G	J	K	L_1	L_2	L_3
A												
B	58·8											
C	8	0										
D	25	26·6	0									
E	19·4	6·6	54	0								
F	30·8	33·8	0	54·5	0							
G	37·5	53·3	0	57	0	54·5						
J	44·5	35·3	8	12·5	19·4	15·4	25					
K	8·3	11·8	0	0	0	10·5	0	0				
L_1	34·5	29·6	17·3	7·7	23·8	8·7	15·4	57·2	5·9			
L_2	25	8·7	20	9·5	16·3	11·1	9·5	43·5	0	54·5		
L_3	15·5	40	8·7	14·3	13·8	18·2	28·6	75	0	30·4	47·6	

is arranged in such a way that the groups with the highest affinities are brought close together. Such an arrangement is termed a "trellis diagram".

This method may be applied not only to indices, such as those of Jaccard and Sørensen, which reflect total similarity between samples or habitats, but also to estimates of the amount of association between pairs of species as determined, for example, by Fager's method (see above). As an illustration of this method, we can use the indices of similarity calculated for the construction of Fig. 2.5, and given in Table 2.3. In this table, the sites A, B, C, D etc. are arranged in a linear sequence in rows and in columns such that the intersects form the squares of a trellis in which the indices of similarity between pairs of sites are located. By re-arranging the linear order of the sites so that those with the highest affinity index are brought close together, it is possible to re-state the trellis diagram in the form shown in Fig. 2.6. It will be noted that the highest indices occur on or near the centre diagonal, and that the lowest are the more remote, in this form. Shading of the top right-hand half shows how the sites considered may be divided into natural groupings corresponding to the basic dichotomy shown in Fig. 2.5.

This is a simple example for the purposes of demonstration. More often, trellis diagrams are larger, with a greater number of sites, samples or species (see, for example, Franz, 1963), and the definition of natural groupings may be more difficult to make. Indeed, one of the criticisms levelled at this method is that the re-ordering of the linear sequence becomes somewhat arbitrary when large numbers of units are considered.

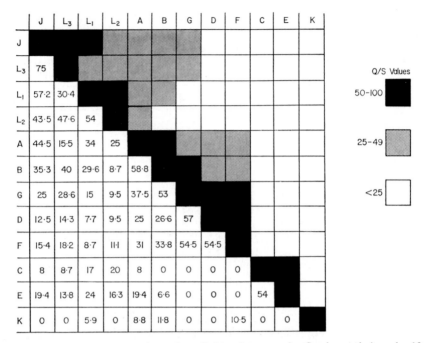

	J	L₃	L₁	L₂	A	B	G	D	F	C	E	K
J												
L₃	75											
L₁	57·2	30·4										
L₂	43·5	47·6	54									
A	44·5	15·5	34	25								
B	35·3	40	29·6	8·7	58·8							
G	25	28·6	15	9·5	37·5	53						
D	12·5	14·3	7·7	9·5	25	26·6	57					
F	15·4	18·2	8·7	11·1	31	33·8	54·5	54·5				
C	8	8·7	17	20	8	0	0	0	0			
E	19·4	13·8	24	16·3	19·4	6·6	0	0	0	54		
K	0	0	5·9	0	8·8	11·8	0	0	10·5	0	0	

Q/S Values

50–100

25–49

<25

Fig. 2.6 Trellis diagram to show the affinities between the floral associations in 12 Pennine sites near Malham Tarn, Yorkshire. *Note:* data presented in Table 2.3 are here re-arranged to bring together the sites with closest affinities. It is evident that a distinction can be made between sites (J, L₃, L₁, L₂, A, B, G, D, F) and sites (C, E). Within the former grouping, the cluster (J, L₃, L₁, L₂) can also be distinguished from (A, B, G, D, F). Site K shows little similarity to any other site.

Concluding remarks

The methods described in this chapter represent only a fraction of the effort that has gone into this question of delimiting natural groupings of species populations. The choice of examples has been determined, to a large extent, by the need to illustrate the different kinds of general approach used. The advantages and limitations of these various methods can now be summarized briefly.

Firstly, it must be apparent that while some methods are more efficient than others, no single one emerges as a complete answer to the problem of delimiting natural associations of soil organisms. There are several reasons for this. Some methods, such as those of Jaccard for example, have been adopted from vegetational analyses in which the distribution patterns and sampling techniques are often quite different from those of soil faunal investigations. Moreover, a method developed to suit one particular set of field data may not be completely satisfactory in dealing with a different set of data, even though these relate to the same taxonomic group of animals. This is illustrated by the work of Crossley and Bohnsack (1960) who applied Fager's method to their own data, but could not operate within the critical limits suggested by that author.

Secondly, a distinction has to be made between methods employing presence or absence (frequency of occurrence), and those using relative abundance as the criterion for comparing distribution patterns. Although several of the former (for example, those of Sørensen and Fager) have the necessary flexibility which permits their use both in intra-group and inter-group analyses, their disregard of relative abundance may lead to biological mis-interpretations. Thus, although a given species may occur in two natural groupings, its relative abundance (and biological importance) may vary in the two, and yet no provision is made for this differential in the calculation of the index of affinity. It is perhaps worth recalling the statement made by Hairston (1959) that " . . . there is no justification other than convenience for omitting any species from an analysis of a community. The importance of knowing the relative abundance of the species is shown both in the demonstration that it provides a valid method of showing the amount of organization present and in the comparison between the two communities, where the same species assume very different degrees of importance." Methods which take no account of species not present in both samples, localities or habitats are open to this criticism.

The use of relative abundance to determine an index of diversity moves closer to a solution of the problem, although difficulties are raised by the fact that such indices are not completely independent of sample size. No really satisfactory way of countering this effect has yet been devised, although Mountford's method demonstrates the greater reliability of this "diversity index" approach in inter-group analyses. Further, the suggestion that the distribution of common and rare species may not be included satisfactorily within the same mathematical model (Hairston and Byers, 1954) is worthy of consideration. The use of any method of comparison which involves the ratios of numbers of species to numbers of individuals has considerable biological significance, for a relationship may then be established between the structural and functional aspects of natural groupings. This idea has been developed by MacArthur (1957) who considered the natural grouping ("community") as consisting of a number of niches (represented by the species content) completely occupied by the total number of individuals present, such that the whole organization could be depicted as a series of contiguous, non-overlapping niches. Although the assumption of random distribution of niches in McArthur's model has been criticized (Hairston, 1959), the implication that food material (the environmental factor most

3

obviously not shared, but completely utilized) determines the species abundance is of great biological significance. This approach has been developed by Lloyd and Ghelardi (1964) who recognize that species diversity has an "equitability" component, a measurable quantity which reflects the extent to which the constituent species have numerical equality. This can be used to indicate the extent to which the observed species diversity realizes the maximum amount of ecological diversity permitted by the habitat. These approaches, which have been discussed in detail by Southwood (1966), remind us that the justification for any ecological model must be based firmly on its biological value, rather than on its mathematical appeal.

One final point must be made. Pre-occupation with analytical techniques should not obscure the possibility that one basic assumption may not be completely valid. This assumption is that natural groupings of species have an objective reality. The available evidence, at least as far as the soil fauna is concerned, suggests that the pattern of species distribution is a mosaic of overlapping ranges which makes it difficult to draw boundaries around natural groupings which have any ecological significance. It is now time to consider this evidence in more detail, and we will begin by considering patterns of distribution in grassland soils.

REFERENCES

AGRELL, I. (1963). A sociological analysis of soil Collembola, *Oikos*, **14**, 237–47.

ARCHEBALD, E. A. A. (1949). The specific character of plant communities. I. A quantitative approach, *J. Ecol.*, **37**, 274–88.

BALOGH, J. (1958). "Lebensgemeinschaften der Landtiere", Akademie-Verlag, Berlin.

BLOWER, J. G. (1970). The millipedes of a Cheshire wood, *J. Zool., Lond.*, **160**, 455–96.

BRAUN-BLANQUET, J. (1932). "Plant Sociology" (English edn., trans. Fuller, G. D. and Conrad, H. S.), McGraw-Hill, New York.

CROSSLEY, D. A. and BOHNSACK, K. K. (1960). Long-term ecological study in the Oak Ridge area. III. The oribatid mite fauna in pine litter, *Ecology*, **41**, 785–90.

DEBAUCHE, H. R. (1962). The structural analysis of animal communities of the soil. *In* "Progress in Soil Zoology", 10–25 (Murphy. P. W., ed.), Butterworth, London.

FAGER, E. W. (1957). Determination and analysis of recurrent groups, *Ecology*, **38**, 586–95.

FISHER, R. A., CORBET, A. S. and WILLIAMS, C. B. (1943). The relation between the number of species and the number of individuals in a random sample of an animal population, *J. Anim. Ecol.*, **12**, 42–58.

FRANZ, H. (1963). Biozönotische und synökologische Untersuchungen über die Bodenfauna und ihre Beziehungen zur Mikro- und Makrofauna. *In* "Soil Organisms", 345–67 (Docksen, J. and van der Drift, J., eds.), North Holland, Amsterdam.

GABBUTT, P. D. (1967). Quantitative sampling of the pseudoscorpion *Chthonius ischnocheles* from beech litter, *J. Zool., Lond.*, **151**, 469–78.

GOODALL, D. W. (1953). Objective methods for the classification of vegetation, *Aust. J. Bot.*, **1**, 434–56.

GREIG-SMITH, P. (1964). "Quantitative Plant Ecology" (2nd end.), Butterworths, London.

HAARLØV, N. (1960). Microarthropods from Danish soils. Ecology, phenology, *Oikos*, *suppl.* **3**, 176 pp.

HAIRSTON, N. G. (1959). Species abundance and community organisation, *Ecology*, **40**, 404–16.

HAIRSTON, N. G. and BYERS, G. W. (1954). The soil arthropods of a field in southern Michigan: a study in community ecology, *Contrib. Lab. Vert. Biol. Univ. Mich.*, **46**, 1–37.

HALE, W. G. (1967). Collembola. *In* "Soil Biology", 397–411 (Burges, N. and Raw, F., eds.), Academic Press, London and New York.

HEAL, O. W. (1962). The abundance and microdistribution of testate Amoebae (Rhizopoda, Testacea) in *Sphagnum*, *Oikos*, **13**, 35–47.

IWAO, S. (1972). Application of the m̂-m method to the analysis of spatial patterns by changing the quadrat size, *Res. Popul. Ecol.*, **14**, 97–128.

JACCARD, P. (1902). Lois de distribution florale dans la zone alpine, *Bull. Soc. vaud. Sci. nat.*, **38**, 69–130.

KERSHAW, K. A. (1964). "Quantitative and Dynamic Ecology", Arnold, London.

KONTKANEN, P. (1950). Quantitative and seasonal studies on the leafhopper fauna of the field stratum on open areas in North Karelia, *Ann. Zool. Soc. "Vanamo"*, **13**, 1–91.

LLOYD, M. (1967). Mean crowding, *J. Anim. Ecol.*, **36**, 1–30.

LLOYD, M. and GHELARDI, R. J. (1964). A table for calculating the "equitability" component of species diversity, *J. Anim. Ecol.*, **33**, 217–25.

MACARTHUR, R. H. (1957). On the relative abundance of bird species, *Proc. natn. Acad. Sci. U.S.A.*, 293–5.

MACARTHUR, R. H. (1965). Patterns of species diversity, *Biol. Rev.*, **40**, 510–33.

MACFADYEN, A. (1963). "Animal Ecology. Aims and Methods" (2nd. edn.), Pitman, London.

MARGALEF, R. (1957). La teoria de la informacion en ecologia, *Mem. R. Acad. Barcelona*, **32**, 373–449.

MORISITA, M. (1962). Iσ – index, a measure of dispersion of individuals, *Res. Popul. Ecol.*, **4**, 1–7.

MORISITA, M. (1971). Composition of the Iσ – index, *Res. Popul. Ecol.*, **13**, 1–27.

MOUNTFORD, M. D. (1962). An index of similarity and its application to classificatory problems. *In* "Progress in Soil Zoology", 43–50 (Murphy, P. W., ed.), Butterworths, London.

MURPHY, P. W. (1962). "Progress in Soil Zoology", Butterworths, London.

NIELSEN, C. O. (1949). Studies on the soil microfauna II. The soil-inhabiting nematodes, *Natura jutl.*, **2**, 1–131.

O'CONNOR, F. B. (1957). An ecological study of the enchytraeid worm population of a coniferous forest soil, *Oikos*, **8**, 161–99.

PIELOU, E. C. (1969). "An Introduction to Mathematical Ecology", Wiley-Interscience New York.

PRESTON, F. W. (1948). The commonness and rarity of species, *Ecology*, **29**, 254–83.

REYNOLDSON, T. B. (1955). Observations on the earthworms of North Wales, *NWest. Nat.*, **3**, 291–304.

SMITH, R. L. (1966). "Ecology and Field Biology", Harper and Row, New York, London.

SØRENSEN, T. (1948). A method of establishing groups of equal amplitude in plant sociology based on the similarity of species content and its application to analyses of the vegetation on Danish commons, *Biol. Skr.*, **5**, 1–34.

SOUTHWOOD, T. R. E. (1966). "Ecological Methods", Methuen, London.

TISCHLER, W. (1949). "Grundzüge der terristrischen Tierökologie", Braunschweig.

USHER, M. B. (1969). Some properties of the aggregations of soil arthropods: Collembola, *J. Anim. Ecol.*, **38**, 607–22.

USHER, M. B. (1971). Seasonal and vertical distribution of a population of soil arthropods: Mesostigmata, *Pedobiologia*, **11**, 27–39.

WALLWORK, J. A. (1959). The distribution and dynamics of some forest soil mites, *Ecology*, **40**, 557–63.

WALLWORK, J. A. (1970). "The Ecology of Soil Animals", McGraw-Hill, London.

WALLWORK, J. A. (1972). Distribution patterns and population dynamics of the micro-arthropods of a desert soil in southern California, *J. Anim. Ecol.*, **41**, 291–310.

WHITTAKER, R. H. and FAIRBANKS, C. W. (1958). A study of plankton copepod communities in the Columbia basin, south-eastern Washington, *Ecology*, **39**, 46–65.

WILLIAMS, C. B. (1964). "Patterns in the Balance of Nature", Academic Press, London, New York and San Francisco.

Chapter 3

Grassland Habitats:
the Soil Fauna and its Food

The term "grassland" covers a broad spectrum of habitat types, within which can be included such diverse forms as heathland and hay meadows, savannah and fen, steppe and prairie. The reason for such diversity lies in the fact that no one single agency is responsible for the formation of grasslands on a world-wide scale. Again, permanent pastures may be dominated by plant species other than grasses. Such sites, which include moorland, heath and fen, have their own special features, as we will see later.

Grassland, defined in its broadest sense, comprises about 17% of the earth's land surface and is responsible for about 15% of the total world net primary production of organic matter (see Whittaker, 1970; Balogh, 1970). In these respects it has a lower ranking than tropical and temperate forests, but it has a special significance in providing habitats in which large herbivorous animals are most successful, and in which the short food chain:

Grass ⟶ Animal

provides a rapid production of protein. It is perhaps no coincidence that most of the great human civilizations have had their origins in this zone, where food material in the form of animal protein can be produced so quickly.

The rate of energy turn-over in grassland soil is high when compared, for example, with that of a forest. This means that organic material, derived from herbaceous vegetation, is broken down very quickly by large saprophages, such as earthworms, molluscs, millipedes and woodlice, and by microfloral decomposers, such as bacteria. Small-sized saprophages, such as mites, collembolans, nematodes and enchytraeids, are evidently of lesser significance in this process. However, in "managed" systems where fertilizers are added to the soil to promote plant growth, these smaller animals become much more important and their contribution to energy turn-over may exceed that of the larger forms (Macfadyen, 1963).

Types of grassland

Natural and Managed Grasslands

Natural grasslands can be regarded as climatic climaxes, developed under natural conditions in regions where the annual rainfall is insufficient to support forest growth, but adequate enough to prevent the development of desert. Such vegetation is to be found in parts of most of the major land masses, for example in North America (prairie), South America (pampas and llano), tropical Africa (savannah) and Eurasia (steppe). In contrast, managed grasslands are produced when interference by man causes natural vegetational succession to be arrested before the climax (usually woodland or forest) is reached. Such interference may involve intensive grazing of domestic animals, clearing and burning to remove woodland, draining of marshes and fens, and subsequent ploughing, fertilizing and seeding to promote the growth of grasses.

Meadows and Rough Pastures

Included within the category of managed grasslands is the cultivated meadow, or ley. This type is often artifically seeded with *Poa pratensis* in Europe and North America to produce a grass crop which is subsequently removed from the site as hay. Such meadows are frequently treated with chemical fertilizers and insecticides to improve crop yield. In contrast, we can also recognize "natural" meadows, often supporting grass-herb associations which, because of certain properties such as high silica content, are not subject to heavy grazing pressure or harvesting. This tall vegetation usually develops a tussock form of growth, particularly when such grasses as *Deschampsia flexuosa*, *Dactylis glomerata* and *Molinia caerulia* are present.

Rough pastures are "natural" in the sense that they are subject to the minimum of human interference; a natural colonization of grasses is permitted, and a large proportion of the crop which is removed by herbivorous vertebrates is returned to the soil in a partly decomposed form as faeces. In this category belong also moorland, heath and fen sites where *Calluna* and *Vaccinium* are grazed by rabbits, sheep and grouse. These are examples of permanent pastures which are common in northern Europe. In other parts of the world, different kinds of permanent pastures occur, and these demonstrate quite clearly the wide variability inherent in this habitat type.

Permanent Pastures

The variability of permanent pastures on a world-wide scale is a reflection of the fact that different environmental factors are responsible for their existence. In Europe, many permanent pastures are maintained as grasslands only by the activities of grazing herbivores, notably sheep and cattle. In the absence of these herbivores, such grasslands would revert to thicket or woodland in all probability. On the other hand, the prairies of North America and the steppes of Eurasia may be controlled by climatic factors. Tropical savannah grasslands are maintained, for the most part, by extensive and regular burning, as are some fen grasslands in Britain. Both of these types often contain species of woody plants and, indeed, would be replaced by a woodland or forest climax if their natural succession were allowed to proceed. It is important to recognize that permanent pastures are only as permanent as the environmental factors which control them. As these factors change with time, whether as a result of long-term climatic fluctuations or changing patterns of land use, so will they alter the characteristic features and distribution patterns of permanent pastures. The operation of the time factor is well illustrated by the dynamic character of grasslands which develop as intermediate stages in a vegetational succession which originates, for example, from a hydrosere or from abandoned arable land. In view of these ecological and historical variations, it is hardly surprising that few generalizations can be made about the grassland type or its soil fauna. This is also true of the soils themselves.

Grassland soils

There is no one type of soil that is typical of all grasslands. Variations occur from place to place in the colour, texture, moisture content, pH, mineral and organic content of the soils which support grassy vegetation. In addition, it must be remembered that grasslands may develop on disturbed sites where natural succession has been halted or diverted, and the soils occurring here may owe their character, very largely, to formative processes which are no longer present. These variables influence the character and distribution of the soil fauna, and we must consider them briefly before going on to look at this fauna in detail.

Prairie Soils

These are fertile soils, conforming to the loam type, which are common in the Great Plains area of the American mid-West and in southern Russia. They are developed from unconsolidated deposits of silt and clay of Pleistocene age. The profile is moderately well drained, with a good crumb structure and a mull type of humus formation. This organic material is dispersed through the profile to an appreciable depth and finely mixed with mineral soil. This dispersion is effected, partly, by the action of rainwater washing

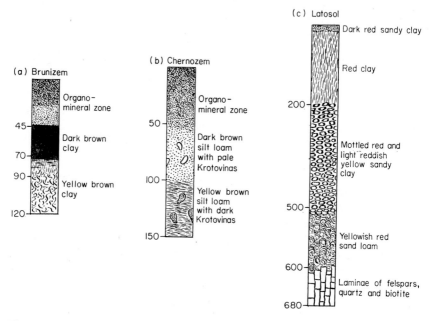

Fig. 3.1 Three grassland soil profiles. (a) Brunizem, (b) chernozem, (c) latosol. All depths are in cm. (After FitzPatrick, 1971.)

organic material and clay from the surface down the profile. It is also the result of the feeding and burrowing activities of some of the larger soil animals, such as earthworms and molluscs.

The term "prairie soil" covers a number of types, of which the brown soils, the brunizems (Fig. 3.1a) and burozems, are the most common. The brunizems are highly fertile soils occurring in rather humid areas. In North America, these soils support tall grasses such as big bluestem *(Andropogon furcatus)* at the present time. In the past, they may have been the sites for spruce and hardwood forests and, indeed, support forest vegetation, mainly oak, in Europe today. These soils are characterized by a low C/N ratio (10–12) and are often rich in exchangeable calcium.

Burozems develop under more arid conditions and support short grass vegetation. In contrast to the brunizems, which are slightly acid soils, burozems are neutral or slightly alkaline. The organic material in the profile is well humified, with a low C/N ratio, and the base saturation is high, with calcium and magnesium predominating. At the dry end of their range, these soils support various desert plants, such as succulents and cacti.

Chernozems

These are the black earth soils which develop in rather dry areas where the summers are hot and the winters cold (Fig. 3.1b). They are characterized by the black colouration of the profile which is due in part to the extensive distribution of organic material and, in part, to the black montmorillonite clay which develops from the parent material (loess). Loess is a wind-borne deposit which imparts a fine texture to the soil, and it is the commonest type of parent material from which a chernozem develops. These soils often have calcium carbonate accumulations below the dark-coloured zone, and the pH becomes more alkaline with depth. Leaching of the surface layers may render these rather acid (pH 5·5), but at a depth of 100 cm the pH rises to 8·0 and over. Organic material is decomposed and incorporated into the soil rapidly, and the resulting profile is well aerated and has a large water-holding capacity. Decomposition of organic material is brought about, in some chernozems, by the activities of earthworms, in others by the burrowing of small vertebrates, such as the mole rat *(Spalax typhluon)* and gopher *(Citellus suslyka)*. The filled-in burrows of these animals, known as krotovinas, are a characteristic feature of chernozem profiles. This type of soil is colonized by tall grasses, almost exclusively, and occurs as steppe grassland in south-western Russia and in parts of the American mid-West.

Chestnut Soils

These grassland soils develop in the drier parts of the American mid-West and southern Russia, and contrast with the chernozems, with which they sometimes intergrade, in several respects. Because of the scarcity of moisture, the profile is not subject to severe leaching. Lack of water also restricts the production of organic material, and that which is incorporated into the soil is rapidly broken down, giving the profile a dark brown colouration. Krotovinas occur in these soils, but they are less abundant than in the chernozems. At the surface of the profile, the soil crumbs are small and unstable, and wind erosion often occurs. The grass vegetation which colonizes chestnut soils has a shorter rooting system than that occurring on chernozems, and annual production is lower. As a consequence, the distribution of organic material down the profile is more restricted, and the zone of calcium carbonate deposition is nearer the surface.

The three soil categories considered above represent the main classes of

3*

soils which support natural grassland. However, they are not the only soils on which grassy vegetation occurs. Soil type is influenced by a number of variables, of which vegetation is only one. It is not uncommon to find the same kind of soil profile supporting forest in one place or at one time, and grassland in another place or another time. Examples of such soil types are the rendzinas, podzols and laterites.

Rendzina

These are shallow soils developed where parent material is rich in calcium carbonate. They are neutral or slightly alkaline in reaction and often occur on chalk (Fig. 3.2), as in the Chiltern Hills of southern England, or on lime-

Fig. 3.2 Soil profile of the rendzina type in chalk grassland, Chiltern Hills, Buckinghamshire. Note the sharp distinction between the upper, dark-coloured zone of mull humus (10 cm thick), and the lower, light-coloured zone of parent chalk. (Photo. by author.)

stone, as in parts of the Pennines. The profile usually shows a distinct demarcation between the deeper, unweathered parent material and the upper, dark-coloured, unstructured zone in which mull humus is intimately mixed with weathered parent material. In arid regions, calcium carbonate may be concentrated near the surface of the profile, although some of this may be leached during the rainy season, and redeposited lower down the profile. Since the characteristics of rendzina soils are determined, for the most part, by the nature of the parent material, they may occur under a variety of climatic conditions. They are particularly common in the Mediterranean region. Because of their high fertility, they are often cultivated for crops, pasture and forests, although the type of vegetation they can support is often limited by their lack of depth.

Podzol

In contrast to the rendzina and chernozem, podzols are acidic soils which develop on alluvial deposits of sand, silt or clay. These soils occur in regions of moderate to heavy rainfall, and are well drained. Rainwater percolating through the profile leaches out soluble bases from the upper layers, leaving an ash-coloured horizon, above which is a distinct zone of mor humus with a relatively high C/N ratio (25–30). Soluble iron and aluminium compounds and humus material, in particular, are translocated down the profile and are redeposited, often as distinct bands, at greater depths. Such acid grassland soils are common on glacial drift in northern Britain where they often occur alongside the wetter, peaty soils of moorland. They also occur in the tropics where they may intergrade with lateritic soils (see below), but they are most strongly developed under vegetation which produces an acid layer of leaf litter, such as conifer forest and heather moorland.

Laterite

The term "laterite" is used to cover a range of soil types occurring in tropical grassland and forest areas. The strictly limited rainfall and high temperatures of the tropics produce soils with a low humus content (Fig. 3.1c). During the rainy season, soluble bases are leached out; during the dry season iron and aluminium compounds are oxidized and give the soil a characteristic red or yellow colour. These soils are more correctly called latosols, or feral-litic soils, rather than laterite, since this term has been mis-interpreted over the years.

Latosols are extensively developed in West Africa, the central part of South America, southern India and south-east Asia. Where rainfall is sufficient, tropical forests are developed on these soils, although they also support thicket or savannah vegetation, where the forest has been cleared, or where conditions of lower rainfall prevail. There are many variants of the latosol profile, and local conditions, such as moisture content and the nature of the parent material, may have a strong influence on soil type. However, in tropical grasslands the most commonly-occurring variants are the ochro-sol and the oxysol. These two types are distinguished by the amount of leaching to which the profile is subjected. Oxysols occur in regions of high rainfall and, during the wet season are strongly leached. As a result, these soils are poor in bases, and the profile is rather acid throughout. Ochrosols occur in drier areas and are subject to less intense leaching. The surface horizon is neutral or only slightly acid, although acidity increases with depth in the profile.

Tropical Black Earths

In much of India, central Australia and East Africa, tropical grasslands develop on soils which are quite distinct from the latosols we have just been discussing. These are the tropical black earths, or vertisols, typically dark in colour, fine in texture and with a low organic content. Like latosols, the black earths are clay soils, but in the former kaolinite is the predominant clay fraction, whereas montmorillonite is the main constituent of the latter. They are formed mainly from alluvial or lacustrine deposits in arid and semi-arid regions where marked wet and dry seasons occur. When dry, black earths often crack to a depth of 1 m, and may experience strong erosion. Their nutrient status is low, and their characteristic vegetation consists of tall grasses.

Faunal characteristics of various grassland soils

The types of grassland we have been considering differ not only in their vegetational composition but also, on a broad scale, in their faunal character. These faunal differences are reflections of geographical and ecological factors. The former can be illustrated by comparing the faunal characteristics of three contrasting grassland habitats, namely temperate grassland, semi-arid steppe and tropical savannah. The influence of ecological factors on the grassland soil fauna will be considered later in this chapter and in the next.

Temperate Grassland

The most conspicuous component of the fauna of neutral or slightly alkaline grassland soils in the north temperate region are earthworms of the family Lumbricidae. Densities of the order of 500–700 individuals per m^2 have been recorded in Europe (Satchell, 1967). It is generally believed that this group of soil animals makes a major contribution to the decomposition of organic material and to the dispersion of this material through the soil profile in the form of organo-mineral complexes. However, the high density and biomass of earthworms are offset, to some extent, by their low metabolic rates. Their contribution to energy flow through the soil/litter subsystem is rather less than their abundance would suggest (Satchell, 1967).

On the other hand, the burrowing activities of *Lumbricus terrestris*, *Allolobophora nocturna*, *A. longa* and *Octolasion cyaneum* undoubtedly promote good soil aeration and drainage and a wide distribution of organic matter in the profile. In acid soils, earthworm numbers are lower, and surface-dwelling species such as *Bimastos eiseni* and members of the genus *Dendrobaena*, which are tolerant of low pH conditions, predominate.

The wide distribution and abundance of earthworms in grassland soils often throw into low relief the ecological significance of other groups, notably the detritus-feeding arthropods. However, it is worth noting that arthropods flourish in grassland soils, and their activities complement those of the earthworms in the breakdown of organic litter. In a comprehensive survey carried out in a Cambridgeshire pasture by Salt *et al.* (1948), a total of 19 Orders of insects, arachnids and myriapods was recorded in densities ranging from hundreds per m², in the case of the larger insects and myriapods, to more than 100,000 per m², in the case of the mites. Conspicuous among the detritus-feeding macroarthropods in grassland soils are various carabid beetles,

Fig. 3.3 The woodlouse *Armadillidium vulgare* (Latreille) from chalk grassland in the Chiltern Hills. (Photo. by A. Newman-Coburn.)

ants, woodlice such as *Armadillidium vulgare* (Fig. 3.3), larval Diptera and Coleoptera, and the pill millipede *Glomeris marginata* (Fig. 3.4). The importance of these groups of animals should not be under-rated. For example, Nielsen (quoted in Elton, 1966) found that *Glomeris marginata* consumed almost the entire quantity of leaf litter produced in a *Brachypodium* grassland

Fig. 3.4 The pill millipede *Glomeris marginata* (Villers) collected from the same site as *A. vulgare* illustrated in Fig. 3.3. (Photo. by A. Newman-Coburn.)

Fig. 3.5 A crab spider belonging to the genus *Xysticus* from chalk grassland (length: 5 mm). (Photo. by A. Newman-Coburn.)

in the Wytham area of Berkshire. Predators are also well represented in this community, frequently by a variety of arachnids, such as spiders and harvestmen (Figs. 3.5–6).

Fig. 3.6 The harvestman *Homalenotus quadridentatus* (Cuvier) from chalk grassland (length: 5 mm). (Photo. by A. Newman-Coburn.)

Steppe Grassland

The steppe habitat varies in its character quite considerably from place to place, and may grade into savannah (for example, in South America and South Africa), or into forest steppe, on the one hand, and semi-desert on the other, as in southern Russia. Wherever it occurs, steppe grassland experiences marked seasonal fluctuations in microclimate, and for at least a part of the year the surface layers of the soil are subject to drought.

Ghilarov (1968) has noted that Russian steppe soils often have a well-developed humus horizon of considerable thickness. This author considers that the deep distribution of humus is a direct result of the continued activity of the soil fauna deep in the profile during the summer season. The information available indicates that this fauna is strongly arthropodan in character, with Coleoptera, Collembola and Acari numerically dominant. Alejnikova (1965) provided data on the faunal composition of forest, meadow and pasture soils in the Middle Volga region of Russia, among which were included a range of black earth sites, in steppe forest and steppe grassland.

These data point to the relatively low densities of lumbricid earthworms compared with coleopteran Carabidae and Curculionidae (ground beetles and weevils, respectively). These two insect families, together with the Elateridae, evidently flourish in a variety of steppe soils. Further, Collembola averaged densities of 22,138 per m² and are generally less abundant than the mites (Acari), averaging 32,900 per m² in black earths. In general, the steppe habitat supports an impoverished soil fauna, compared with many other grasslands in temperate regions, and this paucity is more pronounced in rough pasture than in meadow steppe which, in turn, supports a poorer and less varied fauna than the soil of forest steppe.

The character of the steppe grasslands in South Africa and South America evidently differs markedly from that of southern Russia, particularly in that the organic content is much lower. Loots and Ryke (1967) stress the organic content of the soil as an important factor in determining the composition of the fauna. These workers investigated several steppe and savannah soils in South Africa and noted, with regard to the mites, that the ratio of Cryptostigmata to Prostigmata was low in sites where the organic content of the soil was low, and high when the reverse was the case; their results are given

Table 3.1

The ratio of Cryptostigmata:Prostigmata in relation to soil organic matter content in various steppe habitats in South Africa (from Loots and Ryke, 1967).

Habitat	Ratio: Cryptostigmata : Prostigmata	% Organic Matter
Acacia karoo (thorn-tree community)	0·161	13·7
Themeda grassland	0·185	5·3
Arable soil under maize	0·023	7·9
Bare ground on river bank	0·333	6·3
Salix capensis (willow community on river bank)	1·300	36·0
Populus deltoides (poplar community on river bank)	1·200	39·8

in Table 3.1. The preference shown by Cryptostigmata for highly organic soils may be a reflection of their preferred habit of feeding on decaying organic material and fungi. Little is known of the feeding habits of the small-sized Prostigmata which are common in mineral grassland soils, although it is possible that they feed mainly on bacteria and Protozoa which are plentiful in these sites.

The organic content of the soil influences its water-holding capacity, and highly organic soils tend to be more moist than mineral soils. Many crypto-stigmatid mites are susceptible to desiccation, and the group as a whole is poorly represented in dry soils. di Castri (1963) demonstrated that the ratio of Cryptostigmata to Prostigmata was low in such sites in Chile. In the steppes of southern Russia, the black earth soils have a high organic content, and in these the Cryptostigmata predominate over other groups of mites. Clearly, the broad category of steppe grassland covers a range of environmental conditions, and there are considerable difficulties in formulating definitions of species associations which characterize the steppe habitat in general.

Savannah

Prominent among the soil fauna of tropical grasslands are earthworms and arthropods. As we have seen, these two groups also dominate the fauna of temperate grassland soils. The similarity stops here, however, for the earthworms and arthropods found in savannah soils are quite different, taxonomically and ecologically, from those encountered in temperate regions.

The sharply defined wet and dry seasons in the tropics evidently eliminate those groups of soil animals which either cannot tolerate, or are not able to avoid, a period of extreme drought. During this period, the ferallitic soils which contain no appreciable amount of humus, become hard and compacted. Earthworms of the family Lumbricidae are often replaced by larger forms, such as the Microchaetidae in South Africa and the Megascolecidae in Australia, which live more or less permanently at greater depths in the soil. The deep-burrowing habits of these worms make them relatively independent of climatic conditions at the surface, although they may emerge from their

Table 3.2

A comparison between the earthworm fauna of temperate and tropical grasslands, on the basis of population sizes and surface casting activity.

Habitat	Family	Mean Density	Mean Biomass	Mean Weight of Surface Casts
		Nos./acre	kg/acre	kg/acre
Temperate grassland in Britain	Lumbricidae[1]	2×10^6	450	11,000
Shaded tropical grassland in West Africa	Eudrilidae[2]	$1 \cdot 2 \times 10^5$	50	70,000
Sub-tropical grassland in South Africa	Microchaetidae[3]	3×10^5	315	20,000–100,000

[1] Evans (1948).   [2] Based on data from Madge (1969).   [3] Ljungström and Reinecke (1969).

burrows during the rainy season. A similar pattern of seasonal activity occurs in the eudrilids, *Hyperiodrilus africanus* and *Eudrilus eugeniae* in Nigeria, which produce surface casts during the wet season but not during the dry (Madge, 1969).

Population densities of earthworms in tropical grasslands are rather lower than in temperate sites, but the amount of surface castings is considerably higher (Table 3.2). For example, Madge (1969) calculated that surface castings by *Hyperiodrilus africanus* in shaded grassland in Nigeria could produce a surface layer 1·5 to 2 cm thick each year. This compares with a surface deposit of 0·25 cm thick in temperate grasslands (Evans, 1948). Further, there is some evidence that the earthworms, as a group, can be at least as active in tropical forest soils as in grasslands (Nye, 1955). This extension of activity from grassland to forest is much less marked in the lumbricids of temperate soils.

The arthropod fauna of savannah soils consists mainly of the relatively large-sized termites and ants. The microarthropods are poorly represented, compared with temperate grassland (Strickland, 1947; Salt, 1952, 1955; Belfield, 1956, 1967; Burnett, 1968). Estimates of the size of arthropod populations in tropical grassland soils range from about 7,000 to 87,000 per m², depending on the site, the intensity of grazing and the period of the year when the samples were taken. In temperate grasslands, the density of mites and Collembola alone often exceed 100,000 per m² (see Wood, 1966). In savannah soils, as in temperate grasslands, mites and Collembola are often numerically dominant. However, some of the other groups of arthropods occurring in temperate soils are either absent or present in very low numbers in savannah; these include millipedes, centipedes, Protura, pseudoscorpions, spiders, harvestmen and beetles.

The general distribution of termites in the tropics parallels that of the earthworms, to some extent. The surface mounds of these insects are conspicuous in open savannah, but they are also encountered in tropical forest, particularly where this has been partly cleared (see Chapter 8). The wide distribution of termites in tropical grasslands, and their habit of transporting underground fragments of vegetation, have often prompted the opinion that these insects have a role in the soil/litter sub-system very similar to that of earthworms in temperate grasslands. This implies that termites replace earthworms in the community structure in tropical grasslands. As we have seen, this is not strictly true, since earthworms do occur in such areas. It is true that some termites, such as members of the genus *Macrotermes* in West Africa, which construct large, tower-like mounds each weighing several tons, may produce surface castings at the rate of about 500 kg/acre each year. At any one time, the amount of deposited material may compare favourably with that produced by earthworms, and may be measured in

thousands, even tens of thousands, of kilograms per acre (Nye, 1955). However, these deposits have a low organic content, unlike those of earthworms
and, as Harris (1955) has pointed out, the digestive processes of the termites
are so efficient, and the utilization of ingested food material so complete,
that they make little contribution to the liberation of nutrients into the soil.
The ants, which also form an important component of the savannah ground
fauna, resemble the termites in this respect, although they are probably
responsible for less extensive surface deposits.

The surface fauna of savannah soils includes a number of larger-sized
predators not found in temperate grasslands, such as scorpions and solifugids. These arachnids are nocturnal or crepuscular in habit, living during
the day in soil burrows or moist crevices under stones, and emerging at dusk
to feed on small surface-dwelling insects and other arachnids. Scorpions
and solifugids are normally associated with arid conditions, and there will
be an opportunity to discuss them further in a later chapter devoted to the
desert fauna. However, it is perhaps relevant to mention here that some
scorpions, such as the giant West African *Pandinus imperator*, are distributed
both in open grassland and in wooded localities. Here is yet another example
of the way in which the fauna of grassland and forest overlaps in tropical
regions.

Ecological factors: food

In the preceding pages we have looked at the general character of the grassland habitat and its soil fauna. It will be evident that there are distinct
differences between the fauna in different regions of the world. These differences may be due in part to geographical factors which have provided
barriers to the dispersal of certain groups of animals. They may also be due
to ecological factors which have prevented the establishment of similar
faunas in all types of grassland. These ecological factors operate in a more
restricted sense to produce faunal heterogeneity within a particular locality,
be it temperate, semi-arid or tropical in character. We now go on to consider the effects of some of these factors on the composition and distribution
patterns of the soil fauna. The remainder of this chapter is devoted to a treatment of the biotic interactions associated with food resources. In the next
chapter, the influence of abiotic factors is considered.

The food factor can be approached from several different standpoints.
Grasslands vary from place to place in their vegetational composition, and
the kind of plant material and organic litter may influence the kinds of animals, herbivores and saprophages, which feed directly on this material.
Dung pats deposited by large grazing herbivorous mammals are a common
feature of many temperate grasslands. These deposits have their own

particular associations of animals in which members of the soil fauna participate. Many of these animals feed directly on the dung, others on the bacteria which also flourish here. Finally, predator/prey interactions must be considered, for these shed light on the nature of the grassland ecosystem.

Primary Consumers: Herbivores and Saprophages

Relatively few members of the soil fauna can be regarded strictly as herbivores, that is, feeders on living plant material. The plant-parasitic nematodes, slugs, snails, certain groups of ants and root-feeding insect larvae belonging to the Diptera and Coleoptera fall into this category. It is true that these members of the soil fauna are more strongly represented in open formation types than in woodland and forest. However, the influence of food supply on their distribution is most clearly demonstrated in cultivated soils, and they will be discussed in Chapter 5 when the effects of agricultural practice are dealt with.

The majority of soil animals which ingest plant material feed on litter formed from leaves and stems which die back, and root material which decays *in situ*. In grasslands an appreciable proportion of organic material entering the soil is produced by the decomposition of root systems. Roots, and the channels they create in the soil form a particular kind of microhabitat termed the rhizosphere, with its own associated fauna. We look first at this fauna before going on to deal with that of the surface litter and the humus layers.

The Rhizosphere. The roots of grasses penetrate to various depths in the soil. As they descend they pass through soil which consists mainly of organo-mineral complexes, and often extend into a zone which consists mainly of mineral particles derived from the parent material. These roots can be regarded as vertical extensions of the organic horizons, and the animals which feed on them or their products are thereby able to establish themselves in the deeper layers of the profile.

The rhizosphere is an ill-defined zone although it can be described, in general terms, as the area around a root which is influenced by the presence of the root. Active roots excrete carbon dioxide, and produce various exudates such as amino acids and the growth factors aneurin and biotin (Russell, 1961). Dead tissue is sloughed off from time to time, and the amount of this material increases as the root ages and becomes senescent. These substances stimulate the growth of populations of bacteria, fungi and Actinomycetes. These micro-organisms occur in significantly higher densities in the rhizosphere than in soil where there is no active root growth (Rouatt *et al.*, 1960). The effect of the rhizosphere on the soil fauna is not well documented, although it has been shown that Protozoa are twice as numerous here as in soil where there is no rhizosphere effect (Rouatt *et al.*, 1960). Protozoa are bacterial-feeders, in the main, and their aggregation in the rhizosphere may be seen as a response to the presence of food. The same may be true of many nematodes which feed on bacteria, although certain members of this animal group are parasitic on the roots themselves (see Chapter 5). Other root-feeders are found among the larvae of various Coleoptera and Diptera, and there is some evidence (see Kühnelt, 1961) that these juvenile insects may be attracted to the rhizosphere by a positive behavioural response to the higher concentration of carbon dioxide occurring here. Collembola have also been shown to respond positively to higher than normal levels of carbon dioxide under experimental conditions (Moursi, 1962).

The community of micro-organisms and animals associated with the rhizosphere is not a static one. Roots undergo a succession of changes from a young, healthy condition, through maturity to senescence and, eventually, disease and death. The rhizosphere community undergoes a parallel succession. In the early stages of root development, this community is characterized by a predominance of specialized root feeders and by micro-organisms which feed exclusively on the exudates produced by the young roots, for example Gram-negative, non-sporing bacteria. As the root begins to decay, less specialized forms take over, such as saprophytic fungi, nematodes, Collembola and other members of the soil fauna which feed on decaying root tissue. This succession of organisms in which the specialists are supplanted by forms which occur more widely throughout the soil has its parallels in the communities associated with decaying wood, as we will see in Chapter 9.

Carbon/Nitrogen Ratios and Palatability. The suitability of organic litter as food for saprophagous soil animals depends on a number of factors. The source of the material is important, i.e. whether it is derived from leaves, stems or roots. In the case of leaf material, a distinction can be made between "soft" and "hard" leaves, the former originating from sheltered conditions, the latter from exposed positions on the plant (Heath and Arnold, 1966). Soft leaves evidently have a lower polyphenol content, and decompose more rapidly, than leaves of the "hard" variety. The type of vegetation is also important. Plant material varies considerably in the amounts of carbon and nitrogen in the tissues. The herbaceous litter produced in grasslands usually contains relatively large amounts of nitrogen, i.e. it has a low carbon/nitrogen ratio, compared with that of woody species. It is a well-established fact that there is a relationship between the carbon/nitrogen ratio and the rapidity with which leaves disappear from the litter. Thus, herbaceous material with a low carbon/nitrogen ratio usually breaks down more or less completely during the first year after fall, whereas the leaves of conifers, which have high tannin contents in addition to high carbon/nitrogen ratios, take several years to undergo the major part of their decomposition.

Experiments on food selection by such earthworms as *Lumbricus terrestris* and *L. rubellus* have demonstrated that leaf material with a relatively high content of available nitrogen is more palatable, and therefore more readily accepted, than leaves with a high carbon/nitrogen ratio (Satchell, 1967). Clearly, there is a relationship between carbon/nitrogen ratios, palatability and the rate of litter breakdown. This relationship focuses attention on the role of the soil fauna, particularly the lumbricid earthworms, in the decomposition process. This kind of approach has been taken a stage further by the development of litter-bag experiments (Fig. 3.7).

Fig. 3.7 The litter bag technique. Beech leaf discs (lower right) are cut from entire leaves using a cork borer (size 15) (upper right), and placed in nylon bags of three different mesh sizes (left). The discs are positioned by staples, and each bag is colour coded for ease of identification, before being placed out in the field. (Photo. by A. Newman-Coburn.)

The litter-bag technique involves enclosing fragments of a given type of leaf litter in nylon mesh bags which are then buried in the litter layer of a chosen site. After a period of time, the bags are taken up, their contents examined and the loss of organic material assessed visually or gravimetrically. By using bags of different mesh size, various groups of the soil fauna can be excluded. For example, a mesh size of 0·003 mm will exclude all soil animals but will admit bacteria and fungi; a mesh size of 0·5 mm will allow small soil animals to enter the bag but will exclude earthworms, molluscs and large arthropods, while a mesh size of 7 mm will allow virtually unlimited access for all members of the soil fauna. Using this technique, Heath *et al.* (1966) showed that leaves of maize and various woody species (lime, birch, oak) disappeared at a more rapid rate in bags which did not exclude earthworms and other macrofauna than in bags which did. However, the exclusion of earthworms from bags containing lettuce, beet, kale and bean litter did not affect the rate of disappearance. This suggests that bacteria and fungi have an important role in the decomposition of these materials.

How can we apply this kind of information to the grassland situation? In the first instance, it must be noted that many of the food selection experiments and those using the litter-bag technique have been concerned mainly with studying the effect of the soil fauna on different types of forest litter. However, it is known that carbon/nitrogen ratios of a wide range of grassland litter are fairly consistent around 10:1. This is relatively low by any standard and, of course, is reflected in the rapid rate of disappearance of this litter. However, carbon/nitrogen ratios are probably more closely related to the digestibility of leaf litter, than to its palatability, and it is this latter quality which undoubtedly evokes the feeding response of the saprophagous soil animals. Studies on woody species, notably oak and beech, have shown that palatibility decreases with an increase in polyphenol content (Heath and Arnold, 1966). These observations are relevant to grasslands, for herbaceous species also differ in their polyphenol content. Unfortunately, the extent to which this variation influences the distribution of the soil fauna has not yet been evaluated.

Palatibility is also influenced by the silica content of the vegetation. Silicon is common in grasslands and circulates through the plant/soil system much more readily than, for example, in woodland or forest. Some grass species, notably *Nardus stricta*, have a high silica content in their tissues, and tend to become strongly established on glacial deposits of clay or sand in which silicon is abundant. Because of its high silicon content, *Nardus stricta* is unpalatable to large herbivores, such as sheep, and consequently remains ungrazed. When this grass dies back its relatively slow rate of decomposition results in the formation of an acidic organic horizon near the surface of the soil profile. Such acid conditions favour the growth of fungi which form an important part of the diet of various mites, nematodes, enchytraeids and insect larvae of the dipteran family Tipulidae. These groups of animals usually occur in greater numbers in acid grassland soils than in those of a neutral or alkaline character. The latter develop where the parent material is rich in calcium carbonate, and support such palatable grasses as *Festuca*

ovina, Agrostis tenuis, Anthoxanthum odoratum, Sesleria caerulia and *Helictotrichon pubescens*. Here, the organic material is quickly dispersed through the
profile and mixed with mineral soil mainly by the activities of lumbricid
earthworms which predominate in these situations. The success of lumbricids in neutral or alkaline grasslands is a reflection, in part, of their preference
for the kind of pH conditions which occur here, and also of the plentiful
supply of food in the form of palatable leaf litter and the dung deposited by
grazing herbivores. Dung deposits form a special kind of microhabitat for
soil animals which will be discussed shortly. In a more general sense, however, palatability studies draw attention to the way in which the factors of
food quality and pH conditions act in concert to influence the composition
of the soil fauna. This interaction operates in two ways. Firstly, the chemical
composition of the herbaceous material determines its acceptability as food
to grazing herbivores. This in turn influences the type and rate of decomposition and, subsequently, the pH of the litter. Secondly, pH conditions
in the litter and soil influence the microfloral composition, bacteria favouring
neutral or slightly alkaline soils, fungi flourishing in rather acid profiles.
Consequently, bacterial feeding members of the soil fauna are well represented in the former sites, fungivores predominating in the latter.

Dung Pat Faunas

The dung pat habitat is rich in microbial and animal life, a community in
its own right. It is a rather strictly delimited system and provides a very suitable situation for observing the processes of ecological succession. Common
among the animals encountered in dung pats are Diptera larvae, notably of
the families Scatophagidae, Borboridae, Sepsidae and Muscidae, scarabaeid
dung beetles (*Aphodius* and *Geotrupes* species), earthworms, nematodes and
mesostigmatid mites. It is not our purpose to dwell at length on this community for it has been described in some detail by Elton (1966). Such a
digression would also take us away from the world of the soil since many
dung-dwellers have a special association with the habitat, and are not really
members of the soil fauna. Many Diptera larvae fall into this category, and
their ecological importance, as far as soil organisms are concerned, lies in
the fact that they contribute to the breakdown of organic material in the
dung pat and thereby enrich the soil, notably with nitrogen, in its immediate
vicinity. Other groups, for example certain nematodes and mites, are carried
to the dung by flies and beetles on which they are phoretic. Macrochelid
mites (Order Mesostigmata) are often common in this habitat and they are
predaceous on nematodes and on the eggs and first-instar larvae of the flies
which transported these predators there.

Although the effect of dung pats on the distribution patterns of many groups of soil animals has not been investigated, there is good evidence that these deposits serve as foci for the aggregation of certain groups. Various beetles belonging to the family Scarabaeidae provide good examples of this, and *Geotrupes* species are notable for their habit of constructing burrows in the soil under, and around the periphery of, the dung pat. These insects remove considerable amounts of dung, which is used to nourish their developing larvae, into their underground chambers, and hence perform an important function in distributing this organic material through the soil profile.

Table 3.3

Total collections of earthworms taken from quadrats, each containing one dung pat, and quadrats in the spaces (open soil) between the dung pats (from Boyd, 1958).[1]

Species	Open Soil	Dung Pat
Allolobophora caliginosa	439	277*
A. chlorotica	3	12*
Dendrobaena rubida	5	12
D. octaedra	22	101*
Eiseniella tetraedra	0	0
Lumbricus castaneus	1	65*
L. rubellus	53	456*
Octolasion cyaneum	8	5
O. lacteum	1	0

[1] Data taken from 10 samples (1 m²) per site per month for 12 months.
* Significant ($X^2 > 3.841$).

The attractiveness of dung to certain earthworms is also well known. For example, Boyd (1958) made a comparison between the abundance of 11 lumbricid species in dung pats and in open soil away from the influence of dung, in a maritime grassland. Significant differences occurred in some instances (Table 3.3). In four out of the five comparisons in which such differences occurred, there was a concentration of the populations in the dung or in its vicinity, and this was particularly marked in *Dendrobaena octaedra* and *Lumbricus rubellus*. On the other hand, the abundance of *Allolobophora caliginosa* was significantly lower in dung than in open soil, indicating the possibility of a repulsive effect, although *A. caliginosa* still occurred in appreciable numbers in the dung pats. A further comparison was made between samples taken during April, when the dung pats were fresh and their concentrating effect maximal, and samples taken during October when the pats were old and dry. This showed that the concentrating effect on populations of *D. octaedra* and *L. rubellus* markedly diminished with the age of the dung, for the strongly aggregated pattern of distribution of these two species early in the year changed to a much more scattered pattern at the end of summer. Clearly, the dung pat is a habitat which is subject to relatively short-term changes, and faunal succession takes place fairly rapidly. During this succession specialized groups, which feed only on fresh dung or the rich bacterial component, give way to more generalized feeders derived from the soil fauna, which can utilize decomposing organic material of various kinds. This general trend has its parallel in the rhizosphere community, as we have already seen.

Secondary Consumers: Carnivores

The predatory component of the grassland soil fauna is a distinctive one, quite different in its species composition and its activity from that of, say, a forest soil. In temperate grasslands, vertebrate predators, such as moles, which feed on earthworms and insect larvae, achieve a prominence which they do not enjoy in other situations. Carnivorous ants, such as the turf ant *Tetramorium caespitum*, also have a place in the community structure, particularly in more open, dry sites where the soil is sandy in character. Large-sized arachnids such as spiders and harvestmen and, in the tropics and sub-tropics, scorpions, solifugids and whip-scorpions, are predaceous on many of the smaller arthropods which live on or near the soil surface, in open sites. To this list can be added several groups, such as centipedes, carabid and staphylinid beetles which occur in the grassland habitat, but which probably make a lower overall contribution to the character of this community than they do in the soils of woodlands or forests.

Predation in grassland systems is not a continuous activity. Its level of intensity varies from season to season and, within the diurnal cycle, from day to night. During the winter, low temperatures and snow cover severely restrict animal activity at the soil surface and in the upper layers of the profile. As we will see in the next chapter, many populations shift their centres of density downwards and may become inactive. Food for predators becomes scarce, and they too may become inactive. This is the case with many spiders and carabid beetles which become torpid during the winter months. The harvestman *Phalangium opilio* is a predatory species often encountered in grasslands, and successful here because of its tolerance of rather dry conditions. However, this species is less active in winter than in summer (Williams, 1962).

In open formation types, such as grasslands, there are marked fluctuations in microclimatic conditions during the 24-hour cycle, in temperate regions. Animal activity is most intense during the day, and becomes considerably reduced at night, during the warmer times of the year. As a consequence, the predatory component tends to be restricted to day-active species. This contrasts with woodland systems in which both diurnal and nocturnal components are present. Predation occurs throughout the 24-hour period, in these systems, and this leads to a more efficient exploitation of the prey. A good illustration is provided by the carabid beetles.

Some members of the family Carabidae are more or less restricted in their distribution to one habitat type as, for example, *Notiophilus biguttatus* to woodland, and *N. palustris* to grassland. Other species such as *N. substriatus*, *N. rufipes* and *Nebria brevicollis*, which are mainly woodland forms, can also establish breeding populations in open, grassland sites. Many carabids show clearly defined activity rhythms, being either nocturnal or diurnal. *Nebria brevicollis* is an example of a nocturnal species, at least in the late larval and adult stages, whereas species of *Notiophilus* are essentially diurnal. In sheltered woodland litter, prey organisms are often active throughout the 24-hour cycle, and this system can accommodate nocturnal and diurnal predators. Indeed, the evolution of these two components may be encouraged in order to minimize the amount of interspecific competition for food. In grassland sites, there is much less opportunity for nocturnal predators to find food. As a consequence, the grassland carabid fauna tends to be dominated by species of *Notiophilus* which have intrinsic diurnal activity patterns. So stringent is this requirement for daytime activity, that carabids which are nocturnal in woodlands change their activity pattern and become diurnal when they occur in grasslands. This tendency seems to be developed most strongly in species, such as *Feronia madida*, which have a wide range of feeding habit, and can adapt to whatever is available in the way of food. Incidentally, herbivorous carabids which occur mainly in open sites are markedly diurnal in activity.

Summary

The grassland environment is not a uniform one. The character of the vegetation varies from place to place, and is influenced by climate, the grazing patterns of large herbivores and land management practices. Grassland soils also vary in the content and distribution of organic material, their mineral constituents, pH, and the degree of leaching to which the profile is subjected.

Natural grasslands occur in regions where the rainfall is inadequate for forest growth. Such grasslands are the prairies of North America and the steppes of Eurasia. Some of these grasslands have highly fertile soils, for example the prairie brunizem and the steppe chernozem. However, in drier regions burozems and chestnut soils have a lower nutrient status, and restrictions on plant growth are imposed by the lack of moisture.

Grasslands also develop in regions of moderate to high rainfall where forest or woodland might otherwise occur but for some purely local effect. The development of a rendzina soil is governed by the nature of the parent material which is rich in calcium carbonate. Podzols are acid soils with a strongly leached profile, occurring on alluvial deposits which are well drained. In the tropics, soils with a low nutrient status, such as latosols and tropical black earths, can support grassland, although this is sometimes a secondary vegetation which appears when forest is cleared for cultivation.

In moist temperate grassland, a varied soil fauna occurs in which lumbricid earthworms and arthropods predominate. In rather acid soils, earth worms become less conspicuous and the arthropods more important. Chernozem soils are rather acid, at least in the surface layers, but have accumulations of calcium carbonate at greater depths. In some of these soils, earthworms are virtually absent, and the fauna is strongly arthropodan in character. Other chernozems have an earthworm fauna which, with small burrowing vertebrates, may be responsible for distributing organic material to great depths in the profile. The soils of steppe grassland vary in their organic content, and this is reflected in the composition of their microarthropod fauna. Cryptostigmatid mites are more abundant than Prostigmata in organic profiles, whereas the reverse is the case in mineral soils.

Earthworms and arthropods dominate the fauna of tropical grassland soils, although these two groups of animals differ taxonomically and ecologically from their counterparts in moist temperate regions. The earthworms are deep-burrowing forms, and although they may form surface casts only during the wet season, the amount of material deposited is often considerably higher than that of lumbricid earthworms in temperate regions. Microarthropod populations are generally lower than in temperate grassland, but termites and ants come into greater prominence in tropical grassland. A consistent feature of this tropical soil fauna, earthworms and arthropods alike, is that distribution patterns often extend over a range of habitats from grassland to forest.

The prominence of lumbricid earthworms in the soil fauna of many grasslands is a reflection, in part, of their preference for food material with a low carbon/nitrogen ratio. In sites where the organic litter is relatively rich in nitrogen, earthworms are important agents in mixing this material with mineral soil. Litter which is less palatable, due to its content of polyphenols or silicon, is not dispersed so readily, and may accumulate at the surface of the profile as an acid organic horizon. Such conditions favour the growth of fungi which form the food of various nematodes, enchytraeids and microarthropods, and these groups of animals increase in importance as earthworm populations decline in acid grasslands.

Animal activity in grasslands is not continuous. It is maximal during the day and during the warmer seasons of the year in temperate regions. At night and during the winter, or periods of drought, animal activity on or near the soil surface is reduced to a minimum either because the fauna becomes quiescent or the centres of population density shift down the profile. The predator component of the grassland fauna shows a strong selection for diurnal, as opposed to nocturnal, components, and this is clearly related to the availability of the prey.

Within the grassland ecosystem, discrete microhabitats can be identified such as the rhizosphere and the dung pat. The fauna associated with each of these microhabitats is quite distinct. Bacterial-feeding Protozoa and nematodes are characteristic members of the rhizosphere, while lumbricid earthworms, Diptera larvae and scarabaeid beetles are common in dung deposits. However, in these two contrasting situations there is a fairly rapid faunal succession which has certain common features. In both, early colonizers are mainly represented by specialist feeders which gradually give way to more catholic feeders derived from the true soil fauna.

REFERENCES

ALEJNIKOVA, M. M. (1965). Die Bodenfauna des Mittleren Wolgalandes und ihre regionalen Besonderheiten, *Pedobiologia*, **5**, 17–49.

BALOGH, J. (1970). Biogeographical aspects of soil ecology. *In* "Methods of Study in Soil Ecology", 33–8 (Phillipson, J., ed.), UNESCO, Paris.

BELFIELD, W. (1956). The Arthropoda of the soil in a West African pasture, *J. Anim. Ecol.*, **25**, 275–87.

BELFIELD, W. (1967). The effects of overhead watering on the meiofauna in a West African pasture. *In* "Progress in Soil Biology", 192–210 North Holland, Amsterdam, (Graff, O. and Satchell, J. E., eds.), North Holland, Amsterdam.

BOYD, J. M. (1958). The ecology of earthworms in cattle-grazed machair in Tiree, Argyll, *J. Anim. Ecol.*, **27**, 147–57.

BURNETT, G. F. (1968). The effect of irrigation, cultivation and some insecticides on the soil arthropods of an East African dry grassland, *J. appl. Ecol.*, **5**, 141–56.

DI CASTRI, F. (1963). Etat de nos connaissances sur les biocoenoses édaphiques du Chili. *In* "Soil Organisms", 375–85 (Doeksen, J. and van der Drift, J., eds), North Holland, Amsterdam.

ELTON, C. S. (1966). "The Pattern of Animal Communities", Methuen, London.

EVANS, A. C. (1948). Studies on the relationship between earthworms and soil fertility. II, *Ann. appl. Biol.*, **35**, 1–13.

FITZPATRICK, E. A. (1971). "Pedology", Oliver and Boyd, Edinburgh.

GHILAROV, M. S. (1968). Soil stratum of terrestrial biocenoses, *Pedobiologia*, **8**, 82–96.

HARRIS, W. V. (1955). Termites and the soil. *In* "Soil Zoology", 62–72 (Kevan, D. K. McE., ed.), Butterworths, London.

HEATH, G. W. and ARNOLD, M. K. (1966). Studies in leaf-litter breakdown II. Breakdown rate of "sun" and "shade" leaves., *Pedobiologia*, **6**, 238–43.

HEATH, G. W., ARNOLD, M. K. and EDWARDS, C. A. (1966). Studies in leaf litter breakdown I. Breakdown rates of leaves of different species, *Pedobiologia*, **6**, 1–12.

KÜHNELT, W. (1961). "Soil Biology", Faber and Faber, London.

LJUNGSTRÖM, P. O. and REINECKE, A. J. (1969). Ecology and natural history of the microchaetid earthworms of South Africa, *Pedobiologia*, **9**, 152–57.

LOOTS, G. C. and RYKE, P. A. J. (1967). The ratio Oribatei: Trombidiformes with reference to organic matter content in soils, *Pedobiologia*, **7**, 121–24.

MACFADYEN, A. (1963). The contribution of the microfauna to total soil metabolism. *In* "Soil Organisms", 3–17 (Doeksen, J. and van der Drift, J., eds.), North Holland, Amsterdam.

MADGE, D. S. (1969). Field and laboratory studies on the activities of two species of tropical earthworms, *Pedobiologia*, **9**, 188–214.

MOURSI, A. (1962). The attractiveness of CO_2 and N_2 to soil Arthropoda, *Pedobiologia*, **1**, 299–302.

NYE, P. H. (1955). Some soil-forming processes in the humid tropics. IV. The action of the soil fauna, *J. Soil Sci.*, **6**, 73–83.

ROUATT, J. W., KATZNELSON, H. and PAYNE, T. M. B. (1960). Statistical evaluation of the rhizosphere effect, *Proc. Soil Sci. Soc. Am.*, **24**, 271–73.

RUSSELL, E. J. (1961). "Soil Conditions and Plant Growth" (ninth edn., revised Russell, E. W.), Longmans, London.

SALT, G. (1952). The arthropod population of the soil in some East African pastures, *Bull. ent. Res.*, **43**, 203–20.

SALT, G. (1955). The arthropod population of soil under elephant grass in Uganda, *Bull. ent. Res.*, **46**, 539–45.

SALT, G., HOLLICK, F. S. J., RAW, F. and BRIAN, M. V. (1948). The arthropod population of pasture soil', *J. Anim. Ecol.*, **17**, 139–50.

SATCHELL, J. E. (1967). Lumbricidae. *In* 'Soil Biology", 259–322 (Burges, N. and Raw, F., eds.), Academic Press, London and New York.

STRICKLAND, A. H. (1947). The soil fauna of two contrasted plots of land in Trinidad, British West Indies', *J. Anim. Ecol.*, **16**, 1–10.

WHITTAKER, R. H. (1970). "Communities and Ecosystems", Collier-Macmillan, London.

WILLIAMS, G. (1962). Seasonal and diurnal activity of harvestmen (Phalangida) and spiders (Araneida) in contrasted habitats, *J. Anim. Ecol.*, **31**, 23–42.

WOOD, T. G. (1966). The fauna of grassland soils with special reference to Acari and Collembola, *Proc. N.Z. ecol. Soc.*, **13**, 79–85.

Chapter 4

Grasslands: Physical and Chemical Factors

In the last chapter we looked at the general character of grassland soil faunas and the influence of the food factor on the composition of the fauna. Food, of course, forms a very important part of the environment of an animal population. It is a biotic factor. There is another group of environmental factors, termed abiotic, which are physical and chemical in nature, and which may be just as important as the food factor in determining the species composition and distribution patterns of the soil fauna of grasslands.

Physical factors which form part of the environment of soil animals include the structure of the soil and the vegetation immediately above it, and its microclimate. Chemical influences are exerted mainly through the base content and pH of the soil. In managed systems, chemical fertilizers and pesticides are also important, but since these are applied by man they can be considered as extreme examples of biotic effects. They are dealt with in Chapter 5 where the effects of cultivation are discussed.

Abiotic factors do not always operate in isolation from each other. For example, the structure of the soil and vegetation may have a considerable influence on the microclimate. Similarly, certain abiotic factors may act in combination with biotic influences to determine the composition of the soil fauna in grasslands. For example, in the previous chapter it was noted that the species composition of the carabid beetle fauna was influenced by the availability of prey organisms. In this chapter we will see that the distribution and composition of this fauna is also affected by the structure of the vegetation and also the microclimate.

It is also important to recognize that a particular factor may have a direct influence on the soil fauna, or its effect may be exerted in an indirect manner as part of a complex system of interactions. In the following account, these two operational possibilities are identified as far as possible.

Effect of soil structure

Grassland soils vary from place to place in the relative amounts of sand, silt and clay they contain. They also vary in their organic content. Sandy soils are light and warm, and since the characteristic particle size is relatively large (0·02–2·0 mm diameter), pore spaces are large. These soils are well drained, aerated and loose in texture. In clay soils, there is a predominance of particles with a diameter less than 0·002 mm; these soils are heavy, wet and badly drained. Many grassland soils contain a mixture of particle sizes and can be placed between these two extremes.

Direct Effects

Soil texture has a direct effect on the vertical distribution of many burrowing animals, such as lumbricid earthworms, molluscs, various scarabaeid beetles and millipedes. In compacted soils, these burrowers restrict their activities to the surface layers of the profile, but where the soil has a looser texture they have a more extensive depth distribution. This effect is clearly demonstrated by comparing the vertical distribution of lumbricid earthworms in permanent pasture and ploughed soil. This is discussed in more detail in Chapter 5.

The distribution of certain temporary members of the soil fauna is directly influenced by soil texture. These are the insects and arachnids which live on or above the soil surface as adults, but lay their eggs in the soil. Larvae of the wheat bulb fly, *Leptohylemyia coarctata*, are pests which feed on the roots of grasses and cereals. Eggs are laid in the soil by the adult female fly which selects sites with a rough tilth and an open cavity structure, in preference to uncultivated, compact soils (Raw, 1955, 1967). In this instance, and there are others as we will see in Chapter 5, ploughing the soil facilitates the growth of pest populations. Another example of this direct effect is provided by the British spider *Arctosa perita* which occurs mainly in sites where the soil is loose-textured. This spider constructs a temporary burrow in the soil and will only excavate this where the soil can be easily worked, for example in sand and shales (Duffey, 1968).

We can also find examples of distribution patterns which are related to soil structure, although there is no obvious explanation for this relationship. Members of the beetle family Elateridae illustrate this. Larval elaterids, known as wireworms, feed on roots of grasses. In permanent pastures they can occur in densities of the order of 2–3 millions per acre, and become serious agricultural pests. The commonest elaterid genus in grassland soils is *Agriotes*, often comprising about 75% of the total elaterid fauna in sites where the soil is a medium to heavy loam. In lighter, sandy soils the abundance of agriotids is lower and other genera, such as *Selatosomus* and *Limonius*, become more prominent. At the other extreme, the impoverished fauna of bogs and marshes is often dominated by members of the genus *Corymbites*

(Nadvornyj, 1968). The distribution ranges of various species populations sometimes overlap across this spectrum of habitats, but marked differences in relative abundance usually occur from site to site. In general, the greatest abundance of elaterids occurs in open, agricultural areas, whereas the greatest species diversity is found in forest soils. It is possible that the general pattern of elaterid distribution is governed, not by one environmental factor, but a complex in which soil texture and, perhaps, feeding preferences may be closely inter-related.

Indirect Effects

The examples given above do illustrate that soil structure does influence the distribution patterns of certain groups of soil animals in a direct way. For many other groups, however, the physical structure of the soil, its texture and particle size, are probably only minor influences in determining habitat selection. Nielsen (1949), for example, found a few species of nematodes, such as *Acrobeles ciliatus* and *Tripyla setifera*, which are more or less restricted to sandy soils where, presumably, they have greater freedom of movement. However, such species are in a minority, for most nematodes are distributed over a range of soil types from sand to peat. Many nematodes feed on bacteria, and soil structure may be indirectly involved in this feeding relationship in the kind of substrate provided for micro-organisms. It is perhaps worth recalling here that the nature and distribution of the organic component of the soil may influence the character of the soil fauna through the kinds of micro-organisms that are present. Thus, bacteria and bacteriophages predominate in rather alkaline mull humus formations, whereas fungi and fungivores achieve their highest densities in acid, unincorporated raw humus.

Fig. **4.1** Quadrat survey in a Yorkshire grassland. (Photo. by author.)

4

Structure of the vegetation

Open habitats, such as grasslands, often differ from other terrestrial formation types, such as woodland and forest, in possessing a well-developed field layer of vegetation (Fig. 4.1). In moist temperate regions this field layer consists of a continuous ground cover of herbaceous plants; in arid regions this cover is broken up to some extent, and a tussock growth form occurs. This field layer, whether it be a continuous carpet or a tussock formation, can influence the species composition and distribution of the soil fauna in direct and indirect ways.

Direct Effects

The structure of the surface vegetation can act directly, as a physical factor in the environment of soil animals, in two ways. Firstly, it can impede the mobility of animals moving over the surface of the soil and, secondly, it can provide a structural framework within which a microhabitat, a territory or a spatial niche can be defined.

The growth form of many grasses produces a tangled mat of aerial vegetation on or just above the surface of the soil, and a similar meshwork of roots just below. These structural features present obstacles to free movement, particularly of some of the fast-moving predators. This can be illustrated by considering the effect on the species composition of the ground beetle fauna (family Carabidae) in grasslands.

Although the majority of Carabidae are predaceous, both as larvae and adults, members of the genera *Bembidion* and *Feronia* may be considered more correctly as scavengers, while the Harpalini and Amarini are mainly phytophagous. Predatory carabids, such as *Nebria brevicollis*, *Notiophilus* spp., *Trechus* spp. and *Agonum* spp., are much more common in woodland litter where their microarthropod prey is abundant, than in open grassland where phytophagous species tend to predominate in a species-poor fauna. In a comparative study of the carabids of woodland and open areas, Williams (1959) found that 63% of all carnivores captured were taken from woodland litter, and only 4% from open, uncultivated land, whereas the percentages for phytophages in these two sites were 2% and 60% respectively. The absence of a rich carabid fauna from uncultivated sites may be attributed to the presence of a dense mat of vegetation lying on the soil surface which impedes the free movement of the predatory species. In an open cultivated site, a newly-seeded ley where no such obstruction was allowed to develop, predatory carabids were more abundant, and although their numbers were still lower than in woodland, this group was numerically dominant over the scavengers and phytophages. Here we have an example of a direct effect of the structural features of the environment on the composition of the ground fauna. However, this effect does not operate in isolation. The food factor is obviously important also.

Vegetation provides a structural framework for the habitats of various spiders, which use plants for the support of their webs. Strictly speaking, these animals are not true members of the soil fauna although they frequently move between the vegetation and the soil and some, such as the funnel-web spiders of the family Agelenidae, have extensions of the web which enter soil crevices. Some of the commonest grassland spiders in Britain belong to the family Linyphiidae, the familiar "money-spiders", which construct a hammock-shaped web which is slung among the leaves and stems of tall grasses. These spiders prefer a tussock growth form with a large number of structural components. Duffey (1962) investigated the spider fauna of three limestone sites at Wytham, Berkshire, namely a *Brachypodium pinnatum* site, a *Festuca rubra* turf, and a coral outcrop supporting a sparse, but diverse, vegetation of moss and taller herbs. One of the main conclusions emerging from this study was that the highest density of spiders occurred in the dense vegetation of the *Festuca rubra* site, and the lowest number of individuals on the coral outcrop. The latter site contained the largest number of species, however, and this could be attributed to its floristic diversity.

Vegetational structure is only one of a number of factors which influence the distribution of spiders in grassland. Indeed, the importance of this factor to a particular species of spider can vary from site to site. For example, Riechert *et al.* (1973) found that different habitat factors were responsible for the distribution patterns of the funnel-web spider *Agelenopsis aperta* in three desert grassland sites in New Mexico. In mixed grassland, this spider selected shaded depressions in the ground rather than a particular type of vegetation. There was, however, a correlation between the distribution of *Agelenopsis* and vegetation on a lava flow colonized by shrubs. Here, the spider selected three or four species of shrub with low branching systems for web building sites. At a third site, a rangeland with tall grass vegetation, there was a positive correlation between the distribution of the spider and that of tall grass. All three of the sites studied had similar habitat features, but despite this, the spider chose different features in each of the three sites. This suggests that factors other than vegetational structure are operating directly to influence habitat selection. These factors may include abiotic effects such as microclimate, with which the structure of the vegetation may be indirectly associated. They may also include biotic effects, such as the distribution of prey animals.

Indirect Effects: Tussock Faunas

The presence of a field layer of vegetation covering the surface of the soil provides a measure of protection against disturbance. The erosive effects of wind and water are reduced, and the soil fauna is provided with a relatively stable environment. The microclimate of the soil is affected, and this will be discussed in more detail in the next section. The stability of the soil environment is enhanced when the vegetation is of the tussock type, and it is this effect with which we are immediately concerned.

In rough pasture, certain grass species tend to produce a tussock growth form, a dense tuft of vegetation differing in its structural and microclimatic features from the surrounding, more open, areas. Tussocks are common in moorland sites where clumps of sedges, cotton grass, rushes and *Sphagnum* moss are surrounded by water-filled drainage channels, at least during the wetter parts of the year. The fauna of these tussocks will be considered in a later chapter. For the moment, we are concerned more particularly with tussocks formed in drier sites, mainly by grass species such as *Deschampsia flexuosa* and *Dactylis glomerata*. These formations, like those occurring on wetter sites, can be regarded as refuges to which populations can retreat to

Table 4.1

Total numbers of arthropods in *Dactylis* tussocks and intervening grasses in winter and summer compared (from Luff, 1966).

	Winter			Summer		
	Dactylis tussock	*Holcus*	*Agrostis*	*Dactylis* tussock	*Holcus*	*Agrostis*
Sample area (cm²)	194·0	170·0	142·0	242·0	225·0	259·0
Isopoda	59	2	0	91	4	4
Myriapoda	7	6	0	30	13	3
Collembola	1165	623	87	1790	864	922
Dictyoptera	5	0	0	0	0	0
Psocoptera	0	1	0	0	0	0
Hemiptera	34	4	1	16	9	4
Thysanoptera	147	25	4	10	0	3
Lepidoptera (larvae)	14	19	3	1	0	0
Diptera (adults)	4	0	1	19	4	1
Diptera (larvae)	131	38	15	175	4	8
Hymenoptera	44	1	0	5	1	4
Coleoptera (adults)	111	10	3	15	9	8
Coleoptera (larvae)	6	1	2	78	8	3
Phalangida	0	0	0	23	0	0
Araneida	59	30	12	28	11	11
Acarina	3545	1711	805	3672	1989	1387
Totals	5334	2471	933	5953	2916	2358

shelter from extremes of temperature, flooding and drought. However, they have much more than an occasional function as far as the soil and ground-dwelling fauna are concerned. One of the early, detailed studies of the tussock fauna was made by Ford (1937) who drew attention to the rich representation of microarthropods in clumps of *Bromus erectus* in Oxfordshire. More recently, Luff (1966) recorded 14 Orders of arthropods from *Dactylis* tussocks at Silwood Park, Berkshire, of which the Acari, Coleoptera, Diptera and isopod woodlice were numerically dominant throughout the year. Comparisons between the fauna of these tussocks and that of the intervening grassy areas colonized by *Festuca*, *Agrostis* and *Holcus*, showed that the tussocks supported at least twice as many arthropods as the intervening areas, and that the disparity between the tussock fauna and that of other grasses was greater in winter than in summer (Table 4.1). This finding supports the suggestion, made earlier by Pearce (1948), that these vegetational tufts are seasonal refuges, at least, in some cases.

Luff's study concentrated mainly on the beetle fauna of the tussocks, and some idea of the richness of this fauna can be gained from the fact that 198 species belonging to 23 families were recorded from the Silwood site, with the Staphylinidae, Curculionidae and Carabidae most strongly represented. The tussock beetle fauna comprised a very few species which were represented by a large number of individuals, and a relatively large number of species represented by only a few individuals. Analyses of seasonal variations revealed that the winter increase in size of the populations was brought about by two factors, namely an increase in the number of species present, and an increase in the number of individuals per species. Thus, two elements may be defined in this tussock fauna, one being a temporary component comprising species represented by a few individuals, and the second consisting of the permanent inhabitants of the tussock, and occurring very commonly there. This conclusion was borne out by a comparison of the commoner beetles living between and within the tussocks. Only two genera were equally frequent in both situations, indicating that the common tussock genera were permanently associated with this habitat. These permanent inhabitants belonged to one of three trophic categories, namely (1) predators, such as the carabids *Dromius melanocephalus*, *Trechus obtusus*, *Agonum obscurum* and staphylinids of the genus *Stenus*, (2) omnivorous or saprophagous species, for example the carabid *Bradycellus harpalinus* and the staphylinids *Tachyporus* spp., *Amischia analis* and *Atheta fungi*, and (3) fungus-feeding members of the family Lathridiidae. This is a balanced micro-community, although it should be noted that exclusively phytophagous species are rare here, and none appears to be associated, specifically, with *Dactylis* or *Deschampsia*.

The tussock vegetation form clearly represents a distinct habitat. It has a characteristic fauna formed from a combination of a temporary element, comprising many rare species, and a permanent element consisting of a few common species. Members of the permanent fauna differ from one another in the type of tussock they inhabit, in their life histories, feeding habits and the kind of habitats selected by their larvae. For example, the staphylinid *Stenus impressus* occurs as the adult in tussocks during the summer, whereas the overwintering larvae are found mainly outside these tufts. *Stenus clavicornis*, on the other hand, overwinters as the newly-emerged adult in the tussocks, and its larvae develop around the bases of these during the summer. Both of these *Stenus* species feed mainly on Collembola. In contrast, the diet of *Dromius melanocephalus* consists mainly of mites, and this carabid overwinters as the adult which develops to sexual maturity during this period. It lays eggs in the tussocks during early summer, early enough to allow post-embryonic development to the next adult generation to be completed by late summer or early autumn. These differences may be interpreted as adaptations to varying microclimatic conditions occurring within the tussock type. They may also be a reflection of the distribution of prey animals, which again may be influenced by microclimatic conditions.

Indirect Effects: Grazing

Considerable attention has been paid, in recent years, to the effect of grazing activity of large herbivores on the fauna living in, or near the surface of, the soil. Although the results obtained by different workers are not strictly comparable, because of the disparity between the habitats selected in each case, they demonstrate quantitative and qualitative differences between the fauna of grazed and ungrazed sites. Some of the data obtained by Morris (1968) from chalk grassland in Bedfordshire, England, are presented in Table 4.2 and these show a markedly richer fauna in ungrazed, compared with grazed, sites. The author suggested that this situation may be a reflection of the greater amount of plant food and cover, higher humidity and more equable microclimate prevailing where the ground vegetation is not cut short by grazing. In contrast to these findings, Southwood and van Emden (1967) reported a relatively richer fauna, particularly of spiders and beetles, on cut grassland, compared with undisturbed sites. There are several possible explanations for the discrepancy between the results of these two investigations. Morris' sites were on chalk, whereas those of Southwood and van Emden were on sand, and in the former the grass was cut much closer to the ground than in the latter. Provided the grass is not cut too short, the ground layer may retain enough of the structural features so necessary for spiders, particularly the web-builders, and hence this group may be present in abundance in sites which are grazed, but not heavily so. As already mentioned, carabid beetles tend to avoid uncultivated sites where rank grass

Table 4.2

A comparison between the invertebrate faunas of grazed and ungrazed chalk grassland in Bedfordshire, England (after Morris, 1968).[1,2]

Animal Group	Grazed Sites	Ungrazed Sites
Gastropoda	83	171
Isopoda	22	594
Geophilomorpha	345	432
Lithobiomorpha	601	2664
Polydesmidae	178	910
Iulidae	0	2
Opiliones	1	1
Araneae	177	2023
Pseudoscorpiones	1	4
Heteroptera	184	1236
Homoptera:		
Auchenorrhyncha	80	764
Aphidoidea	17	73
Lepidoptera larvae	82	76
Hymenoptera:		
Parasitica	65	117
Formicidae	33	149
Coleoptera:		
Adults	654	1949
Larvae	612	566

[1] Data expressed as numbers extracted from 4 turves (1/14 m² each) per plot on 14 occasions, 1966–67.

[2] Counts for Acari, Collembola and Diptera not included.

obstructs their free movement. The extent to which this effect may have contributed to the difference between the finding of Morris and those of Southwood and van Emden is not known, but it is obvious that detailed comparisons of this kind must be approached with a great deal of caution and a knowledge of all possible limiting effects.

The striking diminution of the Araneae on grazed plots recorded in Table 4.2 confirms Duffey's (1963) findings on a Yorkshire grassland at Malham Tarn. In this locality, which is in fact a moorland and will be discussed further in this context in a later chapter, sites supporting *Festuca* are heavily grazed by sheep, whereas sites characterized by the grass *Nardus stricta*, an erect species with a high silica content, are hardly touched by large herbivores (Fig. 4.2). Grazing activity on the *Festuca* destroyed the microhabitat characteristics required by spiders, and densities were much lower here than in the more varied microenvironment of the *Nardus*.

From what we know of the ecology of the ground and soil fauna of grasslands, it seems unwise to suppose that grazing, or the lack of it, will affect all

Fig. 4.2 Upland permanent pasture maintained by sheep grazing in the English Pennines In the foreground is *Nardus stricta* growing on glacial drift and subject to slight grazing pressure. In the background is the more closely cropped limestone grassland consisting of *Sesleria caerulia, Helictotrichon pubescens, Festuca ovina* and *Anthoxanthum odoratum.* (Photo. by author.)

groups of animals in the same way. It might be expected that species which prefer rather dry conditions will prevail in heavily grazed sites where the soil and the atmosphere above it have a low moisture content, whereas moisture-loving species will favour ungrazed sites where the soil is protected by a well-developed ground layer. Looking at the arthropods in general, this is true to some extent, but it is by no means the whole story, for preferences other than those associated with moisture may also be important. For example, Boyd (1960) carried out a study on a cattle-grazed maritime grassland, or *machair*, in Scotland, which demonstrated that certain arthropods preferred grazed sites, others ungrazed areas. Some of the principal findings of this study, which are summarized in Table 4.3, deserve to be considered further. The prevalence of beetles belonging to the genus *Calathus*, and ants of the genus *Myrmica* in grazed areas may be a reflection of their preference for dry sites; both groups are often found in sand dunes (see Chapter 7). However, the distribution of ants varied seasonally, and determinations of the age structure of the population suggested that they nested in ungrazed sites, where presumably they would suffer little disturbance, but invaded grazed sites early in the year, possibly to forage for food. Pontin (1960, 1963)

Table 4.3

Preferences of some common groups of arthropods in grazed and ungrazed grassland in Tiree, Argyll (from observations supplied by Boyd, 1960).

| | Species favouring | |
	Grazed Sites	Ungrazed Sites
Coleoptera	Dyschirius globosus	Carabus granulatus
	Amara sp.	Nebria brevicollis
	Calathus spp.	Notiophilus aquaticus
	Staphylinus aeneoephalus	Xantholinus spp.
	Aphodius spp.	Longitarsus jacobaeae
	Longitarsus spp.	Serica brunnea
	(excl. L. jacobaeae)	Liodes dubia
	Sitona spp.	
	Apion apricans	
Hymenoptera	Myrmica spp.*	
	Phalangium opilio	Lycosa pullata
Arachnida	Lycosa tarsalis*	
	Lycosa monticola*	
	Xysticus cristatus	

* Preferences subject to seasonal changes.

also studied the distribution patterns of ants in British grasslands and found that two of the common species, *Lasius niger* and *L. flavus*, generally inhabited exposed sites where there is little shade and sparse vegetation. The horizontal patterns of these two species frequently overlap, and they can form stable and competitive populations provided the vegetation is kept short by grazing. In the absence of grazing, the growth of ground vegetation produces a shade effect which prevents the foundation of new colonies. However, once established in a site, *L. flavus* can often survive the spread of vegetation, owing to its ability to construct tall mounds which reach upwards above the level of the grass sward. In contrast, *L. niger* builds loose mounds which are susceptible to damage and erosion by trampling and water action. Some habitat separation can be expected to occur between these two species, particularly in sites where grazed and ungrazed areas form a mosaic. However, other factors such as interspecific competition also influence distribution patterns, a fact also noted by Elmes (1974) in his study of the spatial distribution patterns of *L. flavus* and the subterranean *Myrmica rubra* in a limestone valley in Dorset. We will return to this topic in Chapter 7.

Differing responses to the grazing factor are also shown by spiders. The wolf spider *Lycosa pullata*, which is known to prefer damp sites (Vlijm and Kessler-Geschiere, 1967), is not usually encountered in grazed grasslands.

4*

In other members of this genus, a change in habitat preference occurs, particularly by the egg-carrying female which moves from damp to dry sites after the breeding season.

The examples of the ants and the spiders indicate that grazing indirectly influences distribution patterns through changing microclimate, or altering the structural features of the microhabitat. There are other examples which suggest that the preference for a certain site may have more to do with the influence of grazing on the food factor. The presence of scarabaeid beetles belonging to the genus *Aphodius* on grazed sites reflects the habit of this group as frequenters of dung, a commodity in good supply here. This contrasts with the pattern shown by another scarabaeid, *Liodes dubia*, which is commonly associated with the subterranean fungi which flourish in the more moist soils of ungrazed sites. Again, referring to Table 4.3, the chrysomelid beetle *Longitarsus jacobaeae* differs from other members of the same genus in its association with *Senecio jacobaea*.

A discussion of the effects of grazing on the soil fauna must include some reference to the herbivores producing these effects. Sheep, cattle, deer and rabbits, for example, may have different grazing patterns and preferences which selectively eliminate different kinds of plants. Some herbivores feed by cropping the grass close to the soil surface, and this may result in a dense but very short ground cover. Others pull out tufts of grass, rendering the vegetation long, loose and more open in structure. Distribution patterns of soil and surface-dwelling animals may be influenced by these differences. For example, a comparison of the cryptostigmatid mites of two adjacent pasture soils under *Poa pratensis* in Kentucky, one grazed by sheep and the other by cattle, revealed different numerically dominant species in each case (Wallwork and Rodriguez, 1961). The two pastures did not differ to any great extent in their vegetational characteristics, so some less obvious environmental factor must be limiting. This could well be some effect produced by the grazing activities of the large herbivores since this was the only major variable in the system.

Effect of microclimate

The influence of microclimate on the soil fauna is, in the main, a direct one. There are indirect effects, it is true, for example when climate influences the character of the vegetation or, through weathering processes, the soil type. These are more properly considered as macroclimatic effects and need not concern us here, particularly since the primary factors of vegetation and soil type have already been discussed. In addition, microclimatic conditions may influence the distribution of predators indirectly by the way in which they act on the distribution of prey. This provides an example of the way an abiotic effect (microclimate) can interact with a biotic one (food) to determine the character of the soil fauna. Similarly, microclimate can interact with other biotic effects, such as intraspecific competition for shelter or breeding sites. This kind of interaction may provide a mechanism for regulating population size in the woodlouse *Armadillidium vulgare* in grassland where periodic flooding or drought can restrict the number of sites available for colonization (Paris, 1963). For the moment, however, we are concerned only with the direct effects of microclimate.

Temperature and Moisture

One of the most attractive features of the soil habitat, as far as its fauna is concerned, is the equable microclimatic features it provides. Fluctuations in temperature and moisture are much less severe than they are at, and above, the ground surface, and the amplitude of these fluctuations decreases with depth in the soil (Geiger, 1959). At the same time, we can recognize that gradients in these two factors occur horizontally from place to place, and vertically through the soil profile. Again, it is important to remember that the effect of microclimate on the soil fauna is not a static one. Cyclical fluctuations of varying periodicity occur both in moisture and temperature conditions. The commonest of these are seasonal and diurnal cycles. There are, therefore, temporal, as well as spatial, effects to be considered.

In practice, it is often difficult to separate the effects of moisture and temperature on the soil fauna. Both of these factors are so closely linked in the ecology of soil animals that, ultimately, their combined effect must be taken into account. It is also true, however, that ecologists tend to study the effects of these two factors separately, at least in the first instance, and the examples given in the following pages reflect this fact.

Soil animals vary in their preferences for microclimatic conditions. In the case of the moisture factor, for example, certain species occur only in relatively dry soils, some in very wet soils, although the majority of the soil fauna evidently prefers conditions that are neither too dry or too wet. The first-named category is termed xerophile, the second hygrophile, and the third mesophile. Some soil animals are strictly limited in their tolerance of environmental conditions and are termed stenotopic. Others are more plastic and are able to survive equally well over a range of conditions; they are eurytopic. Usually, the soil fauna of a particular grassland site, at any one time, consists of a mixture of stenotopic and eurytopic species, although the proportions of these two ecological groups will vary from place to place.

Horizontal Effects: Moisture

The influence of microclimate on horizontal distribution patterns of the soil fauna can be conveniently illustrated by considering first the effect of the moisture factor. The examples chosen represent two groups with different ecologies, namely surface-dwelling beetles of the family Carabidae, and the subterranean mites and Collembola.

Carabidae. These ground beetles are common in a wide variety of grassland sites in the north temperate region. They are most abundant in moist sites, and although the group also occurs in wet marshlands and semi-arid steppe, the number of species and population densities are reduced at both extremes of the moisture spectrum (Fig. 4.3). As conditions become more arid, and desert develops, carabid beetles are largely replaced by others belonging to the family Tenebrionidae.

Some carabids are stenotopic, others eurytopic. To the former category belong members of the genera *Harpalus*, *Calathus* and *Cymindis* which are restricted to dry sites, and *Agonum* species which occur in wet or very moist sites only. Between these xerophilous and hygrophilous extremes is a mesophilous fauna, including *Clivina fossor* and members of the genera *Carabus*, *Bembidion* and *Feronia*. This mesophilous category is less well defined than

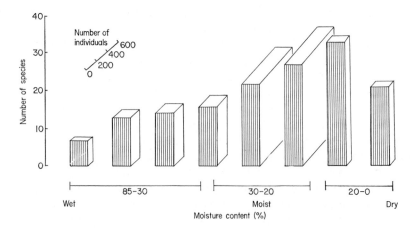

Fig. 4.3 Variations in the number of species and number of individuals of Carabidae along a moisture gradient in grassland soils in Germany. (After Tietze, 1968.)

either of the two extremes. It includes many species which, despite their narrow optima, have distributions which overlap with the xerophiles on the one hand, and the hygrophiles on the other. These species are essentially eurytopic, although they are often classified as "wet" or "non-wet", depending on their preferenda.

This distinction between "wet" and "non-wet" species may seem to be a rather subjective one. It will appear even more so when we examine the distribution of Carabidae in tropical forests (see Chapter 8), for here the pattern changes in response to seasonal flooding and drought. This underlines the fact that the distribution patterns of these soil animals, like those of the rest of the soil fauna, exist in time as well as in space. Another reminder of this is provided by the study of life history patterns. These give a basis for clearer ecological distinctions. For example, Murdoch (1967) found that carabid species associated with wet or moist sites breed in spring or summer in Britain, as a general rule, and overwinter as adults. This pattern could have arisen as a result of the inhibition of breeding and development in autumn due to rising water levels in wet sites. Selection for the overwintering stage would favour the adult, which is more mobile than the larva. Its mobility would enable it to withstand, or avoid, flood conditions more easily than the juvenile. A significant proportion of "non-wet" species, which are not subject to this kind of selection, breed in autumn and overwinter as larvae.

Microarthropods. The general features of carabid distribution are typical of many groups of the soil fauna, particularly of the microarthropod mites and Collembola. In a pioneer study, Weis-Fogh (1948) described variations in the composition of the soil microarthropod fauna along a transect drawn across a sandy plain near Mols laboratory, Denmark. He noted gradual changes in vegetation, soil type and water content such that at one extreme was represented a "dry" association dominated by the herbaceous *Gnaphalium arenarium* (Compositae) and at the other, subject to periodic flooding, was a "wet" association of plant species characterized by the sedge *Carex demisa* and the composite *Hieracium auricula*. Various subdivisions of a predominantly *Agrostis* grassland could be distinguished between these two extremes. Fig. 4.4 shows the picture of the horizontal distribution, expressed as relative abundances, of some of the more conspicuous species of Collembola, mites and dipteran insects (chironomid larvae). It is clear that the Collembola and Prostigmata show a tendency to increase in the drier part of the transect, whereas the cryptostigmatid mites, and also the mesostigmatid mites which are not included in the figure, are relatively more abundant in the wetter parts. Other conclusions may be drawn from this study. For example, the soil fauna of the dry part of the gradient is dominated by a *Folsomia/Variatipes* grouping, and a "wet" association can also be defined in which *Tectocepheus velatus*, *Malaconothrus globiger* and *Scheloribates laevigatus* are numerically dominant. However, these two extremes are connected by a continuum of species, a broad region of intergradation in which there is a considerable

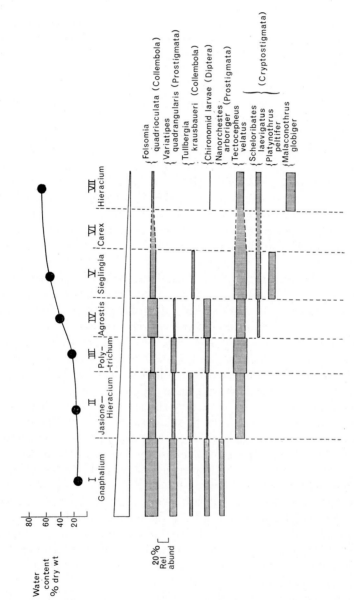

Fig. 4.4 Variations in the species composition of certain groups of soil microarthropods in relation to soil water content and vegetation across a sandy plain in Denmark. (After Weis-Fogh, 1948.)

overlap in species distribution and a gradual change in species composition.

It is but a short step from reaching these conclusions to deducing others which may or may not be correct. For example, it is tempting to suggest that *Folsomia quadrioculata* and *Variatipes quadrangularis* are "indicators" of dry conditions, whereas *Tectocepheus velatus* is an "indicator" of wet conditions. However, all three of these species are found in other situations in both wet and dry localities, and are evidently eurytopic. Such species must be distinguished from those which are associated, habitually, with certain moisture levels. *Malaconothrus globiger* is a good "indicator" of wet conditions, as is the collembolan *Isotomurus palustris*. The latter species, as its name suggests, is often abundant in soils bordering ponds and lakes. Haarløv (1960) found this springtail to be dominant in such a habitat in Jaegersborg Park, Copenhagen. He was also able to identify a "dry" association of soil microarthropods inhabiting exposed ant mounds, and dominated by the mite *Passalozetes perforatus*. Members of the genus *Passalozetes* are xerophilous and stenotopic and, therefore, are good indicators of dry conditions. Clearly, a knowledge of the tolerance limits of individual species is essential before their value as "indicators" of particular moisture conditions can be assessed.

Horizontal Effects: Temperature

Soil moisture gradients in grassland soils occur as a result of variations in topography and soil type, particularly in the extent to which these factors affect drainage characteristics. Slight variations in slope and soil porosity can produce moisture gradients. Horizontal variations in temperature conditions, on the other hand, are determined to a large extent by aspect and the structural features of the vegetation, although the fact that wet soils tend to be colder than dry ones cannot be overlooked. In the north temperate zone, north-facing slopes are generally colder and wetter than south-facing ones, and we may expect to find differences in the composition of the soil fauna in two sites of contrasting aspect. However, in many lowland grassland sites topographical variations are gradual, and the aspect factor is probably of minor significance in influencing the distribution of the soil fauna. It is much more important in alpine regions.

As far as the structure of the vegetation is concerned, we have already noted the effect of the tussock form on faunal distribution, and it is worth recalling that these clumps have a permanent fauna, distinct from that of the intervening, more exposed areas. Tussocks and other vegetational clumps, such as those of heather and bilberry which often occur in permanent pasture, provide shaded microhabitats in which temperature fluctuations are damped down. In clumps of heather, for example, daytime temperatures may be as much as 10°C lower than in more open areas (Delany, 1953). Grasslands can provide a mosaic of temperature conditions which is reflected in the discontinuous distribution of the soil fauna. This is particularly well illustrated by the ants, some of which prefer relatively exposed sites, for example *Tetramorium caespitum*, while others, such as *Lasius niger*, flourish in more shaded microhabitats (Gaspar, 1972). Similarly, termites respond to patterns of sun and shade in a way which facilitates habitat separation. Sands (1965) found different preferences among five species of the genus *Trinervitermes* in open woodland, scrub and grassland sites in northern Nigeria. The most abundant species, *T. ebenerianus*, achieved peak density in cleared grassland, *T. oeconomus* and *T. auriterrae* in shaded locations. *T. suspensus* was mainly restricted in its distribution to mounds of other species, and *T. carbonarius* had a distribution pattern which overlapped that of *T. ebenerianus*.

Animals such as the ants and termites, which spend an appreciable part of their time on the surface of grassland soils are more directly influenced by horizontal temperature gradients than is the subterranean fauna. The latter appears to be more strongly influenced by vertical temperature gradients, and we will be considering the effects of these in a moment. However the distribution of sun and shade does have an effect on the horizontal distribution of certain earthworms, notably the eudrilid *Hyperiodrilus africanus* and *Eudrilus eugeniae* in Nigeria (Madge, 1969) and the Acanthodrilidae in Uganda (Block and Banage, 1968). The distributions of *H. africanus* and *E. eugeniae* overlap in sites exposed to intermittent sun and shade, but the former species confines its activities mainly to shaded localities, whereas *E. eugeniae* predominates in permanently exposed sites. The acanthodrilid *Dichogaster* spp. are characteristic of savannah grassland and cultivated sites, whereas the eudrilid *Eminoscolex* spp. are restricted to the shaded soils of *Eucalyptus* and *Acacia* woodland in Uganda.

The soil provides a vertical, as well as a horizontal, dimension. Similarly, its microclimate varies vertically as well as horizontally. This variation is cyclical, and the particular cycles with which we are concerned are seasonal and diurnal in character. The following account is divided into two sections. The first deals with seasonal and diurnal temperature effects, the second with the seasonal effects of drought in which temperature and moisture factors are so closely inter-related that their effects cannot easily be separated.

Vertical Effects: Temperature

A seasonal response to temperature changes is shown both by vertebrates and invertebrates in their patterns of vertical distribution in the soil. Moles, for example, excavate temporary burrows close to the soil surface during the summer months, but retreat down the profile, often to a depth of 90 cm, where their permanent burrow systems are located, during the colder months. In chalk grasslands in southern England, the pill millipede *Glomeris marginata* and the woodlouse *Armadillidium vulgare* are active in the surface mat of vegetation during the summer, but withdraw into the deeper layers of the soil during winter. This seasonal pattern of vertical movement, which is also shown by lumbricid earthworms and molluscs, is probably typical of many members of the grassland soil fauna which can burrow or make use of existing soil cavities, and provides a means of escaping the freezing temperatures at the soil surface.

Fig. 4.5 A cryptostigmatid mite belonging to the family Scheloribatidae ($\times$ 222). Some members of this family undertake diurnal movements into the aerial parts of grassland vegetation, and are also intermediate hosts of sheep and cattle tapeworms. (Photo. by B. Parry and reproduced through the courtesy of the British Museum, N.H.)

Diurnal changes in vertical distribution patterns are more difficult to detect and, indeed, may not occur entirely within the physical boundaries of the soil. However, it has been established that soil-dwelling mites of the order Cryptostigmata (Fig. 4.5) show vertical movements which take them out of the soil and into the aerial parts of the vegetation (Rajski, 1960; Wallwork and Rodriguez, 1961). Generally speaking, the movement is upwards from the soil to vegetation during the daytime, and downwards at night. However, behaviour patterns are by no means uniform, and some species of mites show a tendency to move downwards into the soil during the midday period, while others maintain high densities in the vegetation throughout the day. The possibility cannot be ruled out that factors other than temperature may stimulate this behaviour. Ticks, for example, undertake similar vertical movements which appear to be governed by relative humidity conditions and the state of hydration of the animals (Lees, 1969). However, cryptostigmatid mites are very sensitive to temperature changes (Wallwork, 1960; Madge, 1965) and they may well be responding, at least in part, to diurnal variations in temperature when undertaking these movements. It is perhaps worth mentioning that these movements are of considerable economic importance, since several species of grassland cryptostigmatids have been shown to act as intermediate hosts in the life cycles of various anoplocephalid tapeworms of sheep and cattle (Rajski, 1959). This behaviour pattern facilitates the transmission of the parasites from the intermediate host to the grazing mammal which is the final host.

Vertical Effects: Drought

In localities where temperatures are high and rainfall is low during the summer months, drought conditions may develop in the upper layers of the grassland soil profile. The effect of these conditions on the soil fauna varies considerably from one group to another. Animals which cannot burrow through the soil, such as worms belonging to the family Enchytraeidae, may suffer severe mortality as a result of desiccation (Nielsen, 1955). The extent of this mortality is, however, a function of the species composition of the fauna, for certain enchytraeids belonging to the genera Enchytraeus, Fridericia and Henlea can produce resistant cocoons (O'Connor, 1967) which can survive dry conditions.

The effect of drought on burrowing members of the soil fauna, for example the lumbricid earthworms, is to drive them deeper down the profile (Gerard, 1967). A similar response may occur among the microarthropods which can utilize soil cavities, and this has been demonstrated in tropical grasslands where drought conditions are the rule, rather than the exception, during the dry season. Belfield (1956) observed a strong decrease in densities of soil arthropods in the top 2 in of a Ghanaian pasture soil during the dry season, while densities at lower levels increased at this time. This author also examined the effect of watering, artificially, grassland plots during the dry season, and found that when a treatment was applied, equivalent to 7 in of rain per month, maximum arthropod densities occurred in the top 2 in of the profile; this pattern was not significantly different from that observed before the start of the experiments. However, in plots which had not been treated, a significant difference occurred in the vertical distribution pattern, with maximum densities shifting to the 2–6 in level. This change was due to a movement of the main population centres and not to differential mortality, for no general decrease occurred in total numbers in the dry plots (Belfield, 1967). Indeed, Athias (1974) found that mite densities reached their maximum in rather dry (pF 4·2–4·7) and warm (33–44°C) soils in savannah sites in the Ivory Coast.

A similar pattern of downward movement has been recorded in steppe soils (Table 4.4). Here, the population density centre moves from the top 20 cm of the profile to the 20–40 cm level during the dry summer season, but reverses this trend during the autumn, both in natural steppe pasture and in semi-desert steppe. However, it is obvious from the data presented in Table 4.4 that the overall population density suffers a marked decline during the summer, followed by a partial recovery during the autumn. This summer

Table 4.4

Seasonal variation in the vertical distribution pattern of soil invertebrates in two steppe habitats in Azerbaydzhanskaya S.S.R., Russia (after Samedov, 1967).

		Density/m^2 at Various Soil Levels					
Habitat	Month	0–10 cm	10–20 cm	20–30 cm	30–40 cm	40–50 cm	Totals
Semi-desert	April	32·3	21·4	5·4	2·0	—	61·1
	July	1·2	3·8	12·4	4·9	1·0	23·2
	October	18·4	19·3	6·1	2·5	—	46·3
Natural steppe pasture	April	28·9	26·1	3·4	1·2	0·8	60·4
	July	—	2·3	9·1	5·2	1·1	17·7
	October	5·9	19·8	13·1	0·6	—	39·4

decline may have been brought about by mortality caused by drought conditions, particularly among the Lumbricidae, but the emergence of adult insects, such as the coleopteran Elateridae, and Tenebrionidae, Diptera and Lepidoptera, and their subsequent departure from the soil, may also have contributed to the decline in total numbers at this time. The effect of such emigration on the total populations obviously is most marked in soils where holometabolous insects have a high relative abundance. In both the semi-desert and natural steppe pasture habitats investigated by Samedov (1967), a group-by-group analysis of faunal density revealed that the Lumbricidae rarely contributed more than 10% of the total invertebrate macrofauna

which was dominated numerically by the coleopteran Carabidae and Curculionidae.

Chemical effects: calcium and pH

Various groups of the soil fauna are affected directly and indirectly by
chemical factors in the environment. Field observations and laboratory
experiments have shown, for example, that some soil animals respond directly
to the presence of calcium, or are intolerant of certain pH conditions. We
are concerned here mainly with these two responses. Indirect effects of the
chemical environment are exerted principally through food supply, for
example when pH conditions determine the abundance of bacteria and fungi
which form the diet of various protozoans, nematodes, enchytraeids and
microarthropods. This food factor was dealt with in the previous chapter.

Calcium

One of the most important elements in the chemical environment of the
soil is calcium. It is important because snails, millipedes, woodlice and
certain mites use calcium carbonate in relatively large amounts to strengthen
their external skeleton. This compound can constitute as much as 50% of
the dry weight of the body of the box mite *Steganacarus magnus* and various
iuloid millipedes. Calcium carbonate is a common constituent of grassland
soils which develop on chalk or limestone, and it is these sites which provide
good examples of the way in which chemical factors can influence the
composition of the soil fauna.

Earlier in this chapter, when microclimatic effects were considered, the
point was made that different groups of soil animals vary in their ability to
tolerate fluctuations in environmental conditions. It was also noted that the
fauna of a particular grassland site usually comprised a mixture of species
with narrow and wide tolerance ranges. These points can be emphasized
further by considering the distribution of snails in relation to the distribution
of calcium in the grassland environment.

Generally speaking, the richest concentrations of snails occur in calcium-
rich grassland soils (Lozek, 1962; Newell, 1967) (Fig. 4.6). Some species,
such as members of the genus *Chondrina*, *Alopia clathrata*, *Pyramidula
rupestris* and *Spelaeodiscus tatricus*, are confined to limestone sites in central
Europe. Others, such as members of the genus *Truncatellina*, *Clausilia
grimmeri*, *Caecilioides petitiana* and *Oxychilus inopinatus*, are associated
habitually with any kind of calcium-rich grassland soil, whether it is developed from limestone or not. A third ecological grouping comprises those
species which can tolerate low calcium concentrations in the soil; this group

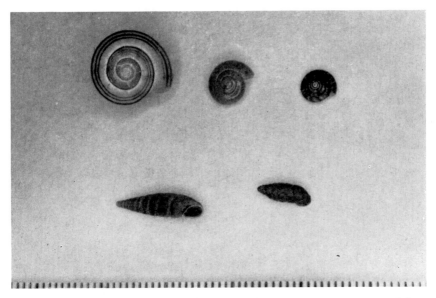

Fig. 4.6 Some common snails associated with calcareous grassland. Top row (left to right): *Cepaea nemoralis* (L.); *Oxychilus helveticus* (Blum); *Discus rotundatus* (Müller). Bottom row (left to right): *Clausilia dubia* (Draparnaud); *Abida secale* (Draparnaud). Scale in mm. (Photo. by A. Newman-Coburn.)

includes various species belonging to the genera *Laciniaria* and *Helicella*. Lozek (1962) found that all three of these ecological groups were represented in compact stony soils of European steppe grasslands, with the "intolerant" species outnumbering the "tolerant" by 4 to 1. In steppe soils formed from looser sediments, however, the intolerant and tolerant groups were equally represented. Clearly, this suggests that, while the presence of calcium is important, the species composition of the snail fauna may also be influenced by soil texture.

Although the millipedes, as a group, are essentially woodland animals, certain species occur commonly in grasslands and other open formation types. These include members of the Iuliformia, the "round-backs", whose calcified integument may account for as much as one-half of the dry weight of the body. It is to be expected, therefore, that this group of animals will be better represented on base-rich soils than on base-deficient ones, and this generally proves to be the case. The term "calcicole" has been applied to such species as *Archiboreoiulus pallidus*, *Cylindroiulus londinensis*, *C. nitidus* *Glomeris marginata* and *Tachypodoiulus niger* (Fig. 4.7), because of their association with calcareous soils (Blower, 1955; 1958). However, not all millipedes are so restricted in their distribution. The common British species

Fig. **4.7** The millipede *Tachypodoiulus niger* (Leach). (Photo. by A. Newman-Coburn.)

Blaniulus guttulatus and *Ophyiulus pilosus* occur in a variety of sites, and although they are infrequent on base-deficient soils they cannot be considered strictly as calcicoles. *Cylindroiulus punctatus*, on the other hand, has a wide distribution in base-deficient soils and is more frequent here than in calcareous soils. Evidently the presence of calcium in the soil is not equally important to all millipede species.

Investigations carried out to determine the reactions of millipedes to certain environmental factors (see Raw, 1967) have shown that moisture preferences are important in determining general patterns of distribution. However, localized patterns of distribution, within a particular moisture regimen, are undoubtedly influenced by the presence of calcium. Here, we have an example of the combined effect of physical (microclimatic) and chemical factors on the distribution of a group of soil animals. Another example is provided by the woodlice.

Terrestrial isopods are, like the millipedes, mainly woodland animals. They also resemble the millipedes in that the exoskeleton is hardened by the deposition of calcium carbonate, rather than by "tanning" as occurs in many other arthropods. However, the cuticular water-proofing mechanism in the woodlice is less efficient than that of the millipedes, and the general distribution of these crustaceans is strongly influenced by the moisture factor. They are best represented in moist calcareous soils of woodlands, although certain species, notably of the family Armadillidiidae, become established in the drier soils of grasslands. The ability of members of this family to colonize open sites is a result of a combination of physiological and behavioural features. These include a facility for reducing the basic transpiration rate which is further enhanced by the ability of these woodlice to conglobulate, i.e. to roll into a ball, thus exposing less surface area to the desiccating effect of the environment. These animals can also avoid severe drought conditions by descending deep in the profile. However, this tolerance of low moisture conditions has not enabled the Armadillidiidae to spread throughout all types of grassland. They rely on an adequate supply of calcium carbonate from the soil and, in Britain at least, are generally restricted to grassland soils developed on chalk or limestone.

pH

Closely related to the base status of the soil is its pH. Base-rich soils are neutral or slightly alkaline in reaction, whereas base-deficient soils tend to be acid. As a consequence, it is often difficult to determine whether soil animals are being influenced by the chemical composition of the soil, i.e. by the extent to which the soil is supplying certain essential elements, or merely by its degree of acidity or alkalinity. We have already looked at some examples of animals whose distribution is related to the presence of calcium in the soil. In contrast to these, there are two important groups of soil animals which appear to be directly influenced by pH conditions. These are the earthworms belonging to the family Lumbricidae, and members of the Phylum Protozoa.

Lumbricid earthworms are familiar examples of grassland soil animals. It has often been suggested that it is the activities of this group which are responsible for the deep mixing of organic material and the mull humus formation so characteristic of all but the most acidic grassland soils. Field observations have suggested that earthworm species vary in their ability to tolerate acid conditions. Thus, *Allolobophora caliginosa, A. nocturna, A. chlorotica, A. longa* and *Eisenia rosea* are considered as "acid-tolerant" since they never occur in soils with a pH lower than 4.5. Species such as *Lumbricus rubellus, L. terrestris, L. castaneus, Octolasion cyaneum* and *Dendrobaena subrubicunda*, on the other hand, can become established in soils where the pH may range from about 7·0 down to 3·8. Other species, for example *Bimastos eiseni, Dendrobaena octaedra* and *D. rubida* usually occur in acid soils where the pH rarely, if ever, exceeds 5·0 (Satchell, 1955, 1967; Nordström and Rundgren, 1974). That these site preferences are a direct response to pH conditions and not to the availability of calcium is suggested by experimental work (Laverack, 1963). Various species of earthworms were subjected to a series of buffer solutions, representing a range of pH, and the electrical activity of the segmental nerves was recorded. The threshold for stimulation varied with the species, but it was always closely correlated with the pH conditions preferred by the species in the field. On the other hand, Piearce (1972) has suggested that factors other than pH may be important in determining distribution patterns of lumbricid earthworms.

The Phylum Protozoa is represented, in the main, by four groups in grass-land soils, namely ciliates, testaceans, naked amoebae and flagellates. Ciliates, amoebae and flagellates, which are usually more abundant than testaceans in open sites, are not particularly sensitive to pH and will grow over a wide range of conditions (Stout and Heal, 1967). The natural distribution of testaceans suggests that pH conditions are important (Bonnet, 1961). Some species, such as *Centropyxis deflandriana*, *Corythion dubium* and *Arcella arenaria* are most frequent in soils with a neutral or slightly alkaline pH, whereas *Centropyxis vandeli* prefers more acid conditions. A preference for more alkaline soils is shown by *Geopyxella sylvicola* and *Centropyxis halophila* (Fig. 4.8). The reasons for these preferences are not obvious for these field observations have not, as yet, been backed up with much experimental work. That pH conditions may directly affect physiological processes is suggested

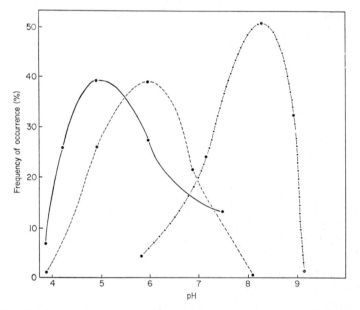

Fig. 4.8 Distribution of three species of testate Protozoa in relation to pH. (After Bonnet, 1961.) *Centropyxis vandeli* ———, *Corythion dubium* ·– – – –·, *Geopyxella sylvicola* ·—·—·—

by the discovery that *Difflugia tuberculata* will not reproduce below pH 4·5 (Heal, 1964). There are, however, other possibilities. The distribution of testaceans may be related, in some cases, to the availability of calcium which may be used to construct the test. Since protozoans are members of the soil-water fauna, the moisture factor is also important in their distribution and may act in conjunction with pH and calcium content. Protozoa feed mainly on soil microflora which, themselves, are influenced by pH conditions. Thus, chemical factors may act indirectly through the food factor.

Summary

The principal abiotic variables affecting the character and distribution of the grassland soil fauna are soil structure, the structure of the vegetation, microclimate and the chemical factors of base content and pH.

Soil structure has a marked influence on the vertical distribution of many of the larger, burrowing soil animals. Lumbricid earthworms, molluscs, millipedes and scarabaeid beetles have more extensive depth distributions in loose soils than in compact ones. Loose textured soils are also selected as oviposition sites by various flies and arachnids.

The structure of the vegetation acts directly on the ground-dwelling fauna in two ways. It presents obstacles to the free movement of surface-active predators, such as certain carabid beetles. It also provides structural components which can be used by various spiders for supporting their webs.

Indirectly, the surface vegetation affects the soil fauna through the way in which it influences the microclimate of the soil. Where growth is of the tussock form, a mosaic of microclimatic conditions is produced, with shaded areas interspersed with more open, exposed areas. These variations in sun and shade have a marked effect on the horizontal distribution patterns of members of the surface fauna, for example, ants, termites, woodlice, Collembola and mites. Among the subterranean fauna, the distribution of some tropical earthworms evidently is related to microclimatic conditions at the surface.

The tussock fauna has two components, namely a permanent element consisting of species which habitually occur here more frequently than they do elsewhere, and a temporary element which seeks refuge in the tussock when conditions in more open habitats become unfavourable for normal activity.

Microclimatic gradients occur not only horizontally but also vertically through the profile and exert a considerable, direct influence on the horizontal and vertical distribution of the fauna. In using these distributions to identify species which are "indicators" of particular microclimatic conditions, account must be taken of ecological tolerances. In particular, species which can tolerate a range of microclimatic conditions (eurytopic) must be distinguished from those with a more restricted tolerance (stenotopic).

Stenotopic species can be divided into "wet" and "non-wet" categories depending on their moisture preferences. These categories are sharply defined in the beetle family Carabidae in which the life cycles are geared to seasonal changes in microclimate. Thus, "wet" carabids breed in spring or summer in Britain, and overwinter as adults, whereas the majority of "non-wet" species breed in autumn and overwinter as larvae.

The grassland soil fauna also responds to seasonal variations in microclimate by altering the pattern of vertical distribution in the soil. During winter and periods of drought, many lumbricid earthworms, molluscs, millipedes, woodlice and microarthropods move from the superficial layers into the deeper horizons, and reverse this trend during the spring and summer. Similar movements, on a diurnal time scale, are shown by certain cryptostigmatid mites.

Chemical factors, notably calcium content and pH, affect some groups of the grassland fauna directly. Snails, millipedes, woodlice and phthiracaroid mites use calcium carbonate in relatively massive quantities to strengthen the exoskeleton. These groups flourish where the parent material is chalk or limestone. Many lumbricid earthworms are intolerant of low pH conditions, and the species commonly encountered in grasslands are those which prefer a pH of 4·5 or higher.

Soil structure, vegetation type and chemical characteristics are relatively static features of the soil environment. Their effects on the soil fauna are essentially qualitative, i.e. they determine the presence or absence of a species in a particular place. Microclimatic factors, on the other hand, are dynamic in character, with diurnal and seasonal periodicities. Their effects are not only qualitative but also quantitative. They promote cyclical shifts, horizontally and vertically, in the centres of population density.

REFERENCES

ATHIAS, F. (1974). Note préliminaire sur l'importance de certains facteurs mésologiques vis à vis de l'abondance des Acariens d'une Savane de Côte d'Ivoire, *Rev. Ecol. Biol. Sol.*, **11**, 99–125.

BELFIELD, W. (1956). The Arthropoda of the soil in a West African pasture, *J. Anim. Ecol.*, **25**, 275–87.

BELFIELD, W. (1967). The effects of overhead watering on the meiofauna in a West African pasture. *In* "Progress in Soil Biology", 192–210 (Graff, O. and Satchell, J. E., eds.), North Holland, Amsterdam.

BLOCK, W. C. and BANAGE, W. B. (1968). Population density and biomass of earthworms in some Uganda soils, *Rev. Ecol. Biol. Sol.*, **5**, 515–21.

BLOWER, J. G. (1955). Millipedes and centipedes as soil animals. *In* "Soil Zoology", 138–51 (Kevan, D. K. McE., ed.), Butterworths, London.

BLOWER, J. G. (1958). British millipedes (Diplopoda), *Linnean Society of London Synopses of the British Fauna*, no. 11.

BONNET, L. (1961). Caractères généraux des populations thécamoebiennes endogées, *Pedobiologia*, **1**, 6–24.

BOYD, J. M. (1960). Studies on the differences between the fauna of grazed and ungrazed grassland in Tiree, Argyll, *Proc. zool. Soc. Lond.*, **135**, 33–54.

DELANY, M. J. (1953). Studies on the microclimate of *Calluna* heathland, *J. Anim. Ecol.*, **22**, 227–39.

DUFFEY, E. (1962). A population study of spiders in limestone grassland, *J. Anim. Ecol.*, **31**, 571–99.

DUFFEY, E. (1963). Ecological studies on the spider fauna of the Malham Tarn area, *Field Studies Council Rep.*, **1**, 23 pp.

DUFFEY, E. (1968). An ecological analysis of the spider fauna of sand dunes, *J. Anim. Ecol.*, **37**, 641–74.

ELMES, G. W. (1974). The spatial distribution of a population of two ant species living in limestone grassland, *Pedobiologia*, **14**, 412–18.

FORD, J. (1937). Fluctuations in natural populations of Collembola and Acarina, *J. Anim. Ecol.*, **6**, 98–111.

GASPAR, C. (1972). Les Fourmis de la Famenne. III. Une étude écologique, *Rev. Ecol. Biol. Sol.*, **9**, 99–125.

GEIGER, R. (1959). "The Climate Near the Ground". Harvard Univ. Press, Cambridge, Mass. U.S.A.

GERARD, B. M. (1967). Factors affecting earthworms in pastures, *J. Anim. Ecol.*, **36**, 235–52.

HAARLØV, N. (1960). Microarthropods from Danish soils. Ecology, phenology, *Oikos*, suppl. **3**, 176 pp.

HEAL, O. W. (1964). Observations on the seasonal and spatial distribution of Testacea (Protozoa: Rhizopoda) in *Sphagnum*, *J. Anim. Ecol.*, **33**, 395–412.

LAVERACK, M. S. (1963). "The Physiology of Earthworms". Pergamon Press, London.

LEES, A. D. (1969). The behaviour and physiology of ticks, *Acarologia*, **9**, 397–410.

LOZEK, V. (1962). Soil conditions and their influence on terrestrial Gasteropoda in central Europe. *In* "Progress in Soil Zoology", 334–42 (Murphy, P. W., ed.), Butterworths, London.

LUFF, M. L. (1966). The abundance and diversity of the beetle fauna of grass tussocks, *J. Anim. Ecol.*, **35**, 189–208.

MADGE, D. S. (1965). The behaviour of *Belba geniculosa* Oudms. and certain other species of oribatid mites in controlled temperature gradients, *Acarologia*, **7**, 389–406.

MADGE, D. S. (1969). Field and laboratory studies on the activities of two species of tropical earthworms, *Pedobiologia*, **9**, 188–214.

MORRIS, M. G. (1968). Differences between the invertebrate faunas of grazed and ungrazed chalk grassland. II. The faunas of sample turves, *J. appl. Ecol.*, **5**, 601–11.

MURDOCH, W. W. (1967). Life history patterns of some British Carabidae (Coleoptera) and their ecological significance, *Oikos*, **18**, 25–32.

NADVORNYJ, V. G. (1968). Wireworms (Coleoptera, Elateridae) of the Smolensk region, their distribution and incidence in different soil types, *Pedobiologia*, **8**, 296–305.

NEWELL, P. F. (1967). Mollusca. *In* "Soil Biology", 413–33 (Burges, N. and Raw, F., eds.). Academic Press, London and New York.

NIELSEN, C. O. (1949). Studies on the soil microfauna II. The soil-inhabiting nematodes, *Natura jutl.*, **2**, 1–131.

NIELSEN, C. O. (1955). Survey of a year's results obtained by a recent method for the extraction of soil-inhabiting enchytraeid worms. *In* "Soil Zoology", 202–14 (Kevan, D. K. McE., ed.), Butterworths, London.

NORDSTRÖM, S. and RUNDGREN, S. (1974). Environmental factors and lumbricid associations in southern Sweden, *Pedobiologia*, **14**, 1–27.

O'Connor, F. B. (1967). The Enchytraeidae. *In* "Soil Biology", 213–17 (Burges, N. and Raw, F., eds.), Academic Press, London and New York.

Paris, O. H. (1963). The ecology of *Armadillidium vulgare* (Isopoda: Oniscoidea) in California grassland: food, enemies and weather, *Ecol. Monogr.*, **33**, 1–22.

Pearce, E. J. (1948). The invertebrate fauna of grass tussocks: a suggested line for ecological study, *Ent. mon. Mag.*, **84**, 169–74.

Pierce, T. G. (1972). Acid intolerant and ubiquitous Lumbricidae in selected habitats in North Wales, *J. Anim. Ecol.*, **41**, 397–410.

Pontin, A. J. (1960). Field experiments on colony foundation by *Lasius niger* (L.) and *L. flavus* (F.) (Hym. Formicidae), *Insectes sociaux*, **7**, 227–30.

Pontin, A. J. (1963). Further considerations of competition and the ecology of the ants *Lasius flavus* (F.) and *L. niger* (L.), *J. Anim. Ecol.*, **32**, 565–74.

Rajski, A. (1959). Moss-mites (Acari: Oribatei) as intermediate hosts of Anoplocephalata (Review), *Zesz. nauk. Uniw. Mick. Biol.*, **2**, 163–92.

Rajski, A. (1960). Quantitative occurrence of the chief intermediate hosts of *Moniezia* (*M.*) *expansa* (Rud.) in the vicinity of Poznan, (in Polish), *Wiad. Parazyt.*, **7**. 39–42.

Raw, F. (1955). The effect of soil conditions on wheat bulb fly oviposition, *Pl. Path.*, **4**, 114–17.

Raw, F. (1967). Arthropoda (except Acari and Collembola). *In* "Soil Biology", 323–62 (Burges, N. and Raw, F., eds.), Academic Press, London and New York.

Riechert, S. E., Reeder, W. G. and Allen, T. A. (1973). Patterns of spider distribution (*Agelenopsis aperta* [Gertsch]) in desert grassland and recent lava bed habitats, south-central New Mexico, *J. Anim. Ecol.*, **42**, 19–35.

Samedov, N. G. (1967). Uber den charakter der Verteilung einiger Wirbellosengruppen in den Grauerden der Sirvan-Steppe in Aserbaidjan, *Pedobiologia*, **7**, 239–46.

Sands, W. A. (1965). Termite distribution in man-modified habitats in West Africa, with special reference to species segregation in the genus *Trinervitermes* (Isoptera, Termitidae Nasutitermitinae), *J. Anim. Ecol.*, **34**, 557–71.

Satchell, J. E. (1955). Some aspects of earthworm ecology. *In* "Soil Zoology", 180–201 (Kevan, D. K. McE., ed.), Butterworths, London.

Satchell, J. E. (1967). Lumbricidae. *In* "Soil Biology", 259–322 (Burges, N. and Raw, F., eds.), Academic Press, London and New York.

Southwood, T. R. E. and van Emden, H. F. (1967). A comparison of the fauna of cut and uncut grasslands, *Z. angew. Ent.*, **60**, 188–98.

Stout, J. D. and Heal, O. W. (1967). Protozoa. *In* "Soil Biology", 149–95, (Burges, N. and Raw, F., eds.), Academic Press, London and New Y rk.

Tietze, F. (1968). Untersuchungen über die Beziehungen zwischen Bodenfeuchte und Carabiden-besiedlung in Wiesengesellschaften, *Pedobiologia*, **8**, 50–8.

Vlijm, L. and Kessler-Geschiere, A. M. (1967). The phenology and habitat of *Pardosa monticola*, *P. nigriceps* and *P. pullata* (Araneae, Lycosidae), *J. Anim. Ecol.*, **36**, 31–56.

Wallwork, J. A. (1960). Observations on the behaviour of some oribatid mites in experimentally-controlled temperature gradients, *Proc. zool. Soc. Lond.*, **135**, 619–29.

Wallwork, J. A. and Rodriguez, J. G. (1961). Ecological studies on oribatid mites with particular reference to their role as intermediate hosts of anoplocephalid cestodes, *J. econ. Ent.*, **54**, 701–5.

Weis-Fogh, T. (1948). Ecological investigations on mites and collemboles in the soil, *Natura jutl.*, **1**, 135–270.

Williams, G. (1959). Seasonal and diurnal activity of Carabidae, with particular reference to *Nebria*, *Notiophilus* and *Feronia*, *J. Anim. Ecol.*, **28**, 309–30.

Chapter 5

Agricultural Practice and the Soil Fauna

Natural systems, such as the grasslands discussed in the last two chapters, are usually very complex, ecologically speaking. The interpretation of distribution patterns and the identification of environmental factors which influence these patterns are often difficult, and, to some extent, speculative. In the present chapter we consider the more clearly defined effects brought about by Man's intervention in a natural system; such effects are usually quite dramatic. It is particularly important to determine the extent of the disturbance such intervention may cause, for a great many human activities damage or destroy, unwittingly or deliberately, natural ecosystems and resources on which Man ultimately relies. Too often in the past, the conversion of natural sites into man-managed agricultural systems has been carried out without a sufficient knowledge of the total ecological effect involved; where short-term gains have led to long-term destruction, for example by promoting soil erosion or the over-exploitation of plant nutrients. It is now recognized that an ecological approach, through conservation, is the only way Man can exploit his natural resources without endangering their potential for future generations. To this end, the United States Soil Conservation Service devised a system of classifying land according to its agricultural capability; this system has subsequently been adopted in Great Britain.

Land capability

The scheme recognizes eight land capability classes, the first four of which are suitable for some form of cultivation (Table 5.1). The basic principle involved here is to allow an assessment of the quality of land, from an agricultural point of view; management practices can then be introduced to ensure that soil erosion is avoided, that adequate drainage or irrigation facilities are provided, and that essential nutrients are not depleted. Such practices undoubtedly change the character of the soil environment, and their effects on the soil community must be examined. For example, the

Table 5.1

A classification of arable soils according to their capability.

Class	Characteristics	Recommended Management
I	Very good land; soils on level ground; deep and fertile; good drainage; suitable for permanent cultivation with few restrictions.	Can be cultivated intensively; organic and inorganic additives required periodically.
II	Good land; soils on slight slopes; subject to slight erosion and slight nutrient deficiency; imperfectly drained; suitable for permanent cultivation with some minor conservation.	Installation of drainage systems; some irrigation may be required; cover crops to reduce erosion.
III	Moderately good land; suffers from either deficient drainage or drought; deficiencies of lime, potassium and manganese common; shallow soils suitable for cultivation only after intensive conservation.	Strip cropping, cover crops and crop rotation needed; terracing and fertilization also necessary.
IV	Fairly good land; soils on steep slopes; deficient drainage; subject to erosion; not suitable for intensive cultivation.	Cultivation only once every 5 or 6 years.

very act of ploughing the soil prior to planting provides suitable conditions for the development of insect pests which will, in time, attack the crop. Again, plant-parasitic nematodes can be encouraged or controlled, depending on the kind of crop rotation used. Management practices can have their undesirable side effects with which the farmer may have to live. On the other hand, a more detailed knowledge of the way in which these practices influence the character of the soil community may suggest modifications which would eliminate or minimize such unfavourable effects.

The agricultural practices we will be considering fall broadly into two categories; those that are detrimental to the soil fauna, and those that are beneficial. They are listed as follows;

Detrimental	Beneficial
Ploughing	Fertilizer application
Crop culture	Drainage
Pesticide application	Irrigation
Inorganic pollutants	Maintenance of hedgerows

It will be recognized that these two categories are not mutually exclusive. As we will see, ploughing has detrimental effects on certain members of the soil fauna, beneficial effects on others. Similarly, some types of irrigation can affect this fauna adversely.

The impact of these practices on the soil fauna varies from place to place and, in this respect, the land capability classification outlined in Table 5.1 is instructive. It can be appreciated that the effect of ploughing, for example, will be most intensive in soils of Class I, and least important in those of Class IV. The effect of fertilizers will be minimal in Class I soils, but increasing in soils of Classes III and IV. Again, the effect of crop culture can be expected to be an important variable in soil Classes I, II and III, where monocultures are repeated annually or seasonally and, alternatively, where crop rotations are practised.

Ploughing

Ploughing has several important effects on the soil environment. Some of these are of benefit to the soil fauna, others are detrimental. On balance, the adverse effects over-ride the beneficial ones, and lead to a reduction in diversity and a lowering of population densities of individual species. Among the detrimental effects may be cited:

(1) The repeated mechanical disturbance of the substratum causes unstable microclimatic conditions in the profile, and also abrasive effects which can produce high mortality among soil animals.

(2) The removal of permanent plant cover or a protective surface of leaf litter exposes the upper soil layers to marked fluctuations in temperature and moisture, and to the possibility of erosion. Compared with soils under grassland or forest cover, arable soils freeze in winter and dry out in summer to a greater depth.

The beneficial effects derive from the fact that:

(1) There is a thorough mixing of mineral soil and surface organic material, producing a relatively homogeneous profile with appreciable organic and mineral fractions throughout.

(2) The substratum is loosened down to a depth of 6–9 in (15–23 cm), and aeration is improved.

Both of these beneficial effects combine to increase the amount of suitable living space for soil organisms. Organic food material is distributed to a greater depth than in undisturbed profiles, and the loose soil is more easily worked by burrowing animals.

Detrimental Effects

The rather harsh microclimatic conditions occurring in the upper parts of the arable soil profile may impose severe restrictions on the richness and variety of the permanent soil fauna. Many non-burrowing microarthropods, slugs and even some shallow-burrowing earthworms show seasonal declines in population size which may be related to drought or frost conditions

Fig. 5.1 Arable landscape in Buckinghamshire, England. Soil type is clay with flints, subject to some erosion on steeper slopes. Main crops are cereals. (Photo. by author.)

(Satchell, 1967; Hunter, 1966; Wallwork, 1970). In arable soils, exposure to such conditions will be maximal, and the composition of the soil fauna may reflect this. A good illustration is provided by the comparison made by Raw (1967) of the arthropod composition and densities in two similar soil types (clay with flints), one being grassland, the other permanent arable land of the type shown in Fig. 5.1. This comparison is given in Fig. 5.2, from which it can be seen that the total arthropod density in the arable soil is less than one-tenth of that in the grassland site, and that several groups present in the latter, notably Araneida, Thysanura, Protura and Pauropoda, are absent from the former. However, the major part of the difference in density between the two sites is accounted for by the strong reduction in numbers of Collembola and Acari in arable soil. In this type of habitat, the most successful groups will be those which can produce a resistant stage in the life cycle. These will include the cyst-forming Protozoa, Rotifera and Nematoda, together with certain enchytraeid worms, such as species of *Enchytraeus*, *Fridericia* and *Henlea*, which cover their cocoons with a protective layer of sand grains and organic material (O'Connor, 1967). Among the arthropods will be representatives of the Astigmata, a group of mites which can include a "resting" stage (hypopus) in the life cycle, and also the holometabolous insects which have a resistant pupal stage. The latter are well represented by the carabid beetles; Greenslade (1964) recorded a density of $80/m^2$ in arable sites, compared with $28/m^2$ and $1.33/m^2$ mean densities in grass heath and beech litter, respectively. Basedow (1973) has also reported high densities of the carabids *Agonum dorsale* and *Feronia vulgaris* in fields cultivated with cereals. However, some holometabolous insects are affected adversely by cultivation. Wireworms, for example, flourish in uncultivated grassland soils, but when cultivation is introduced the species

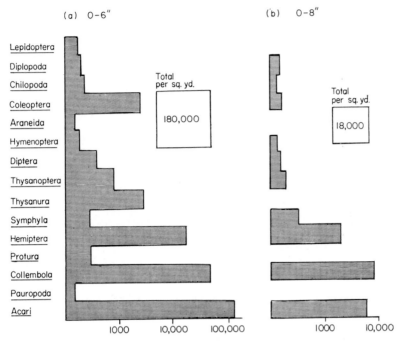

Fig. 5.2 A comparison between soil arthropod densities (numbers per sq. yd.) of (a) permanent grassland to a depth of 6 in (15 cm), and (b) permanent arable land to a depth of 8 in (20 cm). (From Raw, 1967.)

diversity and population densities are reduced. Members of the genera *Selatosomus*, *Adelocera*, *Athous* and *Melanotus* are particularly intolerant of cultivation, members of the genus *Agriotes* less so (Dirlbek *et al.*, 1973).

Beneficial Effects

Ploughing often results in an expansion of the vertical distribution range of permanent members of the soil fauna. A good example is the common slug *Agriolimax reticulatus* (Fig. 5.3) which is often encountered in cultivated and uncultivated soils in Britain. The bulk of the population of this species usually occurs in the top 3 in (8 cm) of the soil in uncultivated sites. However, a comparison of the depth distribution before and after ploughing has shown that the population can extend its vertical range down to 9 in (23 cm) after ploughing (Hunter, 1966). Similarly, the distinction between "deep-burrowing" and "surface-burrowing" lumbricid earthworms may largely disappear in ploughed soils. Species such as *Lumbricus rubellus* and members of the genus *Dendrobaena*, which live in the upper part of the

5

Fig. **5.3** The slug *Agriolimax reticulatus* (Müller). (Photo. by A. Newman-Coburn.)

profile in undisturbed grassland soils, may penetrate into the deeper layers after ploughing (Guild, 1955). Soil arthropods are non-burrowers, for the most part, and their concentration in the upper part of an undisturbed profile is related to the greater amount of living space there. Since ploughing usually creates more inhabitable space in the deeper layers, it is to be expected that these arthropods would show an extension of their vertical range in cultivated soils, comparable with that shown by the slugs and the lumbricids. However, population densities of microarthropods are low in cultivated soils, possibly because the mechanical disturbance affects the stability of their environment, and changes in vertical distribution patterns are difficult to detect.

Effect on Pest Populations

Ploughing encourages the establishment of populations of various agricultural pests, such as larval Coleoptera, Diptera and Lepidoptera which feed extensively on the roots of crops. Although the distribution of these animals is influenced by the kinds of crops present, and we will be considering this effect later, they also favour the loose-textured soils which ploughing produces. Among the Coleoptera, phytophagous larvae are common in the Scarabaeidae, the familiar "white grubs". Especially important are larvae of the Melolonthinae, a group of chafer beetles (Fig. 5.4) widely distributed throughout the world and infesting the roots of grasses, legumes, fruit bushes, shrubs and trees (Ritcher, 1958). Another group of scarabaeids, the Ruteli-

Fig. 5.4 The common cockchafer beetle *Melolontha melolontha* (L.). The larvae of this beetle are serious pests on the roots of cultivated crops. (Photo. by A. Newman-Coburn.)

nae, also includes a number of genera which have economically important larvae because of their habit of feeding on the roots of pasture grasses. Many scarabaeid larvae prefer to feed on dung or organic detritus, but when population densities are high, or the normal food supply scarce, members of the Aphodinae and Dynastinae will attack living root material, particularly of grasses. Phytophagous scarabaeid larvae, which are common in arable soils, have established a symbiotic relationship with gut flagellates and bacteria which are capable of digesting the cellulose present in their diet.

Phytophagous Diptera larvae, common in ploughed soils and consequently of economic importance in Britain, include the cabbage root fly, *Erioischia brassicae*, St. Mark's fly, *Bibio marci*, crane flies of the genus *Tipula*, the carrot root fly, *Psila rosae*, and various pests of grasses and cereals, such as the frit fly, *Oscinella frit*, the wheat bulb fly, *Leptohylemyia coarctata*, and *Opomyza florum*. The biology of the wheat bulb fly has been studied in detail by Raw (1955, 1967). Oviposition in this species occurs in the soil and, on hatching, the larvae migrate through the substratum in search of the roots of suitable host plants, which are grasses, winter wheat, barley and rye. The degree of larval infestation apparently varies with the soil conditions during the oviposition period, for eggs were more abundant in plots with rough tilth, and in cultivated rather than in uncultivated soils (Table 5.2). Evidently, ploughing the soil provides more suitable oviposition

Table 5.2

Effect of soil conditions on wheat bulb fly oviposition (number of wheat bulb fly eggs per ft²) (from Raw, 1967).

| | Plots with | |
	Rough Tilth	Smooth Tilth
Cultivated plots	94	77
Uncultivated plots	78	39

sites, probably because of the better cavity structure produced, and this affords protection for the eggs from the attentions of predatory carabid and staphylinid beetles.

Although phytophagous caterpillars of the Lepidoptera are most commonly associated with the aerial parts of the vegetation, a few species of noctuid moths, such as *Porosagrotis orthogonia* and, possibly, *Agrotis ypsilon* have subterranean larvae which feed on living root material. Larval noctuids are familiarly known as "cutworms", from their habit of severing completely the stem or root on which they are feeding. The densities of some soil-dwelling members of this group are influenced by the physical character of the soil (Singh, 1955), for they prefer the lighter, loose soils of ploughed sites, and are poorly represented in wet, marshy soils and in compacted clay or sand.

Crop culture

The influence of crop type on horizontal distribution patterns of many groups of soil animals, notably mites and Collembola, is much less marked than the effects of ploughing and soil type. The ecosystem of which these saprophages, fungivores and predators are a part is based on the nutrients and energy present in a wide variety of decaying organic material. Within limits, it is not so much the kind of food material available that is the determining factor, but rather its quality. For example, nitrogen-rich accumulations of organic litter provide more suitable food sources than those with a high carbon/nitrogen ratio. Any effect that a crop species may have on the species composition of the soil fauna and on population densities will, most probably, be a reflection of the characteristic carbon/nitrogen ratio of the particular plant type. Although broad distinctions can be made between different groups of plant species, on the basis of this criterion, differences of C/N ratios between species within a particular group, for example between grasses and cereals, are too small to produce any marked effect on the soil fauna. Coupled with this is the fact that many soil animals can adapt to whatever is available in the way of food. In these circumstances a close relationship between crop type and faunal distribution is not to be expected and has not been demonstrated.

Crop type may influence the distribution of those members of the soil fauna which are specifically associated with particular food plants. Monocultures, for example, will eliminate those animal species which are associated with other plants, and it has been suggested (Edwards and Lofty, 1969) that crop rotation decreases species diversity to an even greater extent. To illustrate this phenomenon, examples are chosen from nematode root parasites and root-feeding insect larvae belonging to the orders Coleoptera, Diptera and Lepidoptera, and various molluscs. However, even in these groups absolute specificity is uncommon.

Root-Feeders: Nematodes

In Britain, plant-parasitic nematodes can be divided into two ecological types, namely the migratory and the sedentary forms. Migratory species may be ectoparasitic (the dorylaimoid *Longidorus*, *Xiphinema* and *Trichodorus* groups and the tylenchid *Tylenchorhynchus*), or endoparasitic (*Pratylenchus* spp.). In these groups, both sexes are vermiform in shape and free-moving throughout the life cycle, migrating from one root to another as larvae and adults. Sedentary forms are highly specialized root parasites, some of which form cysts (*Heterodera* spp.) and others galls (*Meloidogyne* spp. and *Ditylenchus radicicola*).

These plant parasites can be arranged in a graded series to reflect an increasing degree of specialization of feeding habit, an increasing dependence on a particular host plant and, consequently, an increasing influence of plant type on distribution. At one extreme can be placed many of the migratory forms which, by virtue of their mobility can move from the roots of one plant to another. This mobility provides them with an opportunity to select from a variety of potential hosts. This might militate against any tendency to develop a strong dependence on one particular host plant. This suggestion is confirmed by the findings of Winslow (1964), who investigated the migratory nematode fauna in the soil of a six-course rotation involving sugar beet, spring barley, red clover, winter wheat, maincrop potatoes and winter rye on two different soil types. In general, soil type had a much greater effect on nematode distribution than did crop type. For example, the spiral nematodes (Hoplolaiminae) were prominent in flinty loam/clay soil, associated with all crops in the rotation, whereas this group was very infrequent in sandy loam. The same was true of members of the genus *Trophurus*. Exceptions to this general rule were few, notably members of the genera *Pratylenchus* and *Tylenchorhynchus* which were much more commonly associated with cereals and clover than with beet and potatoes in both sites. However, this restriction may be an indication of the preference of these nematodes for the undisturbed profiles under wheat and clover, rather than a specialized feeding habit.

A much greater degree of specialization to particular crops is shown by eel-worms (*Heterodera* and *Meloidogyne* spp.), although this specialization is rarely, if ever, absolute. Common pest species in Britain include *Heterodera rostochiensis* (potato root eel-worm), *H. gottingiana* (pea root eel-worm) and root-knot eel-worms belonging to the genus *Meloidogyne* which are often associated with glasshouse tomatoes and cucumbers (Brown, 1955). Members of the latter genus apparently have a wider range of host species than *Heterodera* species which, however, are able to persist in the soil in the absence of a suitable host crop for several years. When a suitable host crop is not available for the root-knot eel-worm, various weed species are attacked, although grasses are not particularly susceptible to infestation. This fact is utilized in the cultivation of tobacco, a crop which is very prone to root-knot eel-worm attack, and rotations involving various grass species are developed to control the numbers of this pest.

Root-Feeders: Insect Larvae

Root-feeding beetle larvae belong mainly to the families Elateridae (wireworms) and Scarabaeidae (white grubs). Wireworms often occur in large numbers in old grassland soils. When these soils are ploughed over and planted with potatoes, onions, lettuce, celery, beet, mangolds, swedes, grass, oats or wheat, populations of *Agriotes* spp. continue to flourish as pests of these crops. These wireworms are less successful in fields planted with lucerne, beans, peas, flax, mustard, barley or rye (Edwards and Heath, 1964). In view of the wide range of host plants susceptible to these larvae, it is probable that crop type will not have a marked effect on distribution. Indeed, the largest populations of wireworms, of the order of 1–2 million per acre, usually occur in undisturbed grassland. Ploughing decreases the numbers, and this suggests that quantitative variations in the horizontal distribution pattern are influenced by mechanical disturbance of the soil. Soil texture was cited by Nadvornyj (1968) as the most important factor influencing the species composition of the elaterid fauna in the Smolensk region of Russia. Here, *Agriotes* spp. predominate in heavy soils, whereas members of the genera *Athous*, *Selatosomus* and *Limonius* come into greater prominence on lighter loam or sandy loam soils.

Much of what has been said about elaterid larvae applies equally well to larval scarabaeids. White grubs are most plentiful in the soils of grasslands, where they feed on root systems. They can also become serious pests of crops when the grassland is ploughed. Larvae of the common cockchafer, *Melolontha melolontha* (Fig. 5.4), attack a wide range of hosts (p. 116–17), and a similar habit is shown by the larvae of *Phyllopertha horticola*, the garden chafer, which is particularly numerous on light calcareous soils in Britain.

The lack of host specificity shown by the soil-dwelling phytophagous Coleoptera larvae is also a characteristic of many of the Diptera larvae which feed on cultivated crops. Larval tipulids, of which *Tipula paludosa* and *Nephrotoma maculata* are common British pests, often build up large populations in grassland soils, but also cause considerable damage to cereals, root crops, lettuce and brassicas. The host range for the stem-boring larvae of *Opomyza florum* and *Oscinella frit* includes various grasses and cereals, whereas the distribution of the larvae of *Psila rosae* appears to be more closely correlated with the distribution of its principal host, the carrot.

The most important Lepidoptera larvae, from the economic point of view, in arable soils belong to the family Noctuidae, the cutworms. As already noted, (p. 118) the horizontal distribution of cutworms, as a group, appears to be influenced by soil texture, for they are more commonly encountered in light soils than in heavy ones. Members of the genus *Agrotis* generally have a wide host range including beans, various root crops, such as potatoes, beet, swedes and turnips and, less commonly, grasses and cereals. Larvae of the Hepialidae (swift moths), on the other hand, normally live in grassland soils, but they often become pests of cereals, root crops and lettuce.

This short survey of phytophagous insect larvae does little to suggest that crop type has an over-riding influence on the horizontal distribution patterns. Many of the species concerned have grasses and various weeds as their natural food, and evidently have been able to adapt very readily to the additional food supplies provided by cultivated crops, without sacrificing their ecological plasticity. Indeed, it is very probable that their ecological niches have been extended as a result of this adaptation, and we can find additional examples of this among the soil-dwelling Mollusca, particularly the slugs.

Foliage-Feeders: Molluscs

In Britain, economically important slugs are *Agriolimax reticulatus* (Fig. 5.3), a pest of cereals, *Arion hortensis*, a slug which causes damage to a wide range of host crops, particularly potatoes, and *Milax budapestensis*, a species common in gardens and potato fields (Edwards and Heath, 1964).

Generally speaking, all three species prefer heavy clay soils and are more prolific here than on sandy soils. Within this preferendum, there is a certain amount of specificity, for *A. reticulatus* feeds mainly on the aerial parts of crop plants, whereas the other two feed on roots and tubers. These preferences are clearly related to differences in vertical distribution patterns, for *A. reticulatus* normally inhabits the surface layers of the soil, whereas the others are more abundant at deeper levels in the profile. On the other hand, none of these species is restricted to arable soils; all three may be encountered in accumulations of moist litter under hedgerows and woodland.

Erosion

There is a general decrease in the diversity and abundance of the soil fauna when soil is ploughed and planted with crops. The reasons for this decrease have been discussed at some length, but perhaps we can return to one particular aspect which deserves to be emphasized, namely the increased danger of soil erosion. When soil is mixed until it is loose and friable, and when a permanent plant cover is replaced by a temporary one, the chances of topsoil being removed by wind and water action are greatly enhanced. Such action may also be directly responsible for removing a large part of the surface dwelling fauna. Many factors may contribute to the erosion process, and among these, climate and topography are undoubtedly of the greatest importance. The degree of erosion of arable soil may also vary with the type of crop under cultivation and is greater, other factors being equal, under seasonal crops than under perennials (Atlavinite, 1965). The diffuse, fibrous rooting systems of monocotyledons such as maize, oats and wheat, serve to counter the erosion process by binding together the soil particles to a much greater extent than do the more compact, tap root systems of dicotyledons. Thus, erosion may be more severe in soils planted with potatoes or carrots, than under cereals. Here then, we see a possible indirect effect of crop type on the character and abundance of the soil fauna, expressed through the extent of the erosion process.

Pesticides

The pesticide problem is not only controversial, it is highly complex. The short list of examples discussed below does not do justice to the problem in its entirety, but it does illustrate a growing recognition of the need to approach it from an ecological point of view, and it provides some idea of the ways in which this approach is being made. More detailed reviews of this problem, as it applies to the soil/litter subsystem, can be obtained from Satchell (1955), Kevan (1962) and Edwards and Lofty (1969), while the accounts given by Rudd (1964) and Moore (1967) provide excellent introductions to the general ecological implications of pesticide practice.

The various kinds of pesticides can be classified into a number of major groupings. Three of these, carbamates, organophosphates and organochlorines, are of special concern since their effects on the soil fauna have been studied in some detail.

Carbamates

Organic carbamates, such as Sevin, apparently have wide-reaching effects on the soil fauna. Voronova (1968) examined the effects of various organophosphate and carbamate pesticides on the invertebrate fauna of soils in the southern taiga region of Russia, and noted that Sevin, applied in suspension at the rate of 0.5 g/m², produced a reduction in the densities of almost all faunal groups; a notable exception being Elateridae larvae. The most severely affected were enchytraeid worms, Carabidae, Staphylinidae and Geotrupinae. Lumbricid earthworms were also affected, but to a lesser degree. In the case of cryptostigmatid mites and Collembola there was a tendency for populations to increase, slightly, immediately after the application, possibly due to a lowering of predation pressure. However, this was succeeded by a gradual decline in numbers over the subsequent 12-month period. All of these effects were more marked when the pesticide was applied as a suspension rather than as a dust, and apparently they persisted for at least a year, for no marked recovery in population densities of the larger soil invertebrates could be detected during this time. Earlier, Edwards (1965a) had noted that Sevin retains its toxicity for about 6 months. Voronova found the pesticide present in litter at a concentration of 100 ppm one year after application. The data provided by Rudd (1964) indicate that this concentration is well below the toxic level; however, population recovery is certain to be slow in animals with a restricted breeding season, as in the Carabidae, Staphylinidae and Geotrupinae, or a long life cycle, as in the cryptostigmatid mites.

Organophosphates

Organophosphate pesticides include parathion and phosdrin which have a high toxicity, and which produce a very marked reduction in lumbricid densities when applied at commercial rates. Also included here is malathion, a pesticide which apparently has little permanent effect on the soil fauna. Hartenstein (1960) investigated the effect of malathion, applied in the form of a spray in doses of 2 lb/acre, on the soil fauna under a mixed hardwood stand and a red pine plantation, and could detect no significant difference between treated and untreated plots. This pesticide is reputed to have

5*

acaricidal properties, but is evidently more effective when applied as a dust than as an emulsion. Voronova (1968) observed that treatments in the form of dust, at a dose of 0·3 g/m², brought about a marked reduction in populations of surface-dwelling carabid, staphylinid and geotrupinid beetles. A diminution in numbers of Diptera larvae, lumbricids and elaterid larvae also occurred immediately after this treatment, but the effect was short-lived, for populations recovered during the subsequent 12-month period. Soil-dwelling mites and Collembola apparently were not affected. These observations are not entirely in accord with those of Edwards and Lofty (1969). These authors noted that carbamate and organophosphate pesticides affected arthropod numbers to a lesser extent than did organochlorines (see below), although these chemicals "consistently change the balance between the mesostigmatid mites and their prey". They also concluded that carbamate and organophosphate pesticides cause a diminution in densities of non-predatory mites and Collembola during the first two months after application, but this is followed by marked increases in numbers to the extent that population densities in treated soil may be several times higher than those in untreated soil. These increases may persist for up to six months after most of the pesticide has disappeared, but they seldom continue beyond this time.

Organochlorines

One of the more serious aspects of the pesticide problem is the fact that repeated applications result in a steady accumulation of toxic compounds in the soil, owing to the slow decay rate of many of the chemicals used. This is particularly true of the chlorinated hydrocarbons, of which DDT, benzene hexachloride (BHC), heptachlor, aldrin and dieldrin are common examples. Considerable attention has been paid to the effects of these substances on the soil fauna, particularly in relation to predator/prey systems. The following examples illustrate this approach.

Effect on predator/prey systems: mites and Collembola. Mites belonging to the Order Mesostigmata include a number of predaceous species which feed mainly on Collembola in the soil. Investigations by Edwards and his co-workers (Edwards, 1963, 1964, 1965b; Edwards and Dennis, 1960; Edwards *et al.*, 1967) have shown that mesostigmatid mites are susceptible to DDT, whereas Collembola are little affected. The extent of this resistance by Collembola has been investigated by Klee *et al.* (1973), who found that the collembolan *Folsomia candida* is apparently unaffected by DDT levels of the order of 100,000 ppm in its food. As a consequence, numbers of Collembola increase in soils treated with DDT, because their predators have been selectively eliminated. Earlier, Sheals (1955, 1956) tested the effects of DDT and BHC applied separately, at the rate of 2 oz per square yard of 2% crude dust, and also in combination. In the plots receiving only BHC, there were marked reductions in populations of mites and Collembola; in plots receiving only DDT, populations of mites suffered a reduction, but Collembola increased significantly to a level 110% higher than the controls. The effect produced by DDT is evidently a reflection of a lowering of the predation pressure exerted by mesostigmatid mites. The greatest reduction in populations of these predators occurred in the plots receiving a mixture of DDT and BHC. Under these conditions, populations of Collembola, and, to a

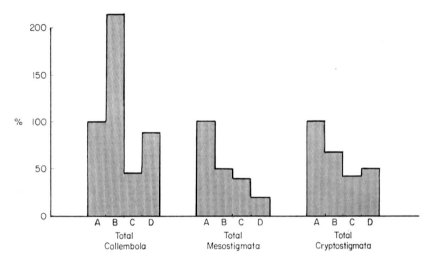

Fig. **5.5** Average effects of insecticides over five occasions, October 1951–52, on microarthropod populations of cultivated soils. (From Sheals, 1955.) A: control; B: DDT; C: BHC; D: DDT and BHC.

lesser extent, cryptostigmatid mites increased (Fig. 5.5), suggesting that the detrimental effect of BHC on both of these groups is offset by the extremely effective suppression of their predators, occasioned by one pesticide reinforcing the other. Heungens (1968) noted a similar disturbance of the predator/prey balance in experiments designed to test the effect of the nematicide Nemagon-20 (19% dibromo-chloropropane), applied at a dosage of 3 kg/acre, on the rhizosphere fauna of azaleas planted in conifer litter. This pesticide has strong acaricidal properties, and the experiments showed that densities of predatory and saprophagous mites were lower on treated plots than on untreated ones. The density of Collembola in treated plots was, however, 84% higher than in untreated controls, and this may be due to a decrease in numbers of predatory mites. On the other hand, Hartenstein (1960) could detect no significant changes in the balance of predator/prey populations of mites and Collembola in soils under red pine and mixed hardwoods sprayed with DDT at the rate of 1 lb/acre, and only an ephemeral effect when this dosage was increased to 10 lb/acre. The discrepancy between these observations and those mentioned earlier may be explained by the different methods used in applying the pesticide. Hartenstein used a surface application whereas Sheals, for example, mixed the pesticide thoroughly with the soil. This mixing would increase, considerably, the chances of the soil fauna coming into contact with it.

Effect on predator/prey systems: control of Pieris rapae. There are many aspects to this problem of the use of persistent organochlorine pesticides. Some of these have been highlighted by Dempster's work (1968a, b) on the faunal changes occurring in agricultural land in the vicinity of Monk's Wood Experimental Station, Huntingdonshire, England, a centre where much important work on the effects of pesticides on terrestrial ecosystems in Britain is being carried out (see also, for example, Davis, 1968). The system investigated by Dempster was that in which crops of Brussels sprouts were sprayed with DDT to control populations of caterpillars of the Cabbage White Butterfly, *Pieris rapae*. Good control of this pest was achieved initially by this treatment, since DDT eradicates the caterpillars *in situ* on the foliage at the time of spraying. However, subsequent invasions of caterpillars are not checked, since they feed mainly on new leaves which are uncontaminated. Further, these populations build up to considerable proportions on treated plots. Dempster found that this increase, which was greater than on untreated plots, was due to the fact that the pesticide produced considerable mortality among two common predators of *Pieris rapae*, namely the carabid beetle *Harpalus rufipes* and the harvestman *Phalangium opilio*. These predators spend much of their time on the surface of the ground, and move up into the vegetation to feed at night. Indeed, the larvae of *H. rufipes* are subterranean forms and may suffer considerable and prolonged mortality, since DDT persists longer in the soil than on the foliage. Thus, the introduction of DDT into this predator/prey system has only a short-term effect on the target organism *(Pieris rapae)*, but a much more sustained effect on predatory animals which, under natural conditions, may exert a considerable controlling influence on the numbers of the pest species. It has been calculated that these predators can inflict a 50–60% mortality on young caterpillars, thus decimating the pest population before it has a chance to produce severe damage on the host crop. Even at sub-lethal concentrations, DDT may alter this pest/predator balance in favour of the pest, for Dempster also showed that the feeding rate of *H. rufipes* is markedly reduced under these conditions.

Effect on predator/prey systems: activity patterns. Carrying this analysis a stage further, Dempster monitored the effects of DDT on populations of other invertebrates inhabiting the soil, its surface and the vegetation. These effects were by no means uniform, for various groups differed in their sensitivity to DDT treatments. Millipedes, root-feeding aphids, Diptera and Coleoptera larvae, mesostigmatid mites and centipedes were, in general, adversely affected, and population sizes decreased progressively with each application of the pesticide (Table 5.3). Collembola and enchytraeids, on

Table 5.3

The response of populations of various members of the soil fauna to successive applications of DDT (0·84 kg/ha) on plots of Brussels sprouts at Monk's Wood, Huntingdonshire (from Dempster, 1968b).[1,2]

| | 1965 | | 1966 | |
Faunal Group	*Unsprayed*	*Sprayed*	*Unsprayed*	*Sprayed*
Collembola	478	11,857†	4538	14,041‡
Acari (total)	281	354	1220	776
Mesostigmata	164	117	379	30‡
Cryptostigmata	13	14	172	304
Astigmata }	104	223	311	133
Prostigmata }			358	309
Diptera larvae	95	36	313	21‡
Coleoptera larvae	55	18‡	62	19‡
Aphids	181	71	155	5‡
Earthworms	18	10	18	25
Enchytraeidae			73	249*
Millipedes			19	0†

* Significant at 0·02 < P < 0·05.   † Significant at 0·001 < P < 0·01.
‡ Significant at P < 0·001.

[1] DDT treatments applied twice each year (July and August) during the period 1964–65.
[2] Numbers for 1965 are totals from 16 cores taken in November; those for 1966 are totals from 25 cores taken in November.

the other hand, showed significant increases in sprayed plots. Mesostigmatid mites and Collembola, as we have already noted, constitute a predator/prey system, and this work of Dempster would appear, at first sight, to confirm the findings of other workers; namely, that the increase in Collembola numbers is a consequence of the relaxation of predator pressure. However, an examination of the data presented in Table 5.3 suggests another explanation. The increase in Collembola numbers occurred before there was any significant decrease in the size of predatory mite populations. Evidently, the presence of the pesticide in some way stimulated the reproduction of the Collembola, although how this is achieved is not known. Experimental studies showed that some insects and arachnids become more active in the

presence of DDT, and such activity must be taken into account when inter-
preting data obtained from pitfall trapping. For example, lycosid spiders
and centipedes suffered a progressive reduction in numbers with successive
applications of DDT, but counts from sprayed plots were generally higher
than from unsprayed plots after the soil had been cultivated at the end of
summer. This difference was attributed to the greater activity of the populations in the presence of DDT. Carabid beetles varied from species to
species in the way they responded to DDT treatments. As we have already
noted, *Harpalus rufipes* is very sensitive and suffered a decline in population
size on sprayed plots. *Feronia melanaria*, on the other hand, apparently was
not affected. Members of the genus *Bembidion* showed significant reductions
in population sizes after the second application of the pesticide each year,
but recovered quickly. *Trechus quadristriatus* and *Nebria brevicollis* populations did not follow the usual pattern (Fig. 5.6), for they showed considerable increases in size following spraying. This was interpreted as a

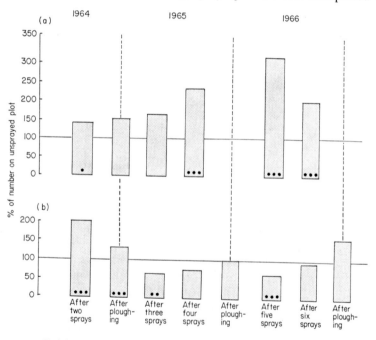

Fig. 5.6 Changes in numbers of (a) *Trechus quadristriatus* and (b) total Carabidae, excluding *Trechus* and *Nebria*, on a plot sprayed with DDT, shown as percentages of the numbers
on an unsprayed plot. (From Dempster, 1968b.)

response to the increased numbers of Collembola on which the adult stages of both these beetles feed.

Pesticides and the soil microflora. Up to now, our attention has been concentrated on what can be considered as direct effects of pesticide applications on the soil fauna. In addition to these, there may be more subtle, indirect effects occurring, for example among microfloral populations upon which many mites and Collembola rely for food material. Indeed, since bacteria and fungi are responsible for the major part of the chemical decomposition of organic material in the soil, their susceptibility to, or tolerance of, pesticides must obviously be taken into account in any assessment of the total ecological effect of such treatments on the soil/litter subsystem.

Although work on this particular problem is only now gathering momentum, some attention has been paid to the effects of chlorinated hydrocarbons on the soil microflora. Dindal *et al.* (1975) found that microbial respiration in an old field community was initially suppressed for a period of one month following a granular application of technical DDT at the rate of 1·12 kg/ha (1 lb/acre). This was succeeded by a period of stimulation which lasted for 17 months. These workers suggested that the DDT caused a high mortality among phytophagous insects, and this would result in increased litter deposition and, consequently, increased microbial substrate. On the other hand, Chandra (1967) studied the action of dieldrin and heptachlor on the nitrification process which is largely promoted in the soil by bacteria. By laboratory studies, this author was able to show that nitrification was influenced not only by the microflora but also by soil type and temperature conditions. The application of the two pesticides at field rates produced an inhibitory effect on nitrification in a clay and three types of loam soils, which was apparent two weeks after treatment at four different temperatures, namely 5°C, 12°C, 19°C and 26°C. After 16 weeks the inhibitory effect of the pesticides remained, although to a lessened extent, in all four soils at 5°C, 12°C and 19°C. However, in two of the soils kept at 26°C (a heavy clay and a mountain loam), detoxification occurred after 16 weeks and nitrification was not inhibited. The results of this work underline the importance of studying rates of detoxification under sub-optimal conditions, since these rates evidently vary with the temperature. They also indicate that caution must be exercised in attempting to apply findings from studies on one particular soil type generally to all soil conditions. Evidence has been presented by Burnett (1968), for example, that organochlorines such as aldrin and dieldrin have much less adverse effects on the soil/litter subsystem in the tropics than in temperate regions, and this may be a reflection of their more rapid loss or decay in warmer latitudes.

Selectivity and Persistence of Pesticide Effects

The extent to which a pesticide affects the soil fauna will depend, as we have already noted, on the degree to which it is mixed in with the soil. Many pesticides are not, of course, applied directly to the soil, but rather to the aerial vegetation, and DDT is one such example. Although a certain proportion of these will find their way into the soil with percolating rainwater, they are less of a threat to the soil fauna than those applied directly to the soil and subsequently mixed in by ploughing. Even compounds sprayed directly on the soil surface may have only a limited effect, and this on those animals which cannot withdraw to the deeper layers of the profile. This has been demonstrated by Anglade (1967) who reported that applications of HCH and aldrin at the rates of 15 kg/MA/ha and 4 kg/MA/ha, respectively, produced an ephemeral decrease in the population of *Scutigerella* sp., a symphylid inhabiting the surface layers of cultivated soil, whereas these treatments had no effect on a population of the deeper-dwelling *Symphylellopsis* sp.

The use of pesticides at the rates they are presently being applied has aroused, and continues to arouse, much controversial discussion, the more so since commercial interests are obviously involved. Clearly, it is difficult to restrict the effect of a pesticide to one "target" organism. Contamination of populations other than that for which the treatment is intended may lead to changes in the community structure, and while pesticides may prove a short-term solution, they may also provide a long-term problem. It may be argued that the natural decay of a pesticide limits its effect, and that those populations which suffer accidentally are soon restored to their normal levels. However, it is hard to sustain this viewpoint with conviction for the information presently available does not answer all of the questions which must be asked before the argument can be settled. For example, laboratory estimates of detoxification rates may have little relevance to field conditions where the initial effect of a pesticide may be sustained and, indeed, re-inforced by re-peated applications over a number of years. Equally important is the fact that certain pesticides break down in the soil into more toxic derivatives which decay at a much slower rate than the original compound. An example of this is the relatively rapid conversion of aldrin into the more toxic dieldrin; similarly, heptachlor breaks down into its more toxic and more persistent epoxide.

Biological Concentrators

An important idea that the ecological approach to the pesticide problem has revealed is that of the possible existence of "biological concentrators".

In essence, it postulates that because an organism does not die when it is exposed to, or assimilates, a pesticide, this is no reason to discount it. There is a growing body of evidence (see Rudd, 1964) to indicate that animals differ in their ability to tolerate concentrations of pesticides in their tissues. Poikilotherms are generally much more tolerant than homiotherms, although differences occur between one group of poikilotherms and another. For example, slugs, snails and lumbricid earthworms can withstand higher levels of pesticide concentration in their tissues than can many arthropods. These tolerant species act as concentrators of the pesticide at various well-defined points in the food web, and they can then deliver to a more susceptible predator a lethal dose of the toxic chemical. A good example of this delayed effect is provided by work done in the United States on high mortalities occurring in robins (Rudd, 1964). This showed that the high death rate of these homiotherms could be attributed to the relatively large amounts of DDT acquired from a diet of earthworms living in soil sprayed with this pesticide.

The fact that these members of the soil fauna could accumulate DDT in their tissues to a level that was greater, by a factor of 10, than that occurring in the sprayed soil, without detrimental effect, emphasizes the importance of considering the total effect of a pesticide on the ecosystem, and of continuing the search for more specific and less persistent compounds to control pests.

Biological Detoxification

There is another aspect of the pesticide problem which deserves to be mentioned although, as yet, little hard fact has been accumulated about it. As we have seen, pesticides are subject to decay or transformation into other compounds under natural conditions. Attention has recently been paid to the chemical changes which occur when DDT is assimilated into the bodies of soil animals. Klee *et al.* (1973) introduced DDT into a forest litter food chain by using a carrier organism, the collembolan *Folsomia candida*. *F. candida* is very resistant to DDT, and was not affected by high levels of this chemical in its food. Its resistance is apparently conferred by the possession of an enzyme, DDT-dehydrochlorinase, which breaks down DDT to its much less toxic metabolite DDE. The liberation of carrier Collembola into forest plots was followed by a regular sampling of the micro- and macro-arthropod fauna for a period of up to 50 days. These samples were analysed for DDT and its metabolites. The investigators reported that almost all of the arthropod groups sampled had metabolized DDT to DDE. The isomer p,p'-DDT was totally converted, within a period of 8 days, into p,p'-DDE by Collembola and cryptostigmatid mites, whereas the isomer o,p'-DDT required about 40 days for its degradation by these two groups of animals. Unfortunately this study did not monitor the fate of DDT in groups such as lumbricid earthworms, slugs and snails which, reportedly, concentrate this pesticide in their bodies (see earlier). Further, no DDT residues of any kind were found in predatory mites and pseudoscorpions, and the authors suggest that this may be an indication, but only such, of the rate at which these residues are eliminated from these predators. However, in view of the well-established reports on the susceptibility of mesostigmatid mites to DDT, it is difficult to accept that this group has an important role in detoxification. It is more likely that the predatory individuals exposed to the pesticide succumb quickly, leaving the uncontaminated individuals to be recovered in the extraction process.

Industrial pollution

Of course, the pesticide problem is only one aspect, admittedly a very important one, of the larger problem of environmental pollution. There is also the increasing need to control the dispersal of industrial waste. Although, here, concern is primarily being directed towards the immediate effects of toxic substances on human health, there is perhaps a tendency to overlook the fact that the entire ecosystem of which Man is a part may be affected. It is convenient to deal with this problem here, for although it is in no sense peculiar to arable situations, it is another example of Man's intervention in natural systems.

It is well established that atmospheric pollution in industrial areas frequently causes changes in the character of the vegetation, promoting the development of ground vegetation at the expense of tree species. This vegetational change is accompanied by changes in the overall density and species composition of the soil fauna. Bassus (1968), for example, noted an overall density decrease in the nematode fauna of soil in areas of Germany where industrial smoke with a high SO_2 content had caused changes in the vegetation. He also established that semi-parasitic nematodes were being replaced by saprophagous and predatory species. Similar changes were noted by Vanek (1967) who studied the effect of industrial chlorine poisoning on the soil mesofauna over a 7-year period in Czechoslovakia. A comparison of the soil fauna of polluted and unpolluted areas revealed that total density of the mesofauna was about 46% lower in the former, and that pollution reduced densities of cryptostigmatid mites by approximately 60%. These reductions were accompanied by a change in the species composition, with different "dominants" in the two areas, and by a decrease in the total number of species present.

Pollution of the soil appears to have an important effect on the soil fauna by decreasing the species diversity. Vanek's work, which reached this conclusion, is supported by evidence obtained from an entirely different kind of polluted system. This is provided by an experimental area in Pennsylvania, U.S.A., in which municipal sewage effluents have been discharged into various terrestrial sites, including a grassland, a mixed oak-hardwood community, a red pine plantation and an old field herbaceous community. The effects of this effluent on the soil fauna have been monitored by Dindal *et al.* (1975). The general conclusions reached by these workers were that wastewater irrigation brought about a general decrease in the species diversity of the soil fauna (Table 5.4), although it caused an increase in

Table 5.4

Diversity characteristics of soil microarthropods from sites treated with sewage effluent (T) and from untreated controls (UT).[,2] (From Dindal *et al*, 1975.)

Group	Mean Total (individuals)		Mean Total (species)		Mean Species Diversity		Mean Species Richness	
	UT	T	UT	T	UT	T	UT	T
Prostigmata	483	443	36	12	1·8397	1·0702	1·64	0·57
Mesostigmata	38	22	7	4	0·4028	0·3053	1·13	0·85
Cryptostigmata	196	24	18	8	1·1094	0·2485	1·28	1·63
Collembola	105	232	17	11	1·1379	0·8421	1·65	0·72

[1] Spring samples from an old field herbaceous community.   [2] Means of 10 sampling units.

densities of the lumbricid *Dendrobaena octaedra* in all sites. Thus, in sites treated with sewage effluent, the ecological emphasis shifts from the saprophagous microarthropods to the lumbricid earthworms. The resultant reduc- in species diversity in this system contrasts with preliminary observations on irrigated systems using uncontaminated water (Edwards and Lofty, 1969). Here, microarthropod populations appear to increase in density under the effect of irrigation, although this discrepancy may be attributable to the absence of a pollutant.

Irrigation and drainage

Irrigation and drainage are simply two different sides of the same coin. They are management practices which are designed to control the soil water content at an optimal level. In arable sites, this means a level which will contribute to maximum yield of crops. At the same time, soil conditions that are neither excessively dry nor excessively wet will encourage a high species diversity among the soil fauna, since the majority of this fauna is mesophilous, rather than xerophilous or hygrophilous, in character. By introducing these topics of irrigation and drainage at this point, we are therefore moving from a discussion of the detrimental effects of agricultural practice on the soil fauna to a consideration of the beneficial effects of such practices. And there are others which will be examined shortly.

It is easier to speculate about the effects of irrigation and drainage than to provide concrete information about these effects on the soil fauna. Irrigation appears to have a beneficial effect on microarthropod populations, promoting increases in population densities, as we have already noted in the previous section. Information on the effects of drainage is provided by studies on the fauna of soils reclaimed from the sea, for example the Dutch polders. Such sites are evidently colonized at an early stage by certain of the more mobile species of lumbricid earthworms, such as *Lumbricus rubellus*, *Allolobophora caliginosa* and *A. chlorotica* (van Rhee, 1969). These earthworms, which probably invade the newly-drained polder from adjacent land, increase the amount of water available to plants by improving the cavity structure of the soil. van Rhee's studies indicate that the Lumbricidae play an important part in the maturation of polder soils, for inoculation experiments have shown that populations of *A. caliginosa* and *A. chlorotica* disperse quickly in such sites.

Fertilizers

The common practice of applying organic and inorganic material in the form of manure, compost or fertilizer to cultivated soils often results in increased densities of soil animals, and promotes a greater degree of activity by soil bacteria and fungi. Raw (1967) was able to show that plots cropped with mangolds receiving inorganic fertilizer (P, K, N and Mg) supported a rather higher total density of arthropods (18,000 per sq yd) than untreated

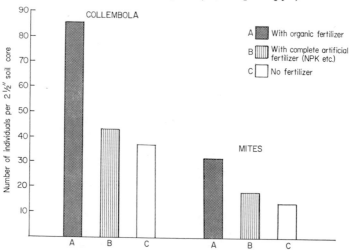

Fig. 5.7 Effect of organic and inorganic fertilizers on soil microarthropod populations. (From Edwards and Lofty, 1969.)

plots (16,000 per sq yd). The addition of farmyard manure at a rate of 14 tons per acre per year almost doubled the density, and permitted groups such as Pauropoda, Diplopoda, Symphyla and Diptera larvae to become established in greater densities than in untreated plots. These findings have been confirmed, more recently, by Edwards and Lofty (1969) (Fig. 5.7). Valiachmedov (1969) has also reported on Russian investigations which show that organic manuring and the application of mineral fertilizers increase population densities of earthworms. The increased organic content of the soil resulting from the application of dung represents an increase in nitrogen-rich food material for earthworms and other soil animals. It also stabilizes soil moisture content at a level higher than is commonly found in mineral soils. Both of these effects will create a more suitable environment for many groups of the permanent soil fauna. However, increases in faunal densities may not be strictly related to the amount of organic material added to the soil. There is some evidence that when abundant energy-rich food material is available, a large proportion of it is dissipated by the respiration of micro-organisms and is not, therefore, available for population growth.

Hedgerows

Although hedgerows cannot really be considered as arable sites, they are often present in the vicinity of the latter so, perhaps, it is pertinent to consider them here, particularly to the extent to which they influence the composition and distribution of crop faunas. Modern methods of farming have tended to reduce or destroy the hedgerow habitat in Britain (Moore et al., 1967), and even when left apparently intact around the margins of arable land, these habitats often suffer from the drift of herbicides and pesticides applied to the neighbouring crop. The role of hedgerows as temporary habitats, refuges or reservoirs for populations of insect pests of crops has been well documented (van Emden, 1965; Lewis, 1965). On the other hand, the hedgerow fauna includes a number of predatory invertebrates such as carabid beetles, spiders, bugs and flies, which may be important in the control of crop pests. Disturbance of the hedgerow habitat may affect the survival of these predators. Pollard (1968a, b, c) noted that removal of the ground flora under a hawthorn hedge reduced the total numbers and biomass of the fauna. This treatment severely reduced the pitfall captures of the carabids *Bembidion guttulata*, *Feronia melanaria*, *F. madida*, *Agonum dorsale* and *Trechus obtusus*. The last-named species is evidently a characteristic member of the hedgerow fauna, being confined to this habitat. The remainder have a much less close association with the hedge, for they are commonly encountered in cultivated soils, and evidently use the hedgerow as overwintering quarters. This study was able to show that hedgerows increase the diversity of the fauna in cultivated areas, and it draws attention to the need for much more ecological work on this important habitat.

Summary

In general, cultivation has the effect of decreasing the species diversity of the soil fauna. However, some species are encouraged and they may increase their population densities to pest proportions. Some agricultural practices have a beneficial effect on the soil community, allowing new species to become established and also promoting general increases in population densities. These factors and their effects are summarized in Fig. 5.8.

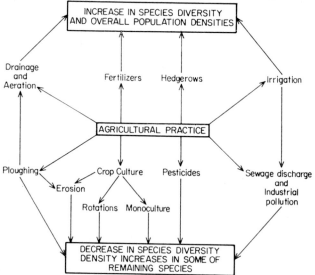

Fig. 5.8 Ecological effects of agricultural practice on the soil fauna. (Adapted from Edwards and Lofty, 1969.)

Ploughing produces conditions which are detrimental to certain groups of soil animals, such as microarthropods, which flourish in undisturbed profiles where equable moisture and temperature conditions prevail. On the other hand, the looser soil structure increases drainage and aeration and encourages colonization by insect larvae. Earthworms, slugs and nematodes can also extend their vertical range in ploughed profiles.

Monocultures and crop rotations decrease species diversity, but favour populations of root-feeding insect larvae and nematodes. However, there is no evidence that crop type has an overwhelming influence on the species composition of this fauna. Other factors, such as soil type, have a marked effect.

Ploughing and crop culture increase the danger of soil erosion and the consequent removal of many surface-dwelling animals. This effect may be pronounced where seasonal cropping is practised, and where root crops rather than cereals are sown.

The application of pesticides, particularly the chlorinated hydrocarbons, often results in a change in the ecological balance. Predator/prey systems are disrupted by the selective elimination of the more susceptible predators. Prey populations may increase to higher levels than normal, partly as a result of the reduced predator pressure, and partly due to increases in activity. Slugs, snails and lumbricid earthworms apparently have the ability to tolerate high levels of organochlorine pesticides in their tissues and, as a consequence, act as biological concentrators. Some microarthropods such as Collembola and cryptostigmatid mites, evidently can metabolize DDT into its much less toxic derivative DDE. Carbamate and organo-phosphate pesticides can produce severe effects on the soil fauna, imme-diately after application, but they do not persist for more than one season.

Drainage and irrigation schemes maintain adequate but not excessive soil moisture levels and, thereby, create suitable microclimatic conditions for the mesophilous element of the soil fauna. However, the use of municipal wastewater and sewage effluent decreases species diversity, favouring acid-tolerant species of lumbricid earthworms at the expense of microarthropods. Similarly, pollution by industrial wastes decreases species diversity and promotes shifts in the trophic structure of the soil community.

The application of fertilizers increases the nutrient content of the soil and, generally, has a beneficial effect on the soil fauna. Enrichment with manure and compost promotes the growth of the soil microflora, and animal popu-lations benefit from the increased amounts of nitrogenous material.

Hedgerows are an important feature of the arable landscape. They serve as seasonal refuges for insect pests of cultivated crops, but they are also habitats for many of the predators which control populations of these pests. Removal of hedgerows may upset the balance of predator/prey relationships in arable sites, and also result in a decrease in the overall species diversity of the fauna.

REFERENCES

ANGLADE, P. (1967). Étude de populations de Symphyles en sol cultivé et d'influence de traitments du sol. *In* "Progress in Soil Biology", 372–81 (Graff, O. and Satchell, J. E., eds.), North Holland, Amsterdam.

ATLAVINITE, O. (1965). The effect of erosion on the population of earthworms (Lumbri-cidae) in the soils under different crops, *Pedobiologia*, **5**, 178–88.

BASEDOW, T. (1973). Der Einfluss epigäischer Raubarthropoden auf die Abundanz phytophager Insekten in der Agrarlandschaft, *Pedobiologia*, **13**, 410–22.

BASSUS, W. (1968). Über Einflüsse von Industrieexhalaten auf den Nematodenbesatz im Boden von Kiefernwäldern, *Pedobiologia*, **8**, 289–95.

BROWN, E. B. (1955). Some current British soil pest problems. *In* "Soil Zoology", 256–68 Kevan, D. K. McE., ed.), Butterworths, London.

BURNETT, G. F. (1968). The effects of irrigation, cultivation and some insecticides on the soil arthropods of an East African dry grassland, *J. appl. Ecol.*, **5**, 141–56.

CHANDRA, P. (1967). Effect of two chlorinated insecticides on soil microflora and nitrification process as influenced by different soil temperatures and textures. *In* "Progress in Soil Biology", 320–30 (Graff, O. and Satchell, J. E., eds.), North Holland, Amsterdam.

DAVIS, B. N. K. (1968). The soil macrofauna and organochlorine residues at twelve agricultural sites near Huntingdon, *Ann. appl. Biol.*, **61**, 29–45.

DEMPSTER, J. P. (1968a). The control of *Pieris rapae* with DDT. II. The survival of young stages of *Pieris* after spraying, *J. appl. Ecol.*, **5**, 451–62.

DEMPSTER, J. P. (1968b). The control of *Pieris rapae* with DDT. III. Some changes in the crop fauna, *J. appl. Ecol.*, **5**, 463–75.

DINDAL, D. L., FOLTS, D. D. and NORTON, R. A. (1975), Effects of DDT on community structure of soil microarthropods in an old field. *In* "Progress in Soil Zoology", 505–13 (Vanek, J., ed.), *Proc. Vth Colloquium Pedobiologiae*, Prague, 1973.

DINDAL, D. L., SCHWERT, D. and NORTON, R. A. (1975). Effects of sewage effluent disposal on community structure of soil invertebrates. *In* "Progress in Soil Zoology" 419–27 (Vanek, J., ed.), *Proc. Vth Colloquium Pedobiologiae*, Prague, 1973.

DIRLBEK, J., BERÁNKOVÁ, J. and BENDLOVÁ, H. (1973). Einfluss der Bodenarbeitung auf die Dichte des Drahtwurmbesatzes (Coleoptera, Elateridae), *Pedobiologia*, **13**, 441–44.

EDWARDS, C. A. (1963). Persistence of insecticides in the soil, *New Scient.*, **19**, 282–84.

EDWARDS, C. A. (1964). Changes in soil faunal populations caused by aldrin and DDT, *Trans. 8th. int. Congr. Soil Sci.*, Bucharest 1964, **3**, 879–86.

EDWARDS, C. A. (1965a). Effects of pesticide residues on soil invertebrates and plants. *In* "Ecology and the Industrial Society", *Symp. Br. ecol. Soc.*, **5**, 239–61.

EDWARDS, C. A. (1965b). Some side effects resulting from the use of persistent insecticides, *Ann. appl. Biol.*, **55**, 329–31.

EDWARDS, C. A. and DENNIS, E. B. (1960). Some effects of aldrin and DDT on the soil fauna of arable land, *Nature, Lond.*, **188**, 767.

EDWARDS, C. A., DENNIS, E. B. and EMPSON, D. W. (1967). Pesticides and the soil fauna. I. Effects of aldrin and DDT in an arable field, *Ann. appl. Biol.*, **59**, 11–22.

EDWARDS, C. A. and HEATH, G. W. (1964). "The Principles of Agricultural Entomology", Chapman Hall, London.

EDWARDS, C. A. and LOFTY, J. R. (1969). The influence of agricultural practice on soil micro-arthropod populations. *In* "The Soil Ecosystem", 237–47 (Sheals, J. G., ed.) Syst. Assoc. London.

GREENSLADE, P. J. M. (1964). Pitfall trapping as a method for studying populations of Carabidae (Coleoptera), *J. Anim. Ecol.*, **33**, 301–10.

GUILD, W. J. McL. (1955). Earthworms and soil structure. *In* "Soil Zoology", 83–98 (Kevan, D. K. McE., ed.), Butterworths, London.

HARTENSTEIN, R. C. (1960). The effects of DDT and Malathion upon forest soil micro-arthropods, *J. econ. Ent.*, **53**, 357–62.

HEUNGENS, A. (1968). The influence of DBCP on the soil fauna in azalea culture, *Pedobiologia*, **8**, 281–88.

HUNTER, P. J. (1966). The distribution and abundance of slugs on an arable plot in Northumberland, *J. Anim. Ecol.*, **35**, 543–57.

KEVAN, D. K. McE. (1962). "Soil Animals". Philosophical Library, New York.

KLEE, G. E., BUTCHER, J. W. and ZABIK, M. (1973). DDT movement and metabolism in forest litter microarthropods, *Pedobiologia*, **13**, 169–85.

LEWIS, T. (1965). The effects of shelter on the distribution of insect pests, *Sci. Hort.*, **17**, 74–84.

MOORE, N. W. (1967). A synopsis of the pesticide problem. *In* "Advances in Ecological Research", **4**, 75–129 (Cragg, J. B., ed.), Academic Press, London and New York.

MOORE, N. W., HOOPER, M. D. and DAVIS, B. N. K. (1967). Hedges I. Introduction and reconnaissance studies, *J. appl. Ecol.*, **4**, 201–20.

NADVORNYJ, V. G. (1968). Wireworms (Coleoptera, Elateridae) of the Smolensk region, their distribution and incidence in different soil types, *Pedobiologia*, **8**, 296–305.

O'CONNOR, F. B. (1967). The Enchytraeidae. *In* "Soil Biology", 213–57 (Burges, N. and Raw, F., eds.), Academic Press, London and New York.

POLLARD, E. (1968a). Hedges II. The effect of the removal of the bottom flora of a hawthorn hedgerow on the fauna of the hawthorn, *J. appl. Ecol.*, **5**, 109–23.

POLLARD, E. (1968b). Hedges III. The effect of removal of the bottom flora of a hawthorn hedgerow on the Carabidae of the hedge bottom, *J. appl. Ecol.*, **5**, 125–39.

POLLARD, E. (1968c). Hedges IV. A comparison between the Carabidae of a hedge and field site and those of a woodland glade, *J. appl. Ecol.*, **5**, 649–57.

RAW, F. (1955). The effect of soil conditions on wheat bulb fly oviposition, *Pl. Path.*, **4**, 114–17.

RAW, F. (1967). Arthropods (except Acari and Collembola). *In* "Soil Biology", 323–62 (Burges, N. and Raw, F., eds.), Academic Press, London and New York.

RITCHER, P. O. (1958). Biology of Scarabaeidae, *A. Rev. Ent.*, **3**, 311–14.

RUDD, R. L. (1964). Pesticides and the living landscape, Faber, London.

SATCHELL, J. E. (1955). The effects of BHC, DDT and parathion on soil fauna, *Soils Fertil.*, **18**, 279–85.

SATCHELL, J. E. (1967). Lumbricidae. *In* "Soil Biology", 259–322 (Burges, N. and Raw, F., eds.), Academic Press, London and New York.

SHEALS, J. G. (1955). The effects of DDT and BHC on soil Collembola and Acarina. *In* "Soil Zoology", 241–52, (Kevan, D. K. McE., ed.), Butterworths, London.

SHEALS, J. G. (1956). Soil population studies. I. The effects of cultivation and treatment with insecticides, *Bull. ent. Res.*, **47**, 803–22.

SINGH, M. P. (1955). A note on cutworms and their rearing. *In* "Soil Zoology", 281–3 (Kevan, D. K., McE., ed.), Butterworths, London.

VALIACHMEDOV, B. (1969). Doždevye červi v osnovnych tipach počv pojasa serozemov i koričnevych počv Tadžikistana, *Pedobiologia*, **9**, 69–75.

VANEK, J. (1967). Industrieexhalate und Moosmilbengemeinschaften in Nordböhmen. *In* "Progress in Soil Biology", 331–39 (Graff, O. and Satchell, J. E., eds.), North Holland, Amsterdam.

VAN EMDEN, H. F. (1965). The role of uncultivated land in the biology of crop pests and beneficial insects, *Sci. Hort.*, **17**, 121–36.

VAN RHEE, J. A. (1969). Development of earthworm populations in polder soils, *Pedobiologia*, **9**, 133–40.

VORONOVA, L. D. (1968). The effect of some pesticides on the soil invertebrate fauna of the south Taiga zone in the Perm region (USSR), *Pedobiologia*, **8**, 507–25.

WALLWORK, J. A. (1970). "Ecology of Soil Animals". McGraw-Hill, Maidenhead.

WINSLOW, R. D. (1964). Soil nematode population studies. I. The migratory root Tylenchida and other nematodes of the Rothamsted and Woburn six-course rotations, *Pedobiologia*, **4**, 65–76.

Chapter 6

Moorland and Fenland

The habitats considered in this chapter can be considered, rather loosely, as sub-divisions of the permanent pasture type dealt with earlier. However, it is convenient to devote a separate section to them for although the vegetational complexes which they represent often include an appreciable number of grass species and serve as food for grazing herbivores, they present some special characteristics which are reflected in the structure of the soil/litter subsystem. These habitats provide good situations for the study of horizontal distribution patterns of soil animals for although the vegetation is often complex, physical factors such as fluctuating water levels, erosion processes and soil type are quite sharply delineated, and many of the faunal patterns can be related to these effects. In addition, moorland and fen often lie in close proximity to more typical grassland pasture, and horizontal distribution patterns can be followed across the vegetational discontinuities so produced.

Soil types

Moorland and fen soils are, characteristically, wet. They develop in areas of high rainfall or, alternatively, where topographical features allow water to accumulate in natural depressions in the bedrock. As a result of this high water content, anaerobic conditions develop which inhibit the decomposition of plant litter, and there occurs an extreme type of raw humus formation, known as peat. Bog soils develop where acidic rainwater or groundwater seeps into the peat. Fenlands occur where groundwater, flowing over limestone or chalk, flushes the peat, rendering it rich in bases and raising the pH.

Bogs can develop in lowland and upland situations. In the former, peat formation usually occurs in natural basins where water can accumulate. Successive layers of peat raise the ground surface above the water table and a dome-shaped bog develops. In uplands, high rainfall produces a constantly wet ground surface. Under these conditions, a continuous blanket of peat occurs.

Peat deposits are characteristic of moorlands, although they are not the only soil type occurring there. In northern England, for example, glacial erosion during the Pleistocene removed the surface layers of shale and grit from the upper slopes of the Pennines, and exposed the underlying Carboniferous limestone. Weathering of this limestone has produced soils of a brown earth type, shallow in depth, with alkaline mull humus and a relatively homogeneous profile. On the floors of dry valleys where glacial drift has been deposited over the limestone, rather acid soils are produced which show some evidences of podzolization and the development of a surface layer of raw humus.

Fens and marshes are primarily freshwater habitats, although it is perhaps more meaningful to regard them as stages in a successional series which may culminate in grassland, woodland or forest. Fenland is characterized

Fig. **6.1** *Sphagnum.* An important peat-forming moss. (Photo. by author.)

Fig. 6.2 Mixed moorland vegetation on blanket bog at Ingleborough, Yorkshire. (Photo. by author.)

by peat accumulation, but unlike the situation occurring in moorland bogs, fen peat is rich in lime, and has an alkaline or neutral pH. In contrast, marshes usually form on silty deposits, and peat accumulation does not occur. In true fen, the water table is at or near the surface, at least in parts of the locality, and drainage from limestone or chalk into the fen produces the characteristic alkaline conditions. In Britain, the most extensive areas of fen occur in the lowlands of East Anglia, where a transition to marshland is not uncommon. Upland fens are much less frequent, although they occur in parts of the Pennines, for example, where they grade into peat bogs and moorland.

Vegetation types

The variations in soil type which occur in moorland are paralleled by differences in the character of the vegetation. The acidic bog peat is mainly formed from the remains of *Sphagnum* moss (Fig. 6.1) which develops extensive hummocks in wet areas. However, in many upland areas of northern Britain, notably along the Pennine chain of hills, blanket bog has been subject to a drying-out process in recent times. This, possibly combined with the trampling effect of sheep and cattle, has caused the disappearance of much of the *Sphagnum* cover. In its stead, a mixed moorland vegetation has become established which includes heather, *Calluna vulgaris*, cotton grass, *Eriophorum vaginatum* and *E. angustifolium*, bell heather, *Erica tetralix*, bilberry, *Vaccinium myrtilus*, and *Empetrum nigrum* (Fig. 6.2). Clumps of *Sphagnum* and *Polytrichum* mosses (Fig. 6.3) persist in wet places where they are often joined by species of *Juncus*.

Fig. 6.3 *Polytrichum.* A common component of the moss vegetation of wet bog soils. (Photo. by author.)

On lower slopes and dry valleys in the Pennines, tracts of grassland are maintained by sheep grazing. These grasslands occur on accumulations of sand or clay glacial drift (alluvial grassland) and on limestone sites, devoid of drift, where a soil profile of the brown earth or rendzina type develops (Fig. 6.4). The alluvial grassland is dominated by *Nardus stricta*, with *Deschampsia flexuosa*, *Potentilla erecta* and *Galium saxatile* also abundant as a general rule. The limestone grassland sites often support a variety of plant species of which the grasses *Festuca ovina*, *Agrostis tenuis*, *Anthoxanthum odoratum*, *Sesleria caerulia* and *Helictotrichon pubescens* are prominent.

Moorlands provide a mosaic of vegetation and soil types, blanket bog, mixed moor, acid and alkaline grasslands, which forms a convenient ecological framework within which to discuss the distribution patterns of the soil fauna. Vegetational mosaics are also characteristic of heathland, as we will see in Chapter 7, and of fenland where plant associations are often described in terms of an ecological succession.

At Malham Tarn, in Yorkshire, is developed a fen which lies in close

Fig. 6.4 Dry valley at Highfolds Scar, Malham, Yorkshire. In the floor of the valley is a deposit of glacial drift colonized mainly by the grasses *Nardus stricta* and *Deschampsia flexuosa*. On the valley sides, a limestone grassland is developed. (Photo by author.)

Fig. 6.5 Wet fen vegetation at Malham Moss (foreground), with fen carr (background). Photo. by author.)

proximity to a raised bog (Fig. 6.5). This fen shows a clear transition from aquatic or semi-aquatic floating vegetation, consisting of *Carex*, *Phragmites*, *Caltha*, *Geum*, *Vaccinium*, *Viola*, *Galium*, *Filipendula* and *Angelica* species (reed-swamp and wet fen), to hummocks of *Sphagnum* and *Polytrichum* and drier tussocks of *Carex appropinquata* (tussock fen). These drier sites are colonized by woody vegetation, notably birch, alder and willow, forming a typical fen carr. Human interference often diverts this succession, for example when *Phragmites* is harvested for thatching, when peat is dug for fuel, and when fen carr is drained and cleared for grazing. As a consequence, fen meadows are developed, consisting mainly of the wavy hair grass *(Deschampsia flexuosa)* and the purple moor grass *(Molinia caerulia)*.

As far as the fenland soil fauna is concerned, the environmental variables are clearly defined. In the succession outlined above, there is a progressive change in pH, from the alkaline reed-swamp and wet fen to the more acid tussock fen, meadow and carr. The introduction of *Sphagnum* into this sequence encourages peat formation and promotes drier conditions by raising the ground surface above the water table. These changes have important repercussions on the character of the soil fauna, as we will see later in the chapter. We must now return to moorland habitats and their faunal characteristics.

Soil fauna of mixed moorland and grassland

The mosaic pattern of vegetation present in moorland influences the distribution of soil animals in a way which varies from group to group. This can be illustrated, very conveniently, by considering two of the most abundant groups of microarthropods, namely the mites and Collembola.

Hale (1966) recorded the following maximum densities for total Collembola from heather moor and grassland sites in the Pennines:

Limestone grassland	77,950 ± 8,600 per m²
Alluvial grassland	55,860 ± 7,330 ,, ,,
Heather moorland	42,350 ± 10,250 ,, ,,
Juncus squarrosus	38,890 ± 6,870 ,, ,,

All six of the collembolan species commonly encountered in this region, namely *Friesea mirabilis*, *Onychiurus* sp., *Tulbergia krausbaueri*, *Folsomia manolachei*, *Isotoma sensibilis* and *Isotomiella minor*, occurred in high densities in the limestone and alluvial grasslands, with all except *T. krausbaueri* and *I. sensibilis* significantly higher on limestone. *T. krausbaueri* was significantly more abundant on the alluvial site compared with the limestone. Generally speaking, the total density estimates for Collembola, as well as the densities for individual species, support the findings of Weis-Fogh

(1948) mentioned in Chapter 4, that the Collembola as a group show a natural preference for relatively dry grassland sites which are not too acid. Of the 38 species recorded from the four sites only two, *Isotoma sensibilis* and *Folsomia brevicauda*, were relatively more abundant on the acid moorland and *Juncus* sites. *Tetracanthella wahlgreni* was almost entirely restricted to the heather moor, but within this locality evidently selected dry microhabitats (see below) where it contributed over 50% to the total collembolan density (Hale, 1963).

The distribution patterns of soil-dwelling mites in this same moorland region have been documented by Block (1965), and the contrast between these and those of the Collembola outlined above is quite marked. The mites achieve their highest density in the alluvial grassland, and their lowest in limestone grassland. Maximum densities for total Acari in each of the four sites are of the same order of magnitude as those already cited for the Collembola, and are as follows:

Alluvial grassland   ..   .. 77,830 ± 4,240 per m²
Mixed moorland   ..   .. 65,790 ± 3,190 ,, ,,
Limestone grassland ..   .. 45,290 ± 1,410 ,, ,,
Juncus squarrosus   ..   .. 43,010 ± 3,090 ,, ,,

This general picture of the horizontal distribution of mites on moorland/ grassland sites has been confirmed in a series of samples taken across a dry

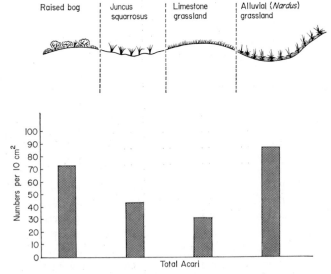

Fig. 6.6 Schematic diagram illustrating the densities of soil mites in four moorland habitats (data for the *Juncus* site taken from Block, 1965.)

valley (Fig. 6.4) in the vicinity of Malham Field Centre, Yorkshire. The floor of this valley is covered with glacial drift supporting *Nardus* grassland, and the sides by a limestone grassland vegetation growing on a brown earth soil. The pattern of distribution of Acari across this valley is compared with that of a raised bog adjacent to Malham Tarn, and with the *Juncus* sites investigated by Block (1965), in Fig. 6.6. These findings underline the preference of mites for soils with a higher moisture content and lower pH than those preferred by Collembola. However, neither of these groups flourishes in the very wet soils supporting *Juncus*.

Other Groups

The distribution patterns of groups other than the microarthropods, such as insect larvae, enchytraeids, lumbricids and nematodes, have also been documented for moorland sites, and it is of interest to compare these with the Collembola and mites. Cragg (1961) has made such a comparison, and the main points can be summarized briefly as follows.

Fig. 6.7 The crane-fly *Tipula maxima* Poda (wing span 6 cm) with clusters of phoretic mesostigmatid mites attached to the sides of the thorax. This specimen was collected from Pennine moorland. (Photo. by A. Newman-Coburn.)

Insect larvae of the dipteran family Tipulidae (Fig. 6.7), enchytraeids and nematodes generally follow the Acari in their distribution patterns, whereas lumbricid earthworms show a greater similarity with the Collembola. The preference of lumbricids for grassland soils, and particularly

6

limestone sites, is partly a reflection of their association with soils which are not too acidic, and also their attraction to cattle dung and sheep dung. Grazing is more intense on limestone *Festuca/Agrostis* grassland than on *Nardus* sites, and dung deposits are more plentiful. According to Svendsen (1957), the species composition of the earthworm fauna varies with different moorland habitats. *Octolasion cyaneum* and *Allolobophora terrestris* are abundant in soils of the brown earth type, whereas *Lumbricus rubellus* and *Dendrobaena rubida* are more prominent in alluvial profiles, and *D. octaedra* and *Bimastos eiseni* are characteristic of soils of mixed moorland.

Enchytraeids and nematodes are often abundant in highly organic soils, and it is not surprising to find them in high densities in peaty moorland. In the English Pennines, at Moor House, Westmorland, the most widely distributed enchytraeid is *Cognettia sphagnetorum*, with *Mesenchytraeus sanguineus* also abundant in *Juncus squarrosus* sites (Peachey, 1959; Cragg, 1961). *Cernosvitoviella briganta* is also common in acid peat soils, but members of the genus *Fridericia* are restricted to mull-type soils on limestone which, according to Springett (1970), harbour more species than acid soils. The soil nematodes are well represented by plant-feeding forms at Moor House, particularly on peaty soils, while miscellaneous feeders achieve more prominence in limestone mineral soils (Banage, 1963). Densities vary from 6,000 to 5,700,000 per m², depending on whether the soil is mineral or peaty. In moist peat soils, miscellaneous feeders show a more marked vertical distribution than in mineral soils, whereas plant feeders tend to be concentrated in the surface layers of both soil types (Banage, 1966). Both the enchytraeids and the nematodes occur in bare peat (Cragg, 1961) and appear to be the most important mesofaunal elements there.

Although there is a considerable degree of overlap between the distribution of individual species populations in moorland sites, there is often a separation of the centres of population density. Among the more abundant species, there is little evidence of severe restriction to one particular microhabitat, but distinctions can be made between populations which flourish best in limestone grassland, for example, and those which occur in highest density in one or other of the more acid sites. The ability to colonize a range of microhabitats has a considerable survival value for the species concerned, and this is clearly illustrated in the case of larval crane-flies (Tipulidae) studied by Coulson (1959, 1962). The centres of population density of the larvae of two species common in moorland, namely *Tipula paludosa* and *T. subnodicornis*, are clearly separated, with the former more abundant in alluvial and limestone grassland soils, the latter in peaty soils of the *Calluna/Sphagnum* moor, *Eriophorum* moor and the *Juncus squarrosus* sites. *T. subnodicornis* occurs in the *Sphagnum* bog habitat in lower, but relatively constant numbers, according to Cragg (1961) who provides records for a 3-year

Table 6.1

Densities per m² of final instar larvae of *Tipula subnodicornis* on different moorland habitats (from Cragg, 1961).

Habitat	1954	1955	1956
Juncus squarrosus moor	168	140	0
Eriophorum moor	70	104	4
Sphagnum bog	17	36	14

period (Table 6.1). During this study, numbers on moorland sites decreased considerably in the dry period, as the table shows, and it was suggested that the horizontal distribution pattern of the larvae of this tipulid forms a mosaic on high moorland, in that a range of habitats may be colonized. The overall effect of this is to provide a safety device for the species, to ensure that it survives even if some habitats become unsuitable. During the dry summer of 1956, the two moorland habitats listed in Table 6.1 dried out and the tipulid population was decimated. The population in the *Sphagnum* bog, which dried out much more slowly, was not seriously affected by this dry period, and although this site normally produced only a low density of crane-flies, this would be sufficient in the normal course of events to enable the species to survive in the area and to become re-established in the moorland sites when wetter conditions returned.

Table 6.2

Parameters involved in the ecological description of Pennine moorland sites (based on Cragg, 1961).

	Limestone Grassland	Nardus Grassland	Juncus Moor	Mixed Moor	Bare Peat
Biomass Rankings	Lumbricidae Long-palped larval Tipulidae Enchytraeidae Collembola	Enchytraeidae Acari Nematoda Collembola	Long-palped larval Tipulidae———— Short-palped larval Tipulidae———— Nematoda	Acari Collembola	
C/N Ratio	6–8	14–18	12–15	14–17	28–34
Soil Respiration (Rate Ratios)	12	6	6	1	0
Biological Activity	LG >	NG >	JM >	MM >	BP

The various studies discussed above provide a comprehensive picture of the horizontal distribution of soil microarthropods and other mesofauna in moorland sites which, in turn, permits us to compare the ecological characteristics of the habitats concerned. This comparison, based on data provided by Cragg (1961), is made in Table 6.2. Evidence from carbon/nitrogen (C/N) ratios and measurements of soil respiration indicate that there is a gradient of metabolic activity, culminating in the lumbricid-dominated grassland. At the lower extreme, bare acid peat, with its impoverished fauna, is characterized by a slow rate of decomposition. The ecological sequence depicted in the table clearly emphasizes the importance of the Lumbricidae and the microarthropods in organic decomposition in the sites concerned. The relative importance of earthworms decreases in peaty moorland soils, while that of the mites and Collembola becomes greater.

Fauna of bog soils

We now turn to a more detailed look at one particular moorland habitat which has attracted the attentions of ecologists for many years. Various kinds of bogs can be recognized, for example blanket bog, valley bog and raised bog, and although we are not particularly concerned here with the way in which these are formed, we are interested in the kinds of environment they provide for their fauna.

Environmental Factors: Moisture and Vegetation Type

The development of a bog, with its characteristic peaty soil, is usually associated with the growth of *Sphagnum* moss, although other plants, such as species of *Eriophorum* and *Carex* may also make a contribution. The development of valley bogs, for example, proceeds from lime-rich, aquatic fenland vegetation which gives way to certain species of *Sphagnum* tolerant of alkaline conditions. As the *Sphagnum* grows, it forms a cushion, or hummock, raising the vegetation above the level of the groundwater, and allowing colonization by less tolerant, acid-loving species. During the course of this succession, which may culminate in a mixed moorland vegetation type dominated by *Calluna*, a gradient of moisture conditions is developed. In this can be distinguished an initial aquatic stage, a semi-aquatic, or mesophile, stage and, thirdly, a dry, or xerophile, stage. This gradient has a profound influence on the distribution of the associated fauna. Two examples will serve to illustrate this.

Distribution of Collembola. The association of various species of Collembola with the moisture gradient produced by bog succession has been documented by Murphy (1955). The initial aquatic stage was characterized by the collembolans *Tetracanthella brachyura*, *Isotomurus* sp. and *Sminthurides malmgreni*. The mesophile fauna was dominated by *Folsomia brevicauda* and also included *Friesea mirabilis* and *Isotoma sensibilis*, while the xerophile association consisted of *Tetracanthella wahlgreni* and *Arrhopalites principalis*.

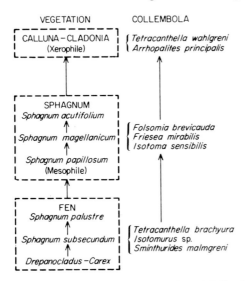

Fig. 6.8 The relationship between horizontal distribution of Collembola and vegetation succession in a Pennine valley bog. (After Murphy, 1955.)

These findings are summarized in Fig. 6.8 and agree, for the most part with those of Hale (1963) who suggested that the distribution of Collembola on eroding blanket bog was influenced to a considerable degree by the soil water content (see below).

Distribution of rhizopod Protozoa. The distribution patterns of rhizopod Protozoa in a Swedish *Sphagnum* bog have been described by Paulson (1952–53), and particular attention was paid in this work to the influence of variations in moisture content of the substratum, vegetation type and pH. Three associations of rhizopod species were defined:

1. A "forest moss" association characterized by the absence of *Amphitrema flavum*, and a poor representation of *Hyalosphenia* species.
2. A *Hyalosphenia papilio/H. elegans* association, common in young bogs and sites at the edge of groundwater supporting a vegetational complex of *Sphagnum apiculatum*, *Dryopteris spinulosa*, *Eriophorum vaginatum* and *Vaccinium oxycoccus*.

3. An *Amphitrema flavum* association characteristic of the wet hollows of old bogs dominated by *Sphagnum parvifolium*, *Pleurozium schreberi*, *Vaccinium oxycoccus* and *Eriophorum vaginatum*.

Although no positive evidence could be provided to establish the factors responsible for these patterns, it appears possible that tolerances for different pH conditions may be related to the presence or absence of some, at least, of the characteristic rhizopods in various parts of the bog. For example, *Amphitrema flavum* only occurred in the more acid sites (pH < 4·0), whereas *Hyalosphenia papilio* was most abundant in sites where the pH exceeded 4·0. However, the restriction of *A. flavum* to the more acid sites in this case may also reflect a lack of opportunity for dispersal to other parts of the bog, for this species is more widely distributed in certain Lapland bogs, for example, which are subject to seasonal flooding (Harnisch, 1938). It is well established (see Stout and Heal, 1967) that rhizopods have a more limited tolerance to variations in pH than other groups of Protozoa, and the work of Bonnet (1961) has shown that the distribution patterns of individual species may vary with the pH conditions. However, local factors such as periodic flooding, which may permit a wide dispersal within an acid bog, must be taken into consideration in any attempt to interpret the overall pattern.

An important feature of *Sphagnum* succession is the development of a hummock of vegetation. This is an integral part of the transition from an aquatic to a terrestrial environment both in moorland and fenland localities, and plant species other than *Sphagnum* may be involved in this succession. Here we have, condensed into a microcosm, many of the ecological events which occur when an aquatic ecosystem grades into a terrestrial one, and these are now considered in more detail.

Environmental Factors: Hummock Formation

In a study of the relationship between the distribution of cryptostigmatid mites and microhabitat factors, Popp (1962a, b) distinguished three types of hummock, namely *Sphagnum*, mixed moor or heathland (Reiser), and *Carex* hummocks. Different parts of the hummock structure have different moisture characteristics, and these can be equated with different ecological zones (Fig. 6.9). This basic arrangement is characteristic of all three hummock types, but beyond this, marked differences occur. In comparing the two extreme hummock formations in an ecological sequence, namely the *Sphagnum* (representing the initial stage) and the mixed moor hummock (terminal stage), it is evident that the former has a structure which is essentially horizontal. Vegetational succession in depth is lacking, i.e. there is no

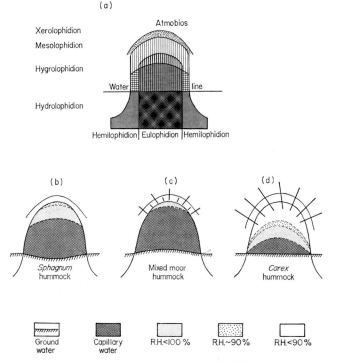

Fig. **6.9** Ecological zones and moisture characteristics of hummock formations. (After Popp, 1962a.) (a) A generalized plan of horizontal and vertical zonation; (b–d) moisture profiles through three different kinds of hummock formations.

higher plant cover. On the other hand, the mixed moor hummock has a vertical structure provided by secondary growth of birch and pine covering the primary shrub vegetation. The extent to which capillary water contributes to the water regimen in each of these three hummock types is illustrated in Fig. 6.9. It is evident that the upper parts of the *Carex* hummock are appreciably drier than those of the other two types, at least in this particular study. The predominance of capillary water in the hummocks of *Sphagnum* and mixed moor vegetation is reflected in their faunal characteristics. The cryptostigmatid mite fauna is poorly represented, and is dominated by species of the genus *Limnozetes* noted for their association with wet habitats. The centres of population density of this fauna show vertical shifts with the seasons, moving with the capillary water to deeper layers of the hummock during hot summer weather, and reversing this trend during wet weather. To this extent, then, the distribution of the mite fauna in wet hummocks is mainly controlled by climatic conditions. In the drier *Carex* hummocks, particularly in lowland areas, no such seasonal vertical shift occurs, since the mite fauna consists, in the main, of "terrestrial" species whose distribution is influenced more by edaphic factors.

In order to define the limits of species associations in this range of hummock types, Popp (1962b) used various stages in vegetational succession as points of reference, and characteristic microarthropod species were identified from abundance and frequency data. The associations of mites, and the stages in succession with which they can be identified are shown in Table 6.3. The sequence shown is essentially similar to that recorded from a

Table 6.3

The relationship between species associations of mites and various types of hummock formation (after Popp, 1962b).

MOORLAND *SPHAGNUM* HUMMOCK SUCCESSION			LOWLAND FEN and MARSH HUMMOCKS
Sphagnum	Sphagnum	Mixed moor plants	Carex
I ────────►II──────────►III			
Sphagnetum fusci association	Sphagnetum magellanici association	Rhodoreto-Vaccinietum association	Magnocaricion association
Limnozetes ciliatus and L. rugosus			
	Pilogalumna tenuiclavus		
		Adoristes ovatus and Scheloribates pallidulus	
Steganacarus striculus– Cyrtolaelaps mixed association			Liebstadia similis and Nanhermannia comitalis
Suctobelba subcornigera– Malaconothrus egregius mixed association			

Swedish mire by Tarras-Wahlberg (1952–53). It suggests that while a definite succession of mite species can be described, ranging from the aquatic or semi–aquatic *Limnozetes* spp. at one extreme, to the more truly terrestrial *Adoristes ovatus* and *Scheloribates pallidulus* at the other, some species occur in high abundance and frequency in all three types of the moorland succession. Similarly, when the comparison is extended to include the lowland *Carex* hummocks, it is evident that the species most frequent and abundant here, such as *Malaconothrus egregius*, *Suctobelba subcornigera*,

Scheloribates confundatus, *S. laevigatus*, *S. latipes*. *Pergamasus crassipes* and *Hydrozetes confervae*, have distribution patterns which extend into moorland sites.

Even greater difficulty was experienced in attempting to characterize the various ecological zones within a particular hummock in terms of its mite fauna. Some species, such as the small *Oppia nova*, *O. falcata* and *O. subpectinata*, showed some preference for the Eulophidion, but this zone was often not sharply delimited from the Hemilophidion, and many species moved between the two in response to changes in environmental temperature conditions. Within the Hemilophidion, the distribution of many species extended over two, three or sometimes four of the sub-divisions, so that again characteristic species are hard to define. The picture is further complicated by the fact that many of the species found in the Hydro- and Hygrolophidion zones of moorland hummocks move up into the Mesolophidion to overwinter, and go down again in the summer, as we have already seen. This phenomenon does not occur in the lowland *Carex* hummocks where the fauna is localized, to a great extent, in the Hygrolophidion and Mesolophidion zones throughout the year. An additional point to be borne in mind when describing these distribution patterns is that juvenile forms of many species occur in, and may be restricted to, lower and wetter levels of the hummock than the adults.

This description of horizontal distribution patterns overlapping and extending into a range of vegetational types is a familiar one, and serves to underline the fact that many species of soil animal which are successful in one particular habitat have the ability, possibly through their great tolerance, to extend their range into other localities of a different character. Of the species listed in Table 6.3 as "characteristic" of a particular successional stage only three, *Adoristes ovatus*, *Liebstadia similis* and *Nanhermannia comitalis*, are restricted to the association with which they have been identified. In dealing with common species, differences in horizontal distribution become differences in numbers, rather than in kind. However, if we include in this comparison species which occur rarely and in low abundance, the distinctions between the various stages in succession become greater.

Table 6.4

The numbers of species of mite and the degree of exclusiveness in four hummock formations compared (based on data provided by Popp, 1962b).

| | Moorland | | | Lowland |
	Sphagnum I	Sphagnum II	Mixed moor	Carex
Number of species	45	80	114	78
Number of restricted species	8	10	41	44
Degree of exclusiveness (%)	17·8	12·5	35·9	56·4

6*

Table 6.4 shows that marked qualitative differences occur between the fauna of the *Carex* hummock and that of the three moorland types. A less marked, but still appreciable, distinction may also be made between the mixed moor fauna and that of the other three sites. These differences may be a reflection of the microhabitat differences illustrated in Fig. 6.9.

Earlier it was suggested that the *Sphagnum* hummock displays an essentially horizontal, rather than a vertical zonation. This is not entirely true, especially in the case of emergent hummocks in which the *Sphagnum* carpet is overgrown with an open vegetation of *Eriophorum*, *Vaccinium* and *Carex*. Even within the *Sphagnum* carpet above water level, two layers may be distinguished, namely the smooth surface layer formed by the crowding together of the heads of the *Sphagnum* plants, and the stalk layer which contains small air spaces suitable for colonization by certain animals. These two layers provide contrasting microhabitats, and distinct faunal differences may occur between them. Heal (1962) noted that the rhizopod Protozoa fauna of *Sphagnum* showed a vertical stratification of species. *Hyalosphenia papilio* and *Amphitrema flavum*, which contain zoochlorellae, are mainly localized in the upper parts of the clump, whereas the majority of species which do not require high light intensities to promote their metabolic processes occur deeper down where the minerals needed for the construction of their shells are in good supply.

Nørgaard (1951) found that distinct microhabitat preferences were shown by two lycosid spiders common in wet *Sphagnum*, namely *Lycosa pullata* and *Pirata piraticus*. The former showed a preference for the smooth surfaces of the carpet, while the latter chose the closed spaces within the stalk layer. This segregation remained fairly constant throughout the year, for even during winter *L. pullata* did not descend too deeply into the stalk layer, preferring instead to hibernate in tussocks of *Carex* and *Polytrichum*. On the other hand, *P. piraticus* could survive the freezing conditions in the *Sphagnum* tussock at this time of year, and thus was able to remain in the same site throughout. The results of experiments on temperature preferences and resistance to desiccation showed clear differences between these two spiders, which could be correlated with their distribution patterns. *L. pullata* could survive a combination of high temperatures and low relative humidities much better than *P. piraticus*. With its lower tolerance, the latter species would be unable to become established on the surface of *Sphagnum* growths. These behavioural responses provide a mechanism for achieving a spatial separation of two species which have broadly similar feeding habits, and otherwise might be directly competitive. This separation is further enhanced by selection of different structural, as well as microclimatic, features. If an analogy can be made with the woodland system, we could regard *L. pullata* as a canopy dweller, while *P. piraticus* lives among the trunks and the branches and, when disturbed, can climb down into the underlying substratum which, in this case, is water.

Environmental Factors: Effect of Erosion

The development of the bog habitat is intimately associated with the growth of *Sphagnum* moss, as we have seen. As a bog matures and ages the *Sphagnum* cover is often replaced by mixed moorland vegetation, as is the case, for example, with upland blanket bog. As it dries out, this bog frequently undergoes erosion, and the complex developed under this process provides a variety of microhabitats for the soil fauna. Hale (1963) and Block (1966) recognize four stages in the erosion and subsequent re-colonization of blanket bog peat, namely:

(1) Non-eroded moor supporting a mixed vegetation in which *Calluna vulgaris*, *Eriophorum vaginatum*, *E. angustifolium*, *Empetrum nigrum*, *Vaccinium myrtilus*, *Rubus chamaemorus* and *Sphagnum rubellum* are abundant.
(2) Persistent peat hummocks covered with *Calluna* and *Empetrum*.
(3) *Eriophorum angustifolium* growing on bare peat.
(4) Tussocks of *Eriophorum vaginatum* on peat.

Fig. 6.10 Eroding blanket bog on Fountains Fell, Yorkshire. On the right is an eroding peat hummock supporting *Vaccinium myrtilus* and *Empetrum nigrum*. An expanse of bare peat (middle distance) extends to a sward of *Eriophorum angustifolium*. Beyond this are tussocks of *Eriophorum vaginatum* and eroding hummocks. (Photo. by author.)

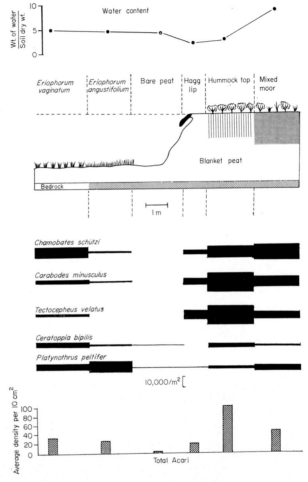

Fig. 6.11 Horizontal distribution of water content, vegetation, five species of Crypto-stigmata and total Acari on an eroding blanket bog at Moor House, Westmorland, England. (Modified after Block, 1966.)

This general sequence is shown in Fig. 6.10, and a diagrammatic representation which includes a fifth site, the overhanging peat of the residual hummock (hagg lip) is given in Fig. 6.11. This figure also summarizes the distribution of the most abundant cryptostigmatid mites, the variations in water content of the soil and the vegetational stages in the erosion sequence. The erosion to, and re-colonization of, bare peat proceeds from right to left in this figure. It will be noted that erosion causes a drying out of the saturated mixed moor, with the hummock top and the hagg lip rather more desiccated than the other sites. The mites in the hummock top zone occurred in significantly higher density than in the other zones. Bare peat (Fig. 6.12) did not support a permanent mite fauna, although *Platynothrus peltifer* and *Ceratoppia bipilis* (Fig. 6.13) occurred locally in crevices and under algal mats where high humidities were maintained. To this extent, then, soil moisture influences horizontal distribution patterns. Further, the greater density of mites in the hummock top zone compared with the mixed moor may be a reflection of the fact that the latter is normally saturated. Free water in the soil lessens the ability of microarthropods to move and forage for food, and also may produce anaerobic conditions which lower the metabolic activity of these animals. However, variations in soil moisture cannot account for all of the faunal differences. Striking differences occur in the densities and species composition of the mite fauna when comparisons are made between hummock top and hagg lip, bare peat and the two *Eriophorum* sites, but these are not paralleled by marked differences in the water content of the soil. Mixed moor and hummock top samples yielded 15 species from each site, despite differences in water content. The two *Eriophorum* sites have almost identical moisture characteristics, and yet 12 species of cryptostigmatid mite occurred in the *E. vaginatum*, compared with only 6 from *E. angustifolium*. It is possible that the changes in plant cover resulting from the erosion of the moor may be the explanation for these horizontal variations. Such changes could alter microfloral and detritus components of the

Fig. **6.12** The bare peat zone of an eroding blanket bog, showing crevice formation which provides microhabitats for soil animals. (Photo. by author.)

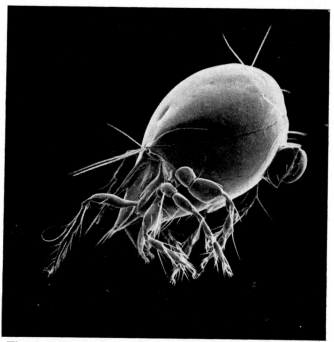

Fig. 6.13 The cryptostigmatid mite *Ceratoppia bipilis* (Hermann), (× 42), a temporary inhabitant of bare beat. (Photo, by B. Parry and reproduced through the courtesy of the British Museum, N.H.)

soil on which these microarthropods feed. Freshly eroded and re-distributed peat offers little in the way of nutrients, and the low densities recorded here and from the *E. angustifolium* zone probably reflect this. A detailed examination of the distribution patterns of individual species (Fig. 6.11) shows that the commonest species, *C. schützi*, *C. minusculus* and *T. velatus*, occur in high densities both in wet and dry sites, indicating that water content does not influence significantly their horizontal distribution pattern. However, it cannot be emphasized too often that "wet" and "dry" categorizations can over-simplify the situation. For example, *Platynothrus peltifer*, often characteristic of "wet" sites as we have noted in Chapter 4, is more abundant in the "dry" zones of *Eriophorum* vegetation than it is in the mixed moor. In these relatively dry localities, however, it appears to be localized in cracks and crevices where conditions of high humidity prevail.

It must be obvious, in considering horizontal distribution patterns of soil animals, that the broad classification of sites into "wet" and "dry" has little relevance, for these patterns must be interpreted in ecological terms commensurate with the microenvironmental factors involved. Haarløv (1942) found this to be the case in an investigation of a typically wet "glumiflores" meadow in north-east Greenland. Here, vegetational tufts of *Carex*, *Eriophorum* and *Juncus*, separated by water-filled drainage channels, were often

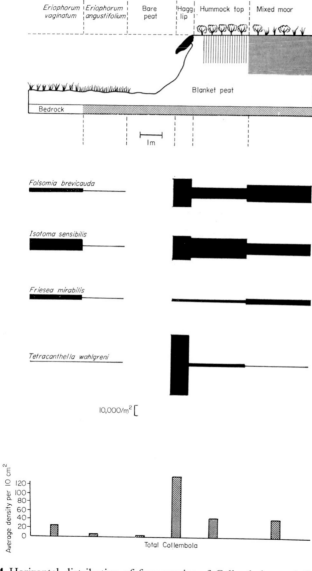

Fig. 6.14 Horizontal distribution of four species of Collembola, total Collembola and vegetation on an eroding blanket bog at Moor House. (Modified after Hale, 1963.)

high enough above the water table to qualify as dry habitats. The distribution of microarthropods confirmed the distinction between the dry (upper) and wet (lower) zones of the tufts.

On the other hand, plant cover and soil water content may be related, and this seems to be a general feature of the erosion complex discussed above. As erosion proceeds, a drying out of the soil occurs, although localized areas of bare peat may have high humidities and even standing water during particularly wet weather. In a study of the Collembola of this complex, Hale (1963) related the horizontal distribution patterns illustrated in Fig. 6.14 to the gradient of soil water content. Comparing this picture with that given in Fig. 6.11, it is evident that the general patterns of the mites and Collembola are similar, with the notable exception that the latter group achieves its highest density in the hagg lip zone, where three species, *F. brevicauda*, *I. sensibilis* and *T. wahlgreni*, collectively make a contribution which far outweighs that of any other species or combination of species. Unlike the mite

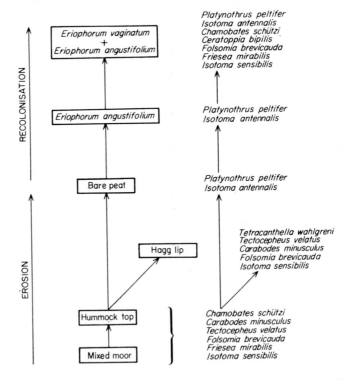

Fig. 6.15 Horizontal distribution of species of Oribatei (Cryptostigmata) and Collembola on an eroding blanket bog. (Compiled from Hale, 1963 and Block, 1966.)

populations, the densities of Collembola on hummock top and mixed moor were not significantly different. Bare peat is not permanently colonized by Collembola, although in peat crevices and regions of high humidity under algal flakes and surface rocks, *Isotoma antennalis* occurs with a distribution pattern closely paralleling that of the mite *Platynothrus peltifer*.

In summary, it is evident that both the mites and Collembola suffer decreases in numbers of species and individuals as erosion produces bare peat surfaces, but within this sequence, Collembola flourish in rather drier habitats than do the mites. Both of these groups are sparsely represented in the early stages of re-colonization of the eroded peat by *Eriophorum angustifolium*, but become more abundant as the vegetation cover increases with the establishment of *E. vaginatum*. This sequence is summarized in Fig. 6.15.

Fauna of tundra soils

The so-called barren lands of the North, in which tundra vegetation predominates, frequently develop podzolic soils similar in character to those of temperate moorland, except for the presence of permafrost. The tundra habitat has many unusual characteristics and is a major ecological system in its own right, distinct from grasslands and moorlands. However, grass species, together with sedges, rushes and low shrubs, make an appreciable contribution to the vegetation, and the soil microarthropods present include many species of the more southerly moorland and alpine soils.

Tundra vegetation is present in the coastal regions of Greenland, northern Canada and northern Eurasia, and the soil microarthropod fauna of a wide variety of vegetational and topographical situations has been examined, Hammer (1937, 1938, 1944, and earlier under her maiden name of Jørgensen, 1934) classified various sites into dry and wet "biotopes", the main distinction between the two being that the former became dry in summer but the latter did not. Included in the dry biotopes were vegetational associations dominated by sedges, willow, saxifrage, *Empetrum*, *Dryas octopetala*, *Silene acaulis*, *Cerastium alpinum* and *Cassiope* species, among others. Wet biotopes, such as lake bank, bog and moss, were characterized by *Eriophorum polystachium*, *Salix arctica*, *Calamagrostis neglecta*, *Equisetum variegatum*, *Sphagnum* spp., *Juncus biglumis* and *Carex subspathacea*.

A recent review of the tundra soil microfauna (Bliss *et al.*, 1973) has pointed out the low levels of diversity which are commonly encountered. The most important components of this fauna are nematodes, mites, Collembola and, in damp sites particularly, insect larvae belonging to the dipteran Chironomidae (midges) and crustacean Copepoda and Ostracoda (Table 6.5). The densities of these populations vary considerably from place to place, and although the estimates given in Table 6.5 are generally lower than those recorded from temperate moorland, especially with regard to the nematodes, mites and Collembola (see earlier in this chapter), they may approximate to the latter in localized situations. A major contribution to total faunal biomass is made by the chironomid larvae in moist sites.

Earlier work on dry and wet biotopes by Hammer, mentioned above,

Table 6.5

Estimated mean populations (seasonal averages) of various invertebrate groups in the arctic tundra of Truelove Lowland Devon Island, Canada. All data are thousands/m² (unpublished data from Bliss *et al*).

					Invertebrate group					
Site	*Nematoda*	*Rotifera*	*Enchytraeidae*	*Tardigrada*	*Copepoda*	*Ostracoda*	*Cladocera*	*Acarina*	*Collembola*	*Diptera Nematocera*
Raised beach	3840	0·2	8·7	5·7	—	—	—	9·8	18·2	0·4
Hummocky sedge meadow	1273	6·4	24·5	2·3	23·1	0·6	0·006	9·7	7·0	10·6

distinguished a "dry community" characterized by the mites *Zygoribatula exilis*, *Calyptozetes sarekensis* and *Camisia horrida*, and a "wet community" dominated by *Platynothrus peltifer* and *Melanozetes mollicomus*. In making this distinction however, the author drew attention to the fact that there was no sharp demarcation between the two in a qualitative sense, i.e. many species were common to both types of habitat. Indeed, *Z. exilis* had a frequency of 19% in wet sites, so that it cannot strictly be considered as an indicator of dry conditions. Knülle (1957) stated that this species finds its optimum environmental conditions in moist habitats provided by the growth of moss on compact soils.

As noted in Chapter 4, the gradation of "wet" associations into "dry" ones is a common feature among soil populations, and in the tundra this involves a sequence of habitats ranging from permanently saturated lake banks, at one extreme, through bog and moss to the drier mixed chamaephyte vegetation, and that of the fell-field which represents the dry extreme. The great impression is one of considerable overlap in the pattern of species distribution, with mites such as *Tectocepheus velatus*, *Zygoribatula exilis*, *Trichoribates trimaculatus* and the collembolan *Folsomia quadrioculata* generally distributed through the habitats, wet and dry alike.

When comparing the faunal characteristics of "wet" and "dry" habitats, it must be borne in mind that the "wet" type has a more stable environment, at least as far as moisture content is concerned. Hence, it is to be expected that species with a preference for moist soils, such as the mites *Platynothrus peltifer* and *Malaconothrus globiger*, can become established and maintain populations which would not be subject to short-term fluctuations in size. On the other hand, the "dry" environment is much less stable, for not only does the moisture content vary between summer and winter, but other environmental changes consequent on this may occur. For example, winter flooding may alter salt concentrations in the soil profile by leaching some ions and depositing others. Short-term waterlogging effects may alter the composition of the soil microflora and, indirectly, the microfauna which feed on these plants. Finally, surface run-off may cause erosion of the litter layer. During the drying-out period in summer, these processes will go into reverse. In view of these environmental changes, which can be quite drastic, we could perhaps expect a microfauna which varies considerably in its species composition, seasonally and from one locality to another. Ecologically tolerant forms would make an appreciable contribution to this fauna, and because these species occur in a wide variety of habitats, the "dry" association becomes difficult to define.

The species composition of a "dry" association at any one place and at any one time will depend not only on the prevailing environmental conditions, but also on the immediate past history of the locality and the ecologically tolerant species available to colonize it. This point is well illustrated by the work of Tarras-Wahlberg (1961) on the microarthropods of a Swedish bog, a dystrophic habitat similar in some respects to the tundra soils. The numerically dominant species in this site was the cryptostigmatid *Tectocepheus velatus*. Now, ecologically speaking, *T. velatus* is a tolerant species, and as a rule in this type of habitat, at least in northern Europe, the dominant is the less tolerant *Parachipteria punctata* (Strenzke, 1952; Knülle, 1957). However, in the situation studied by Tarras-Wahlberg, a previous hot, dry summer had created conditions unfavourable for the survival of less tolerant, more "typical" bog species. The wider tolerance of *T. velatus* permitted it to survive and maintain high population density, while the numbers of other species were declining. Thus, the use of this species as an "indicator" would be entirely misleading under these circumstances.

Fauna of fenland soils

From the point of view of the soil fauna, two of the most important variables in fenland succession are moisture levels and vegetation type. The effect of these variables can be illustrated by the following examples.

Moisture Gradients and Carabid Beetles

The variations in moisture content, which are such common features of fen and marsh soils, influence the distribution patterns of such macroarthropods as the Carabidae, a group of beetles which often occurs in high densities in such sites. Dawson (1965) investigated the distribution of carabids in Wicken Fen, Cambridgeshire, on a site where tussocks of the grass *Molinia* and the fern *Thelypteris palustris* were common. These tussocks evidently serve as overwintering quarters for the beetles, for winter densities here were 4–8 times greater than in summer. This work traced the distribution patterns of eight species of beetle belonging to the genera *Agonum* and *Feronia* in six different sites in the fen. Differences in vertical distribution were noted, with *Agonum* species preferring the surface litter layer, *Feronia* the underlying soil. Horizontal patterns were also evident. Three of the commonest species, *Agonum obscurum*, *A. fuliginosum* and *Feronia diligens* were widely distributed through the fen, although the two *Agonum* species were especially abundant in wet habitats formed by the sedge *Cladium mariscus*. On the other hand, the highest densities of *F. diligens* developed in habitats often dominated by *Molinia*, some of which were inundated in winter, others remaining drier. Rather more clear-cut habitat preferences were shown by the rarer species, such as *Feronia vernalis*, which evidently selected grassy clearings where *Carex* was abundant, *F. minor* which was also found in the soil layer of this type of habitat and in fern tussocks growing in the fen carr, *F. strenua* which apparently avoided peaty sites and was restricted to the edges of the fen, and *Agonum thoreyi* which was strongly associated with the sedge *Cladium* in which it hibernates.

A study rather similar to the one just described was carried out by Murdoch (1966) on the ground-dwelling carabids of a marshy area in Wytham Woods, Berkshire. Here again, the familiar pattern of population movements into overwintering sites in grass tussocks and decaying logs was noted. In the centre of the marsh, where water-logged soil supported growths of *Carex*, *Epilobium* and *Equisetum*, the commonest carabid was *Agonum fuliginosum*, whereas in drier areas near the periphery of the marsh, with *Deschampsia*, *Pteridium* and *Rubus* forming the ground layer under mixed deciduous or conifer stands, this beetle was replaced by *Feronia madida* and *Abax parallelipedus*. This work suggests, as does Dawson's, that distribution patterns of these Carabidae are influenced to a considerable extent by specific preferences for certain microclimatic conditions, and the gradients occurring in fenland and marsh undoubtedly permit colonization by a varied fauna. To some extent at least, the microclimate of the soil/litter subsystem is influenced by the type of ground vegetation present, and it is to this effect that we now turn.

Vegetational Succession and the Microfauna

The fen at Malham Tarn, Yorkshire, mentioned earlier in this chapter, provides a good illustration of the faunal changes occurring as one vegetation type succeeds another. Samples of microarthropods and enchytraeids have been taken from three sites in this fen, namely clumps of *Acrocladium* moss associated with the floating mat in the wet fen, clumps of *Sphagnum squarrosum* growing in the drier tussock fen, and leaf litter at the base of birch trees growing in the carr (Fig. 6.16). These three sites provide a sequence of habitats becoming progressively drier and more terrestrial in character. The counts obtained from these samples are given in Fig. 6.17, from which it can be seen that a much richer, more diverse fauna is present in the more stable habitat provided by the accumulation of organic debris around tree bases in the carr. The microfauna of the *Acrocladium* includes an appreciable number of enchytraeids, as well as cryptostigmatid mites belonging to the genera *Phthiracarus* and *Brachychthonius*. Prominent among the fauna of the *Sphagnum* clumps were Collembola (*Onychiurus* spp.) and representatives of the mite genera *Trimalaconothrus* and *Hermannia*. In the leaf litter of the fen carr, the more varied fauna was dominated by a mixture of semi-aquatic and terrestrial mites belonging to the genera *Oppia*, *Nanhermannia*, *Phthiracarus* and *Trimalaconothrus*, and members of the Astigmata. Collembola were also well in evidence in this site.

In many lowland situations in Britain, human interference has changed considerably the original character of fenland. Examples of such activity include drainage and burning to allow reclamation for arable farming, peat cutting for fuel, and harvesting of *Phragmites* for thatch. Clearly, fenland in its broadest sense is ecologically very complex, and special attention must be paid in soil fauna studies to the effects of microenvironmental factors, the composition of the vegetation, the stage in ecological succession which this represents, and the history of land use in the area. A *Molinia* fen sampled by Macfadyen (1952) may be considered to have reached the grassland stage of development, and evidently was maintained in this condition by harvesting and burning. Analyses of variance carried out on counts of soil microarthropods in this locality revealed that density variations were no greater when different parts of the fen were compared, than they were between pairs of samples taken from adjacent points. The conclusion drawn was that, in general, these mites and Collembola appeared to have a fairly uniform distribution over the fen, and this was particularly true of the cryptostigmatids *Minunthozetes semirufus* and *Oppia nova*. However, significant differences were detected in the general distribution of *Folsomia quadrioculata* and *Nanhermannia elegantula*, indicating aggregation and, in addition, significant differences in densities of microarthropods in different plant types were also

Fig. 6.16 Fen carr at Malham Tarn, Yorkshire. The principal tree species are birch and willow. In the foreground is a *Molinia/Deschampsia* meadow. (Photo. by author.)

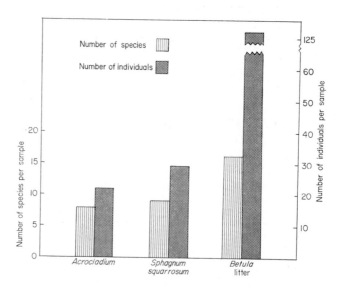

Fig. 6.17 Numbers of species and numbers of individuals of microarthropods from three habitats in Malham Fen, compared.

noted. The species recorded from this fenland in southern England are those normally associated with terrestrial habitats, and the same is true of the soil fauna of Woodwalton Fen, Huntingdonshire. This fen has been partly drained, and its peaty soil supports a thicket vegetation and even oak woodland in parts. Samples taken from oak litter and the underlying peat contained a total of 19 species of cryptostigmatid mites, all of which are typically soil forms, including *Nanhermannia elegantula*, *Adoristes ovatus*, *Achipteria coleoptrata*, *Tectocepheus velatus* and members of the Camisiidae, Oppiidae and Phthiracaridae. Clearly, we are dealing here with a situation which grades into that of the forest soils considered in Chapter 8.

Summary

In this chapter we have been particularly concerned with the fauna of the wet soils of moorland, fen and marsh. In such sites, environmental gradients are often clearly defined, and the most important parameters, as far as the soil fauna is concerned are: (1) soil moisture, (2) vegetation type, (3) pH, and (4) soil type.

Variations in soil moisture levels in moorland and fen produce a mosaic pattern of wet and dry microhabitats which encourages the establishment of a species-rich fauna. A variety of carabid beetles, mites and Collembola are represented in this fauna by xerophilous, mesophilous, as well as hygrophilous, forms. Within this environmental framework, habitat separation is achieved by differing degrees of tolerance to soil moisture levels. This is particularly well illustrated when the distribution patterns of rare species are considered. However, common species are not so severely restricted, and their distributions overlap and extend across a range of moisture conditions. In permanently wet conditions, the terrestrial component of the soil fauna is reduced. Waterlogged moorland soils are characterized by anaerobic conditions which reduce biological activity and promote the development of an extreme form of raw humus, namely peat. The acidic soils which result from this process are colonized mainly by enchytraeids and nematodes; they have high carbon/nitrogen ratios and low levels of biological activity.

The establishment of wet and dry mosaics is also enhanced by the growth form of the vegetation. Such common moorland plants as *Calluna vulgaris* and *Vaccinium myrtilus* form hummocks of vegetation, while the grasses *Nardus stricta* and *Eriophorum vaginatum* have a tussock growth form (Fig. 6.18). In fenland, tussock formation occurs in various grasses and sedges, such as *Molinia caerulea* and *Carex appropinquata*. As the tussock or hummock builds up with the accumulation of organic material around the roots of the plants, the upper regions are progressively removed from the influence of groundwater and increasingly exposed to the drying effects of wind and

Fig. 6.18 Tussocks of *Eriophorum vaginatum* growing on flooded peat on Fountains Fell, Yorkshire. A dense sward of *Eriophorum angustifolium* extends out to eroding peat hummocks in the background. (Photo. by author.)

sun. In this way, a vertical pattern of wet and dry microhabitats is produced. The extent to which habitat separation is achieved by this pattern depends on the type of hummock. In hummocks of *Sphagnum*, the distinction between a dry, upper, zone and a wet, lower, one is only evident during the summer; during the winter, capillary water extends into the upper zones, and the "aquatic" fauna shifts its centres of population density in response to changing water levels. On the other hand, in drier *Carex* hummocks the distinction between dry and wet zones is maintained throughout the year and, as a consequence, the fauna shows a higher degree of exclusiveness, and terrestrialness.

The variety of vegetation type occurring in moorland and fen influences the species diversity of the soil fauna by providing a variety of different kinds of food material for saprophagous species. The leaf litter which accumulates under mixed moorland and grassland vegetation originates from a variety of plant species, and may provide a wider range of food material, a greater number of food niches, than the litter which accumulates, for example, under a pure stand of *Eriophorum angustifolium*. The distribution patterns of cryptostigmatid mites on eroding blanket bog and the later stages in fen succession can be explained, at least partly, in these terms.

Gradients of pH are well marked in the fenland succession and also in

moorland sites where glacial drift and blanket peat are deposited on Carboniferous limestone. As was noted in Chapter 4, the two groups of the soil
fauna most directly affected by this factor are the lumbricid earthworms
and the Protozoa. Most moorland soils are unsuitable for colonization by
Lumbricidae because of their acid, peaty nature, although this faunal group
becomes prominent in rendzina-type soils developed from the weathering
of limestone. Collembola also achieve maximum densities in these soils,
whereas cryptostigmatid mites, tipulid larvae, enchytraeids and nematodes
are more successful in the rather acid, alluvial soils developed from glacial
drift. However, it is likely that factors other than pH are responsible for the
habitat preferences of these microarthropods, enchytraeids and nematodes.
Such factors include soil type, the organic profile and moisture content of
the soil.

The effect of factors such as soil type and pH are often difficult to separate
in moorland situations, for the two are often closely related. However, there
is no evidence that pH conditions, in themselves, have a major effect on

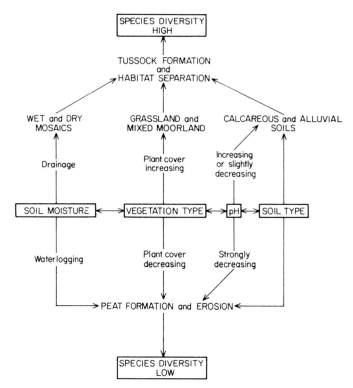

Fig. **6.19** Factors affecting the species diversity of the soil fauna of moorland sites.

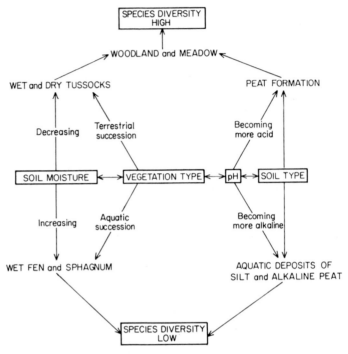

Fig. 6.20 Factors affecting the species diversity of the soil fauna in fenland sites.

distribution patterns, except in the case of the Lumbricidae. In fenland, on the other hand, where the soil water fauna achieves a greater prominence, a varied fauna of rhizopod Protozoa develops and this can be attributed to the direct effect of the pH factor. Rhizopod species vary in their pH tolerance, and the existence of a gradient of pH in the fenland succession, particularly associated with *Sphagnum*, allows the establishment of a species complex in which acid-tolerant and acid-intolerant forms can exist in the same locality.

The factors we have been considering here do not work in isolation. Soil moisture levels are often related to vegetation type and soil type. The composition of the vegetation may be influenced by the pH of the soil and, as we have just seen, pH and soil type are closely related. Despite these interactions, there are certain clear-cut effects which influence the species diversity of the soil fauna in moorland and fen. These effects are summarized in Figs. 6.19–20.

REFERENCES

BANAGE, W. B. (1963). The ecological importance of free-living soil nematodes with special reference to those of moorland soil, *J. Anim. Ecol.*, **32**, 133–40.

BANAGE, W. B. (1966). Nematode distribution on some British upland moor soils with a note on nematode parasitizing fungi, *J. Anim. Ecol.*, **35**, 349–61.

BLISS, L. C., COURTIN, G. M., PATTIE, D. L., RIEWE, R. R., WHITFIELD, D. W. A. and WIDDEN, P. (1973), Arctic tundra ecosystems. *A. Rev. Ecol. Syst.*, **4**, 359–99.

BLOCK, W. C. (1965). Distribution of soil mites (Acarina) on the Moor House National Nature Reserve, Westmorland, with notes on their numerical abundance, *Pedobiologia*, **5**, 244–51.

BLOCK, W. C. (1966). The distribution of soil Acarina on eroding blanket bog, *Pedobiologia*, **6**, 27–34.

BONNET, L. (1961). Caractères généraux des populations thécamoebiennes endogées, *Pedobiologia*, **1**, 6–24.

COULSON, J. C. (1959). Observations on the Tipulidae (Diptera) of the Moor House Reserve, *Trans. R. ent. Soc. Lond.*, **111**, 157–74.

COULSON, J. C. (1962). The biology of *Tipula subnodicornis* Zetterstedt with comparative observations on *Tipula paludosa* Meigen, *J. Anim. Ecol.*, **31**, 1–21.

CRAGG, J. B. (1961). Some aspects of the ecology of moorland animals, *J. Anim. Ecol.*, **30**, 205–34.

DAWSON, N. (1965). A comparative study of the ecology of eight species of fenland Carabidae (Coleoptera), *J. Anim. Ecol.*, **34**, 299–314.

HAARLØV, N. (1942). A morphologic-systematic-ecological investigation of Acarina, *Meddr. Grønland*, **128**, 1–71.

HALE, W. G. (1963). The Collembola of eroding blanket bog. *In* "Soil Organisms", 406–13 (Docksen, J. and Drift, J. van der, eds.), North Holland, Amsterdam.

HALE, W. G. (1966). A population study of moorland Collembola, *Pedobiologia*, **6**, 65–99.

HAMMER, M. (1937). A quantitative and qualitative investigation of the microfauna communities of the soil at Angmagssalik and in Mikisfjord, *Meddr. Grønland*, **108**, Nr. 2.

HAMMER, M. (1938). The zoology of East Greenland. Collemboles, *Meddr. Grønland*, **121**, Nr. 2.

HAMMER, M. (1944). Studies on the oribatids and collemboles of Greenland, *Meddr. Grønland*, **141**, Nr. 3.

HARNISCH, O. (1938). Weiters Daten zur Rhizopoden fauna Lapplands, *Zool. Anz.*, **124**, 138–50.

HEAL, O. W. (1962). The abundance and microdistribution of testate amoebae (Rhizopoda, Testacea) in *Sphagnum*, *Oikos*, **13**, 35–47.

JØRGENSEN, M. (1934). A quantitative investigation of the microfauna communities of the soil in East Greenland, *Meddr. Grønland*, **100**, Nr. 9.

KNÜLLE, W. (1957). Die Verteilung der Acari: Oribatei im Boden, *Z. Morph. Okol. Tiere*, **46**, 397–432.

MACFADYEN, A. (1952). The small arthropods of a *Molinia* fen at Cothill, *J. Anim. Ecol.*, **21**, 87–117.

MURDOCH, W. W. (1966). Aspects of the population dynamics of some marsh Carabidae, *J. Anim. Ecol.*, **35**, 127–56.

MURPHY, D. H. (1955). Long-term changes in collembolan populations with special reference to moorland soils. *In* "Soil Zoology", 157–66 (Kevan, D. K. McE., ed.), Butterworths, London.

NØRGAARD, E. (1951). On the ecology of two lycosid spiders (*Pirata piraticus* and *Lycosa pullata*) from a Danish *Sphagnum* bog, *Oikos*, **3**, 1–21.

PAULSON, B. (1952–53). Some rhizopod associations of a Swedish mire, *Oikos*, **4**, 151–65.

PEACHEY, J. E. (1959). "Studies on the Enchytraeidae of Moorland Soils". Ph.D. Thesis, University of Durham.

POPP, E. (1962a). Semiaquatile Lebensräume (Bülten) im Hoch- und Niedermooren I. Die Standortsfaktoren, *Int. Revue ges. Hydrobiol.*, **47**, 431–64.

POPP, E. (1962b). Semiaquatile Lebensräume (Bülten) im Hoch- und Niedermooren II. Die Milbenfauna, *Int. Revue ges. Hydrobiol.*, **47**, 533–79.

SPRINGETT, J. A. (1970). The distribution and life histories of some moorland Enchytraeidae (Oligochaeta), *J. Anim. Ecol.*, **39**, 725–37.

STOUT, J. D. and HEAL, O. W. (1967). Protozoa. *In* "Soil Biology", 149–95 (Burges, N. and Raw, F., eds.), Academic Press, London and New York.

STRENZKE, K. (1952). Untersuchungen über die Tiergemeinschaften des Bodens: Die Oribatiden und ihre Synusien in den Böden Norddeutschlands, *Zoologica*, **37**, 1–167.

SVENDSEN, J. A. (1957). The distribution of Lumbricidae in an area of Pennine moorland, *J. Anim. Ecol.*, **26**, 411–21.

TARRAS-WAHLBERG, N. (1952–53). Oribatids from Åkhult-Mire, *Oikos*, **4**, 166–71.

TARRAS-WAHLBERG, N. (1961). The Oribatei of a central Swedish bog and their environment, *Oikos*, suppl. **4**, 56 pp.

WEIS-FOGH, T. (1948). Ecological investigations on mites and collemboles in the soil, *Natura jutl.*, **1**, 135–270.

Chapter 7

Heathland and Sand Dunes

In this chapter we look at the fauna of soils which are basically sandy in character. These soils are usually well drained and, as a consequence rather dry, so that they provide an interesting contrast to the wetter, moorland, sites discussed in Chapter 6, which often have similar vegetational features. In particular, the prevalence of heather, *Calluna vulgaris*, is a feature common to both of these types of sites. In upland areas, heather is often found growing on the deep, acid, peaty soils of moorland, as we have seen. In the lowlands, on the other hand, this species can also form a conspicuous component of the vegetation on dry, rather shallow, sandy soils of heathland.

Dry sandy heaths are developed in some parts of northern Europe, such as the Jutland region of Denmark and the north German plains, and in southern Russia. Opinion is divided on the origin of these heathlands. They have been interpreted as an arrested stage in the vegetational succession towards woodland, but recently Gimingham (1972) has suggested that they

Fig. 7.1 An area of heathland in the New Forest colonized by *Calluna*, bracken and grasses. The site is being invaded by neighbouring woodland. (Photo. by author.)

are derived from forested sites from which the forest cover has been removed. In its present form this vegetational type supports, in addition to heather, such plants as bracken, gorse and a variety of grasses (Fig. 7.1).

Heathlands often provide a range of habitats, a successional series from bare sandy soil, at one extreme, to *Calluna* heath, at the other. In tracing patterns of faunal distribution along this succession, the influence of intervening grassland sites must also be taken into consideration. The faunal characteristics of these are included here, not so much as grassland faunas *per se*, since this topic has already been dealt with, but rather as stages in heathland succession.

In coastal regions, the initial stages in this succession are often represented by sand dune formations which form a horizontal sequence of habitats extending from the shore line to the interior. Here, the stages in succession are usually clearly demarcated, and faunal distribution patterns can be analysed in relation to vegetational changes. This topic will be considered later in this chapter.

Heathland succession

As bare, sandy soil becomes colonized by heather, bracken, gorse, grasses and other herbaceous species, a series of ecological events is set in train which results in the development of a variety of microhabitat conditions. First and foremost, there is a progressive increase in the amount of ground cover, and since this cover is provided, at least in part, by woody plants, it has a degree of permanence through the seasons. This cover tends to ameliorate the temperature and moisture conditions at the soil surface, particularly during the day, as the measurements made by Delany (1953) show. This author noted that ground surface temperatures under the shade of *Calluna* were 10°C lower than on exposed soil. Around the base of heather plants, the relative humidity was found to be up to 25% higher than on bare soil, and these moist sites permit the growth of moss and lichens which, in turn, assist in preserving the moist conditions. As Elton (1966) pointed out, such heathland will have a pattern of cool, damp, sheltered microhabitats interspersed with hot, dry, exposed ones. As the vegetation becomes more luxurious with the development of grasses, the organic content of the soil increases, and in many heathlands, particularly in the later stages when *Calluna* predominates, a layer of acid "raw humus" develops at the soil surface.

Looking now in more detail at the soil and its fauna, heathland succession can be regarded, in an idealized fashion, as a transition from an unstructured largely embryonic, sandy soil type, through a more stable enriched grassland soil, to a structured acid profile of the podzol type. An entire picture of the faunal succession paralleling this sequence has not yet been assembled .but a few examples will serve to illustrate the general character of the heathland soil fauna, and the effects of the various factors mentioned above on the patterns of distribution.

Heathland soil fauna

We can start by making a few general statements which appear to be valid, at any rate for the heathland soil fauna studied to date. Firstly, this fauna is a rather poor one, in terms of species diversity, and this is particularly true at the two extremes of the succession. Lumbricid earthworms are poorly represented, if at all, in these soils, possibly because of their shallow depth. It is often suggested that the accumulation of a surface layer of acid humus which characterizes the later stages in succession, is unsuitable for earthworms, and a greater ecological significance attaches to groups which can tolerate these conditions, such as nematodes, enchytraeids, mites and, possibly to a lesser extent, Collembola. In the early stages of heath development, when the mineral soil is exposed to the erosive effects of wind and water, the fauna consists mainly of ants, sand wasps, beetles belonging to the families Carabidae and Cicindelidae, and certain spiders. As far as the microarthropods are concerned, Knülle (1957) noted that cryptostigmatid mites were characteristically represented by forms which are tolerant of rather dry conditions, such as the xerophilous *Humerobates rostrolamellatus*, *Camisia segnis* and *Eupelops* spp. In more moist situations, these species normally prefer habitats above the ground (see Chapter 9). Some of these groups are now considered in more detail.

Distribution Patterns of Ants

The prevalence of the Hymenoptera among the larger fauna of exposed sandy soil has been highlighted in the classical study by Richards (1926) on the succession of communities at Oxshott Heath, Surrey, England. The bulk of this fauna comprised various bees and wasps which have, perhaps, only a marginal significance as far as the true soil community is concerned. Ants, on the other hand, are often more closely associated with the soil and although the species diversity is generally low, members of this group may occupy a dominant position in the heathland community. Various aspects of their ecology have been investigated by Brian and his co-workers in Britain (Brian, 1964; Brian *et al.*, 1965, 1966, 1967). They noted that distribution patterns are influenced, to some extent, by direct competition between and within species populations, and also by the settlement of fertilized females in suitable nesting sites. The mosaic pattern of moist and dry areas in heathland, already referred to, evidently is of considerable importance in the spatial distribution of certain species, such as *Lasius niger*, *L. alienus* and *Tetramorium caespitum*. Fertilized females of *L. niger* tend to settle in rather moist, even shaded places, whereas the other two favour drier sites. In attempting to explain this segregation, Brian *et al.* (1966) suggested that moisture preferences may be less important in the selection of habitat than differences in the seasonal timing of nuptial flights. Both of the *Lasius* species, for example, may show similar microhabitat preferences, but since *L. niger* flies mainly during the hottest part of the summer, whereas *L. alienus* flies later, the two species will tend to settle in different areas. The dry sites occupied by *L. alienus* in September could have microclimatic conditions very similar to those occurring in the moist sites selected by *L. niger* in July and August. Any tendency for these two species to overlap is further prevented by direct competition in which *L. niger* is usually the more aggressive and the more successful. This competition may also have contributed to segregation by encouraging a vertical, as well as horizontal, separation of the populations, with *L. niger* dominant above ground, and *L. alienus* below the surface. Within the predominantly dry areas, *L. alienus* is commoner in arid sites, whereas *T. caespitum* prevails on soil with a moderate amount of organic material and a greater moisture content. In competing for the limited number of nesting sites available in the more favourable zones, *T. caespitum* has a much greater advantage by virtue of its better fighting ability, and its capacity for constructing mound formations. Both of these species make underground nests which consist of "brood zones", more sharply delimited in *T. caespitum* as definite chambers, from which a system of anastomosing galleries, or foraging channels, radiates. This method of colony formation evidently limits the distribution of these species to warmer organic soil beneath grassy vegetation, and in this situation *T. caespitum* forms a species continuum in which pockets of *L. alienus* are embedded (Brian *et al.*, 1965).

All three of the ant species which we have just been considering are predominantly predators, feeding on other ants, beetles, centipedes, flies and bugs, although they may also utilize exudates produced by bracken and various bugs such as aphids and mealy bugs. *T. caespitum* is unusual in that its diet also includes large quantities of seeds of heather and various grasses. Although the differences in food habit between these ant species are not very striking, they may be sufficient to re-inforce the other mechanisms which influence the ecological differentiation of these populations. Thus, *L. niger* feeds mainly at the surface where it preys on other ants and surface-dwelling elaterid, tenebrionid and staphylinid beetles. This species also imbibes exudates from bracken shoots and aphids living on the aerial parts of heather, gorse and the grass *Molinia caerulea*. *L. alienus* and *T. caespitum* will also forage at the surface, but probably feed more commonly underground on soil organisms and the exudates of root aphids. Clearly, the spatial distribution patterns we have been discussing are the end products of a complex series of interactions, biotic and abiotic.

Faunal Succession

As the degree of ground cover, dead organic matter and soil moisture increases during heathland succession, the character of the soil fauna changes in several ways. Firstly, there is a general increase in the abundance of mesophilous forms, such as enchytraeids, nematodes, isopods, beetles and microarthropods. For example, Nielsen (1949, 1955) compared the overall densities of nematodes and enchytraeids in sandy heathland soil and in raw humus, and noted a 3-fold increase in the nematodes as the organic content of the soil increased. An even greater increase (more than 12-fold) occurred in the enchytraeid density. Working in Jutland, Schjøtz-Christensen (1957) found a progressive increase in the biomass of ground-dwelling beetles through four pioneer stages of heath succession characterized by an increasing development of grassy ground cover, followed by a decline in the biomass as an acid grassland yielded to a *Calluna*-dominated heath. Secondly, at least among the nematodes, beetles and cryptostigmatid mites, there is a change in the species composition of the fauna as ground cover and moisture content increase. Again, Nielsen (1949) noted that nematodes largely restricted to open sandy soil include *Tylencholaimus mirabilis*, *Tripyla setifera* and *Acrobeles ciliatus*, whereas in raw humus sites under ground vegetation the species *Acrobeles vexilliger*, *Wilsonema otophorum* and *Bunonema reticulatum* came into greater prominence. Schjøtz-Christensen's work, discussed in detail by Elton (1966), also showed that changes in the species composition of the beetle fauna accompanied the progressive development of heath conditions. In the early stages, where tussocks of the grass *Corynephorus canescens* are separated by areas of bare soil, the elaterid *Cardiophorus asellus* was prominent, along with members of the Carabidae. In later successional stages, when a more complete cover of *Corynephorus*, *Agrostis* and *Anthoxanthum* develops, *C. asellus* was joined by the herbivorous carabid *Harpalus anxius* and scarabaeids such as *Phyllopertha horticola* and *Amphimallon solstitialis*. As *Calluna* becomes a more conspicuous element in the vegetation, the carabids and scarabaeids become increasingly scarce. In the case of the microarthropods, which are probably more closely bound to microhabitats within the soil than are the surface-dwelling insects, xerophilous species characteristic of dry sandy soils are often replaced by less tolerant, more truly subterranean forms, as moisture and organic content increase (see Knülle, 1957). Finally, the pattern of vertical distribution in the soil changes. This can be illustrated by the data on nematode distribution given

7

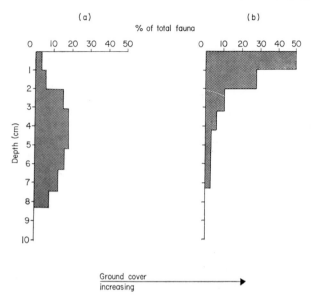

Fig. 7.2 The changing pattern of vertical distribution of the nematode soil fauna in relation to changes in vegetation and the type and amount of plant cover in a sandy heath. (a) Exposed *Corynephorus* slopes, (b) permanent pasture. (From Nielsen, 1949.)

in Fig. 7.2, which show that the bulk of the fauna moves upwards to become concentrated near the surface as this becomes enriched with organic material.

Sand dunes

Dune formation occurs where deposits of sand are subjected to the violent action of wind and water. Such exposed situations are more common in coastal regions than inland, although dune systems occur along the shores of the Great Lakes in North America, and along the margins of rivers as, for example, at Beynes, Haute Camargue, France. These systems have much in common with the heathland type we have just considered, and may intergrade with heathland in the later stages of succession. The following account is concerned with identifying the various habitats present in the sand dune system, and the patterns of faunal distribution associated with these habitats. The process of dune formation has been discussed in detail recently by Ranwell (1972) and will be referred to only briefly here.

The Habitats

Coastal sand dunes provide a complex of habitats which has received

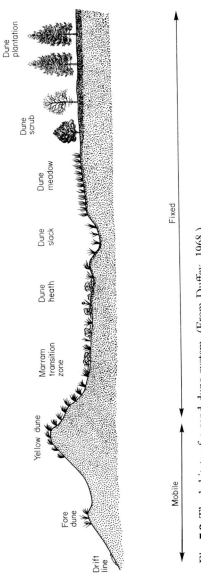

Fig. 7.3 The habitats of a sand dune system. (From Duffey, 1968.)

much attention from biologists over the years (see, for example, the review by Elton, 1966), although little detailed ecological information is available concerning the ground fauna. This complex includes a number of well-defined vegetational types, which can be arranged in a horizontal, chronological sequence extending landwards from the strand line. This sequence ranges from the accumulated debris at the upper tide mark, over a zone of embryonic "fore dunes", often colonized by marram grass, *Ammophila arenaria* or *A. breviligulata*, and by *Agropyron* species, to a higher dune ridge composed of mobile sand supporting marram grass. Beyond this is an area of stable, or "fixed", dunes in which a heath type of vegetation, often consisting of mosses, lichens, heath, wild thyme and a variety of other prostrate herbs, may be distinguished from the scrub or woodland vegetation occurring farther inland. The fixed dune system is sometimes dissected by low-lying channels, seasonally waterlogged and colonized by marsh-type vegetation consisting of a variety of low herbs interspersed with clumps of *Juncus*; these areas are termed dune "slacks". This ecological sequence is illustrated in Figs. 7.3–7.

Composition of the Sand Dune Fauna

The soil fauna of sand dune systems is dominated by the arthropods. Earthworms are relatively uncommon here; the unstable character of the substratum, its dryness and low organic matter content may be responsible for this. Reynoldson (1955) found only two lumbricid species of any consequence on the dunes at Newborough Warren, North Wales, namely *Allolobophora chlorotica* and *Octolasion cyaneum*, and even these were scarce on the mobile dunes.

As far as the microarthropods are concerned, the information is fragmentary, but enough is known about the general ecology of the Collembola and mites to predict that these groups would be poorly represented, at least in the early stages of dune succession. The low organic content of the sandy soil, together with its unstable character, hardly encourages the establishment of these groups, although it could be expected that they would come into greater prominence in the permanent dune habitats, particularly where there is complete ground cover.

The most important groups of macroarthropods in the sand dune system are insects belonging to the Coleoptera, Hymenoptera and Diptera, and the spiders (Arachnida). Krogerus (1932) recorded over 1,000 species of arthropods from a variety of coastal sites in Finland over a 10-year period. He also noted that many dune arthropods are well adapted for digging in the substratum. This is particularly true of various beetles, such as *Dyschirius* spp., *Bembidion argenteolum*, *Amara fulva* and *Liodes ciliaris* among others.

Compared with their relatives in other habitats, these beetles have much more strongly developed fore-legs, often equipped with thick spines, which are used as excavating organs. Other morphological features often associated with the sand dune fauna include weakly pigmented bodies, as for example in the carabid *Eurynebria* spp., *Amara fulva* and *Liodes ciliaris*, and reduction in wing development, such as occurs in *Bembidion argenteolum*, *Eonius bimaculatus* and various Hemiptera, Diptera and Hymenoptera.

Distribution Patterns

Although the general outline of the dune system we have described differs in detail from place to place, major sub-divisions are recognized which are common to many systems. An ecological distinction can be made, for example, between "slack" habitats and those of the higher dune ridges; between "mobile" and "fixed" dunes. In some localities, at least, a characteristic animal association can be identified with these sub-divisions (van Heerdt and Mörzer Bruijns, 1960), although many dune animals do not show marked habitat specificity. Krogerus (1932) found that less than 10% of the arthropods recorded from Finnish sand dunes were obligatory dune-dwellers, and distribution patterns were influenced, to a large extent, by temperature conditions, particle size and relative humidity conditions. Ardö (1957) found that the Diptera fauna of sand dunes comprised a eurytopic group of species, with rather broad habitat preferences, and a stenotopic group whose more restricted distribution was related to temperature preferences. We can now examine these patterns in more detail by selecting a few examples drawn from work on the more conspicuous elements of the sand dune fauna.

Collembola of Haute Camargue. Distribution patterns of the microarthropods of dune systems at Beynes, Haute Camargue, France, have been studied by Poinsot (1966). The localities are river dune formations, associated with the Rhône, and provide a range of habitats paralleling those of coastal dune systems. Basically, a distinction can be made between slack habitats, periodically inundated with brackish water, and supporting a marshy vegetation of *Juncus* and *Scirpus*, and those of the higher dune slopes which included mobile habitats constantly exposed to wind action, and fixed dunes on which oak woodland was developed. Associated with each of these two major divisions was a characteristic group of collembolan species, namely a "wet" fauna dominated by *Isotomurus* and *Isotomodes* species in the slacks, and a "dry" fauna indicated by *Isotomina orientalis* and *Xenylla brevicauda* in habitats not periodically flooded. The identity of these two associations persisted throughout the year, for the "wet" species were prevented from extending into the drier areas by their susceptibility to desiccation, while the

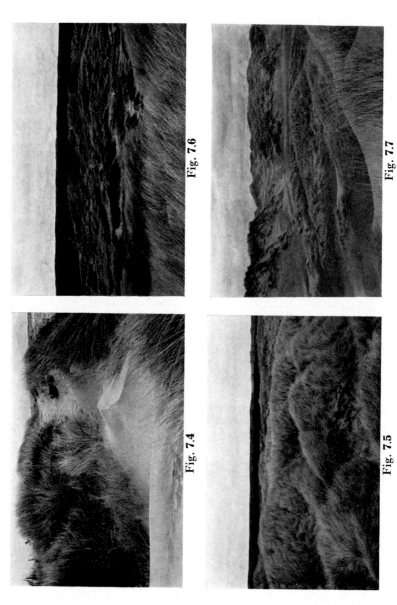

Fig. 7.4   **Fig. 7.6**

Fig. 7.5   **Fig. 7.7**

Figs. 7.4–7 Four stages in sand dune succession at Newborough Warren, Anglesey, North Wales. *Fig. 7.4* Fore dunes showing the effect of wind action and the localized distribution of marram grass. *Fig. 7.5* Yellow dune formation with more extensive colonization by marram grass. *Fig. 7.6* Permanent dunes supporting a heath-like vegetation. *Fig. 7.7* A coastal dune slack showing waterlogging effects. (Photos by author.)

"dry" species did not move into the slacks, even when these were not flooded, on account of the high salt content. Generally speaking, the micro-arthropod fauna was impoverished over the whole of this system, and especially so in the mobile dune habitats. The richest development of this fauna occurred in the more stable habitat under dune woodland.

Carabidae on Newborough Warren. On the south-western corner of the island of Anglesey, North Wales, is located the extensive sand-dune formation of Newborough Warren National Nature Reserve (Figs 7.4–7). The arthropod fauna of this system has been studied in some detail (see Hobart, 1968) and, in particular, attention has been paid to various aspects of the ecology of some of the larger ground-dwelling insects. Gilbert (1956) investigated distribution patterns and life histories of four species of the carabid genus *Calathus*, namely *C. fuscipes*, *C. erratus*, *C. mollis* and *C. melanocephalus*, which are widespread in this dune system. Two sites were selected for detailed study, just over a mile from the shore and about half a mile apart. One site was an area of stable grey dunes in which the main elements of the vegetation consisted of mosses *(Hypnum* and *Brachythecium)*, lichens *(Cladonia)*, *Senecio jacobaea*, *Rhytidiadelphus triquetrus* and *Carex arenaria*, with marram grass sparsely represented. The upper 10 cm of the soil profile showed some accumulation of organic material (organic carbon content 0·7% dry weight). The second site, termed the Seaward area, represented an earlier stage in dune succession, and although the substratum was rela-tively stable and extremely low in organic carbon (0·1% dry weight), it was densely colonized by marram grass.

All four species of beetle occurred in both of these sites but distinct pre-ferences were shown. *C. mollis* and *C. erratus* were predominant in the seaward dunes, whereas *C. fuscipes* and *C. melanocephalus* favoured the grey dunes. In attempting to explain these differences, any limiting effect of food preference can be discounted, for all four species are omnivorous in this area. The grey dune site provided a more open habitat than that of the sea-ward area, and the prevalence of *C. melanocephalus* in the former may be due to the fact that this species is intolerant of shade. The preference shown by *C. fuscipes* for the grey dunes probably reflects its association with a sub-stratum having some accumulation of organic material. The relatively greater abundance of *C. erratus* in the seaward site, which is richer in exchangeable calcium than the grey dune site, can be attributed to this species' predilection for lime-rich soils. Here we have an example of species populations over-lapping in their horizontal distribution patterns but being separated, to some extent at least, by different microhabitat preferences. Differences also occur in the timing of maximum adult activity, which might tend to reduce direct competition between species with coincident distributions. Adults of *C. fuscipes* reach their activity peak in July and August, and are undergoing a decline by the time adults of *C. melanocephalus* are reaching their maximum activity in September and October. A similar separation distinguishes the two species prevailing on the seaward site, with *C. mollis* reaching maximum activity before *C. erratus*. It is difficult to determine to what extent, if any, such differential activity reduces interspecific competition. It may act as a part of a complex of effects in which selection of different microhabitats, even within the same vegetation zone, may also be important. However, it must be pointed out that a parallel can be drawn here with the activity patterns of ants in heathland, discussed earlier in this chapter. Ants form an important component of the sand dune fauna, and it is therefore appropriate to deal with them next.

Ants on Newborough Warren. The influence of microhabitat factors on the distribution of the ground fauna at Newborough is also well illustrated by Stradling's (1968) observations on the ants (see also the discussion in Hobart, 1968). This study surveyed distribution patterns of *Lasius niger*, *L. flavus*, *Formica fusca*, *Myrmica rubra*, *M. scabrinodis* and *M. ruginodis* in a variety of sites ranging from the seashore through permanent dunes to an inshore plantation of Corsican pine. Ant nests were absent from some of the slacks and more mobile dunes, and from the coniferous stand. Unstable sand is unsuitable for permanent colonization by ants, and the low ground temperatures in the wooded site may have been responsible for the lack of nests here. Over the remainder of the dune system, the diversity of the ant fauna increased with an increase in the complexity of the ground vegetation. On semi-permanent dunes supporting patches of dwarf willow, two species of ant occurred, namely *Lasius niger* and *Myrmica rubra*. These two were joined by *M. scabrinodis* in sites where dwarf willow formed a complete ground cover, whereas all six species occurred on fixed dune grassland covered with *Festuca*. Despite the fact that two, three or more species occurred within the same vegetation, or dune type, different preferences for microhabitat conditions may have provided some degree of ecological isolation. For example, the two most abundant species, *L. niger* and *M. rubra*, were always recorded from the same sites, but *L. niger* was generally associated with cooler surfaces than was *M. rubra*. Another feature which undoubtedly served to separate different species populations was the territorial behaviour shown by the commoner species, such as *L. niger*, *M. rubra* and *M. ruginodis*, in which the size of the territory appeared to be related to the size of the worker population.

Spider Fauna of Sand Dunes. In contrast to the situation just described Duffey (1968) could find no correlation between the species diversity of the spider fauna and the complexity of the vegetation in sand dune systems. Spiders are very common in sand dunes, and this author recorded 188 species from two locations, one in South Wales and the other in Scotland. It is impossible to discuss all of the significant findings of this detailed study here, and the following summary is restricted to a few of the more important points. The dune system was sub-divided, for the purposes of this work, into nine sites as shown in Fig. 7.3, and collections were made over a period of four years. Looking at the various sites in terms of the species diversity of the spider fauna, two major groupings could be recognized: (1) the drift line and fore dunes, which showed a low diversity, and (2) the fixed dune sites, comprising the marram transition zone, dune heath, slack and dune meadow, in which the species diversity was higher. Diversity in the yellow dunes varied much more widely than that of any other site and showed an overlap with the two major groupings. Within each of these groupings, there was little difference in the range of diversity from site to site.

In analysing the collection data to determine the existence of natural associations of species, Duffey discovered that the distribution and number of spider species are influenced by two factors, namely habitat and season. It was found that faunal differences from habitat to habitat were more marked during the spring than during the summer in some cases, and this would indicate that the pattern of species associations changes with the season, along with changes in the vegetation and microclimate. The timing of activity peaks and maturation will also influence this pattern, and to illustrate this we can refer to the observations of Vlijm and Kessler-Geschiere (1967) on the spiders of a sand dune system in the Friesian island of Schiermonnikoog. Three lycosid species common in this system demonstrated different microhabitat preferences, such that *Pardosa pullata* selected damp sites, *P. nigriceps* rather drier areas under a scrub vegetation of *Salix repens* and *Carex arenaria*, while *P. monticola* was mainly concentrated in dry, barren localities. However, this pattern was subject to some modifications after the period of reproductive activity, when male spiders dispersed to less densely populated areas, and females carrying eggs moved into warmer, more exposed habitats than they occupied at other times of the year. Furthermore, a shift in habitat preference during the course of post-embryonic development has been recorded for a number of spider species. Duffey (1968) noted that immature stages of *Tibellus* spp. were mainly distributed in yellow dune and dune meadow habitats in July, but became rare in dune meadows and more common in the marram transition zone in September. Again, *Agroeca* spp. were most common, as adults, in the marram transition zone, and, to a lesser extent, in the yellow and fore dune habitats, whereas the majority of the immatures of these species occurring in early summer were found in the dune meadows.

7*

On the other hand, it is also evident from Duffey's work that certain habitats have a characteristic association of spiders which is maintained through the seasons. Although many species encountered in this study were distributed over a range of sites, sharp peaks in numbers of individuals occurred in only one site, in the majority of cases. Examples of this were provided by *Silometopus ambiguus*, a species exclusive to the drift line, *Oedothorax fuscus* occurring mainly in the fore dunes, *Agroeca proxima* and *A. inopina* mainly in the marram transition zone, and *Lycosa monticola*, *Styloctetor romanus*, *Argenna subnigra* and *Zelotes electus* mainly on the dune heath. A smaller group of species showed distribution patterns which extended over two

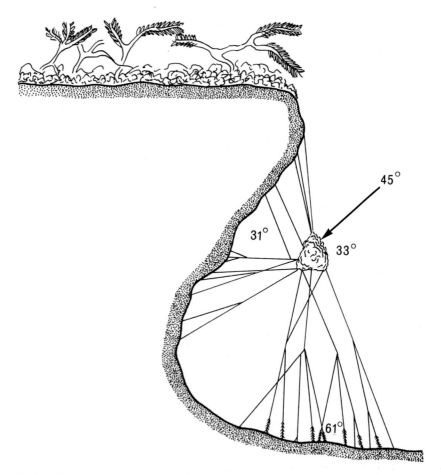

Fig. 7.8 The microenvironmental conditions associated with the nest of the ant-feeding spider *Theridion saxatile*. (From Nørgaard, 1956.)

adjacent habitats, at least at certain times of the year, while a few species showed density peaks in two widely differing habitats. *Theridion bimaculatum*, for example, is a spider commonly encountered in grassland and heath. In the sand dunes of Whiteford Burrows Nature Reserve, Glamorgan, South Wales, this species evidently concentrated in the fore dune and yellow dune habitats, with a subsidiary density peak in the dune meadow. Towards the end of summer, populations of *T. bimaculatum* were greatest in the yellow dunes and the dune meadow. Another example is provided by *Erigone arctica maritima*, a spider commonly associated with the drift line in Britain. In Scandinavia and Iceland, this species is common in inland sites and very frequently occurs above the snow line on mountains. The ability to establish flourishing populations in two or more apparently contrasting habitats is not uncommon among spiders and insects, and such species have been called *diplostenoecious* by Duffey, who also suggested that this phenomenon does not necessarily represent a shift in habitat preference. Under more detailed scrutiny, it can often be demonstrated that apparently dissimilar habitats may have certain microenvironmental features in common and it is these, rather than the general facies of the habitats, that determine the faunal distribution pattern. For example, the lycosid spider *Arctosa perita* is very common in yellow dunes where it constructs a temporary burrow in the sandy soil. This species also occurs inland in sandy areas where microclimatic conditions are quite different from those of the coastal dunes; it also inhabits deposits of shale and coal around collieries where the character of the ground vegetation is far removed from that of the dunes. Evidently the most important factor limiting the distribution of this species is not the microclimate or the vegetation type, but rather the nature of the substratum in which the burrow is constructed. Since sand and coal dust provide a loose-textured, workable material suitable for excavation, both of these unrelated habitats are optimal for *A. perita*.

The particular factor limiting distribution may vary from species to species. In the case of hammock-web spiders of the family Linyphiidae, which are active among the aerial parts of vegetation where their webs are constructed, the growth form of this vegetation may be limiting (Cherrett, 1964). In other cases, it can be shown that microclimate is the determining factor, and similar conditions may occur on a variety of soil types and in a range of vegetation. In evaluating the limiting effect of microclimate on distribution patterns, the topography of the area must also be taken into account. The ant-feeding spider *Theridion saxatile* is common in many inland sandy areas of Europe and the USSR where it constructs its web on the sloping faces of small hummocks and terraces. Such situations provide the warm, dry microclimate preferred by this species (Fig. 7.8). In many parts of the sand dune system the ground is too flat or the sand too mobile to permit *T. saxatile* to become established, but it does occur in localized situations where sandy hummocks are developed.

Drift line

One group of habitats in the coastal sand dune system about which very little has been said so far is that associated with the drift line on the shore. Environmental factors operating on the seashore are less equable than those of the adjacent terrestrial and marine localities, and demand of the animals becoming established there a wide tolerance of changes of salinity, water content, mechanical disturbance, oxygen concentrations, organic content and temperature. Where some degree of environmental stability is achieved, as for example in sheltered rock crevices or in the grassy turf of a salt marsh, an indigenous fauna may develop, consisting of species highly adapted to rigorous conditions. Duffey (1968) recorded 49 species of spider from a permanent drift line habitat formed at the junction of salt marsh and sand dune formations in South Wales, compared with only 22 species from the foreshore. This author also noted that drift material deposited high on the salt marsh was exposed not only to tidal influences but also, at certain times of the year, to fresh water brought in by rainfall and seepage from neighbouring dunes. This was reflected in the presence of spiders usually associated with freshwater marshes, such as *Pirata piraticus* and *Clubiona stagnatilis*. Sudd (1972) was able to divide 48 species of spider occurring in a salt marsh/sand dune system in Yorkshire into five ecological groupings on the basis of their tolerance to immersion in sea water. At one extreme were species such as *Lycosa purbeckensis*, *Erigone longipalpis* and *Silometopus curtus*, which can tolerate high salinity levels. At the other, were the typical sand dune species, such as *Lycosa pullata* and *Agroeca proxima*.

Pseudoscorpions are typical inhabitants of the seashore, and Weygoldt (1969) has described a marked zonation of species on an island site off North Carolina. Wetter parts of the drift line are selected by *Dinocheirus tumidus*, while drier areas in this same vicinity are occupied by *Parachernes litoralis*. Higher up on the shore in grassy sites, *Chthonius tetrachelatus* occurs, but is replaced by *Serianus carolinensis* on high sand dunes. This zonation may well reflect differing degrees of tolerance to submersion in seawater, or differing resistance to desiccation, in the species concerned. Little attention has been paid to this topic in relation to pseudoscorpions. Rather more is known about the factors influencing the distribution of soil mites, as the following example illustrates.

Mites on an Estuarine Salt Marsh

A study of the distribution of soil mites on a highly organic salt marsh on the Gower Peninsula of South Wales by Luxton (1964, 1967a, b) revealed that a zonation of Cryptostigmata and Mesostigmata occurred that paralleled both the zonation of the flora and the various tide levels. The upper limit of the high water spring tides marked the boundary between a Festucetum (above LHWS) and a Puccinellietum (below LHWS). Despite the fact that three of the commonest mites, *Hygroribates schneideri*, *Punctoribates quadrivertex* (Cryptostigmata) and *Cheiroseius necorniger* (Mesostigmata), occurred at all levels on the marsh and often extended into the Juncetum above, and *Spartina* vegetation below, marked differences in abundance occurred. *H. schneideri* achieved its maximum population size in the middle of the *Festuca* turf zone, whereas *P. quadrivertex* reached its peak in the *Puccinellia* zone. The zonation of less common species is more marked (Fig. 7.9).

In attempting to determine which factors were responsible for limiting the distribution of the various species to particular levels on the marsh, Luxton could find no clear correlation between the gradient of salinity and species distribution. However, the ability to tolerate conditions of high osmotic pressure is a pre-requisite for the establishment of populations here, especially at lower levels where tidal flooding occurs periodically. Experiments confirmed that this tolerance was possessed not only by species which extended throughout the marsh, such as *H. schneideri*, but also by *Hermannia pulchella* which is restricted to the upper levels where there is no direct tidal effect. On the other hand, a marked discontinuity in the horizontal distribution of the cryptostigmatids occurred between the *Festuca* and *Puccinellia* zones, i.e. between the zone above or outside the tidal limit, and that below or within this limit. It was suggested that tidal effects in some way prevent species present in the upper levels of the marsh from establishing populations in the lower levels. Clearly, these effects are limiting only for certain species, for others occur throughout the marsh, as we have seen. Two of these ubiquitous forms, *Hygroribates schneideri* and *Cheiroseius necorniger*, are viviparous, at least facultatively, as are *Halolaelaps nodosus*, *Digamasellus halophilus* and *Leioseius salinus*, mesostigmatids that are also relatively more abundant in the lower levels. Possibly the production of living young has a survival value where periodic tidal flooding might dislodge the egg stage. This is not the whole story, however, for the tidal limit provides no barrier for some egg-laying species, such as *Punctoribates quadrivertex*, which develops peak populations within the tidal region. How survival is achieved under these conditions is by no means certain; the selection of sheltered oviposition sites may be important. On the other hand, the viviparous habit

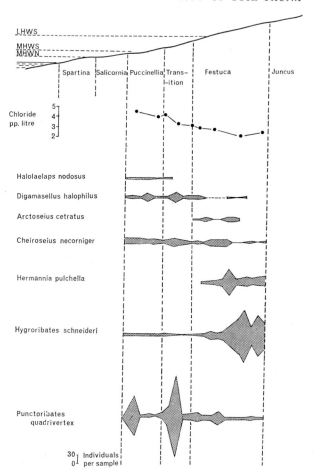

Fig. 7.9 Diagrammatic representation of the zonation of some mites on a salt marsh in South Wales, U.K. (After Luxton, 1964.)

does not necessarily mean that a species can establish populations below the tidal limits. The viviparous *Arctoseius cetratus* is present in the *Puccinellia* zone during the summer, but it does not establish a permanent population here as it does in the Festucetum. Evidently the factor limiting the distribution of this species is seasonal, rather than tidal, and may have little to do with viviparity.

Fauna of Temporary Habitats

The seashore also provides many temporary habitats, consisting of accumulations of organic debris, seaweed and dead animals. These deposits, washed high on the shore, may subsequently either dry out and be dispersed by the wind, or reclaimed by the sea on a high tide. The habitats thus provided may not be stable enough to permit the development of an indigenous fauna. The observations made by Duffey (1968) are again relevant here, for he found that a spider fauna, which in April comprised 19 species, was reduced to three species, each represented by one specimen, in September when the drift material had been dispersed or partly covered by sand. For the most part, the drift line fauna is composed of pioneer species derived from both the marine and the terrestrial fauna. These pioneers are predominantly detritus-feeding or predatory forms (Backlund, 1945), and frequently include carrion-feeding and carnivorous beetles, crustacean isopods and amphipods, Diptera and mites, in addition to the spiders just mentioned. In the north temperate zone at least, cryptostigmatid mites are often well represented in the detritus-feeding component. Inter-tidal species, such as *Punctoribates quadrivertex*, *Haloribatula tenareae*, *Hermannia scabra* and members of the Ameronothridae, occur in the same habitat as the more terrestrial *Punctoribates hexagonus*, *Oribatella arctica litoralis*, *Eupelops phaenotus* and *Passalozetes* spp. (Strenzke, 1950, 1954; Luxton, 1964; Travé, 1958). Clearly, the species composition of the fauna, particularly the terrestrial component, will vary from place to place depending on the species composition of the neighbouring environments from which this component is derived. For example, Travé (1958) noted that Cryptostigmata of the genus *Passalozetes*, which are common in rather dry sandy soils, colonize deposits of dead vegetation washed up by tidal pools in sand dune localities on the French Mediterranean coast. In more highly organic soils, such as the salt marsh, the terrestrial pioneer species are more often the mesophilous *Trichoribates incisellus* and *Platynothrus peltifer* (Luxton, 1964). We have already noted that there may be seasonal changes in the faunal composition of temporary drift deposits. There is also some indication that spiders such as *Arctosa perita* and the carabid beetle *Eurynebria complanata* leave drift deposits and move inland, deeper into the dune complex, to overwinter (Duffey, 1968).

Faunal Succession in Seaweed

In certain situations, for example at extreme highwater mark, accumulations of seaweed may persist for varying periods of time during which they undergo decomposition. This process is accompanied by a faunal succession which has been studied, under both experimental and semi-natural conditions, by Strenzke (1963). The principal faunal elements concerned in this sequence were mites, Collembola, Diptera and Coleoptera, comprising a fauna which could be sub-divided, ecologically, into four groupings, namely thalassobionts, thalassophiles, indifferents and thalassoxenes. The thalassobiont element, represented by the mites *Parasitus kempersi* and *Halolaelaps celticus*, the dipteran *Scatomyza litorea*, *Ceratinostoma ostiorum* and *Fucellia intermedia*, and the coleopteran *Cafius xantholama* and *Cercyon litoralis*, is restricted to the marine littoral zone, whereas the thassalophiles, such as the collembolan *Hypogastrura viatica* and the dipteran *Limosina fuscipennis*, *L. brachystoma* and *L. zosterae*, are not so strictly confined, although they achieve their highest abundance in the marine littoral. The indifferent species, as their name suggests, occur in coastal accumulations of seaweed, but show no particular preference for this habitat, for they also occur in equal abundance in inland deposits of compost and dung. This group includes mites, such as *Macrocheles glaber*, Collembola (*Isotoma olivacea*) and various dipterans of the genera *Psychoda*, *Limosina*, *Eristalis* and *Fannia*. Finally, the thalassoxenes, which include such mites as *Parasitus fimentorum*, *Tyrophagus brauni*, *Caloglyphus berlesei* and *Histiosoma ferroniarum*, the collembolan *Hypogastrura assimilis* and *H. denticulata*, and a variety of Diptera, are forms which rarely occur in seaweed, but are to be found in decaying compost, dung and carrion in sites remote from the coast.

In order to study in detail the ecological events occurring during the decomposition of seaweed, Strenzke placed fresh *Fucus spiralis* in open wire baskets, and allowed this material to decompose in a variety of situations. In this way, samples of the fauna could be taken conveniently every two weeks for the period of a year. By extending the study over four years, he was able to determine the influence of various environmental factors, such as salinity and temperature, on the course of faunal succession. This succession was characterized by four distinct phases which could be defined as follows. An initial phase, occurring during the first two weeks, in which the seaweed showed little evidence of decay but had an increased water and NaCl content, was indicated by the presence of large numbers of thalassobiont species, with Diptera larvae of *Fucellia intermedia* and the mite *Halolaelaps marinus* attaining their peak densities. This phase was succeeded by what Strenzke defined as the "decayed matter phase", extending from the third week to the beginning of the fourth month. During this phase, the amount of decayed organic matter increased, although the NaCl concentration was still high, and peak densities occurred of the fly larvae *Limosina brachystoma* and *L. fuscipennis* and the mites *Halolaelaps celticus* and *Parasitus kempersi*. The third phase in this sequence, termed the "transitional phase", extended from the fourth month to the seventh or eighth, during which time the NaCl concentration fell before 2°/$_{oo}$, and peak densities occurred in the dipteran *Scatopse notata* and Syrphidae. The final phase, from the eighth to the twelfth month, in which the seaweed reaches an advanced stage of decay, is characterized by the predominance of thalassoxenes, such as the winter mosquito *Trichocera regulationis* and the collembolan *Hypogastrura assimilis* and *Isotoma olivacea*.

The general course of this ecological succession is very reminiscent of that occurring in another "minor" community discussed later in this book, namely that of decaying wood (Chapter 9). In both cases, organic material in an early stage of decomposition is invaded by rather specialized animals. In the case of seaweed, these are thalassobionts, mainly Diptera larvae, whereas in wood they are wood-boring beetles. As decomposition proceeds, the faunal character becomes more varied with the invasion, in the case of seaweed, of thalassophilous Diptera larvae. The predominance of these

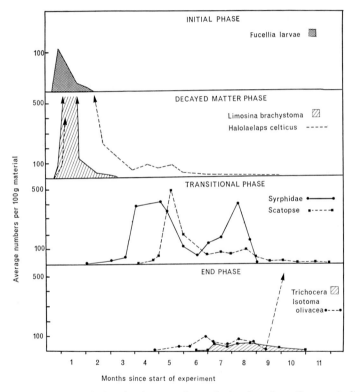

Fig. 7.10 Succession of the dominant arthropods in decaying *Fucus spiralis*. (After Strenzke, 1963.)

larvae and of thalassobiont mites, such as *Halolaelaps celticus*, is maintained as long as the NaCl concentration remains above $10°/_{oo}$, but as the influence of this special factor diminishes during the later stages of decomposition, the less specialized species, the thalassoxenes, come into prominence. These events are illustrated in Fig. 7.10. Following this sequence to its final phase when some, at least, of the decomposed seaweed becomes incorporated into the soil of the littoral zone, we encounter more frequently members of the soil fauna, particularly mites such as *Hermannia subglabra*, *Punctoribates quadrivertex*, *Trichoribates incisellus*, *Scutovertex* spp. and predatory Meso-stigmata, together with various Collembola.

Summary

Despite their different histories of succession, heathlands and sand dunes have several important features in common. In both, the sandy soil is usually well drained and low in organic matter, and in both, the ground cover of vegetation is often discontinuous.

The combined effect of these features has important repercussions on the character of the soil fauna in both of these formation types. Species diversity is generally lower than in the more organic soils of grassland and woodland. Lumbricid earthworms and microarthropods, in particular, are poorly represented in pioneer heathland and the early stages of dune succession. On the other hand, several groups of macroarthropods flourish, notably the ants, carabid beetles, spiders and Diptera larvae. Many members of this fauna are not obligatory heath- or dune-dwellers, but have invaded these sites from neighbouring habitats. During later stages in ecological succession, when a complete ground cover of vegetation develops and the organic content of the soil increases, there is a shift in the character of the soil fauna, with the mesophilous enchytraeids, nematodes, isopods and microarthropods coming into greater prominence.

The discontinuous ground cover of vegetation in heathland provides a mosaic of habitats for the fauna. Clumps of vegetation provide shade, organic material and protection from extremes of temperature and relative humidity, while intervening areas have the opposite characteristics. This heterogeneity encourages a diverse assemblage of ant populations in which habitat separation is achieved by the interaction of a number of factors. Different preferences for sun and shade, and seasonal differences in the timing of nuptial flights ensure a spatial separation of the various species populations which is maintained by direct competition for nesting sites.

In the sand dune system, two broadly contrasting habitat types are provided by mobile dunes, on the one hand, and fixed dunes on the other. By virtue of their unstable nature, mobile dunes provide unsuitable habitats for soil animals, and species diversity is low here compared with that on fixed dunes. The diversity of the ant fauna appears to be directly related to floristic diversity which is higher in fixed dune habitats. Habitat separation also occurs within the latter, however, and here the limiting factors are more complex. Differing degrees of tolerance of shade and soil type are probably responsible for the spatial separation of carabid beetle populations, and a temporal separation is also achieved by differences in the timing of activity peaks of adult beetles. Spider populations are separated by different preferences for microclimatic conditions, although these preferences may vary, within a species, during the reproductive period. In addition, soil texture may have an important influence on the distribution patterns of spiders which deposit their cocoons in the soil.

Perhaps the only general statement one can make about the faunal characteristics of these dry, sandy habitats is that species diversity increases as pioneer stages in succession give way to "climax" communities. The latter have achieved some degree of stability and floristic diversity which allow organic matter to accumulate in the soil. Dry, exposed habitats are succeeded by moist, sheltered ones, and there is a shift in emphasis from a xerophilous fauna, dominated by macroarthropods, to a meosophilous fauna of lumbricid earthworms, nematodes and microarthropods, as succession proceeds towards the "climax". However, the factors which influence the distribution patterns of individual species are varied and complex. They include:

(1) Direct competition for nesting sites, which maintains habitat separation among different species of ants.

(2) Complexity of the ground vegetation which provides a variety of habitats for macroarthropods, notably the ants.

(3) Timing of activity peaks, particularly dispersal activity following the breeding period. Differences in activity patterns may produce a spatial separation even between species populations with similar microclimatic preferences, as in ants, spiders and carabid beetles.

(4) Microclimatic preferences are undoubtedly important in reducing direct competition for living space, and for maintaining species diversity. Differing degrees of tolerance of shade among ants and carabid beetles in particular illustrate this.

(5) Character of the substratum, its texture, base content, pH and organic content are properties which vary from place to place in heathland and sand dunes. All of the major faunal groupings in these sandy habitats, macroarthropods, microarthropods, earthworms and nematodes, include species whose distribution patterns can be related to one or other of these properties. However, a much more general distinction can be drawn between those habitats in which the overall species diversity is low and those in which this diversity is relatively high. The former include pioneer heathland and mobile sand dunes, the latter the more acidic fixed dunes and *Calluna* heath.

A "minor" community system is provided by the decomposing organic material which accumulates along the drift line of sandy shores. A marked faunal succession occurs here which has certain general features in common with those of other minor communities, for example, decaying wood, dung and the rhizosphere. Pioneer organisms in this succession are specialized forms which can tolerate relatively high concentrations of NaCl. Once the salt concentration falls below $10°/_{oo}$, a more varied fauna of less specialized forms enters into the succession. Finally, as the decomposing material becomes incorporated into the soil, its associated fauna is derived, for the most part, from the true soil fauna.

REFERENCES

ARDÖ, P. (1957). Studies in the marine shore dune ecosystem with special reference to the dipterous fauna, *Opusc. ent. Suppl.*, **14**, 1–255.

BACKLUND, H. O. (1945). Wrackfauna of Sweden and Finland, Ecology and Chorology, *Opusc. ent. Suppl.*, **5**, 1–236.

BRIAN, M. V. (1964). Ant distribution in a southern English heath, *J. Anim. Ecol.*, **33**, 451–61.

BRIAN, M. V., ELMES, G. W. and KELLY, A. F. (1967). Populations of the ant *Tetramorium caespitum* Latreille, *J. Anim. Ecol.*, **36**, 337–42.

BRIAN, M. V., HIBBLE, J. and KELLY, A. F. (1966). The dispersion of ant species in a southern English heath, *J. Anim. Ecol.*, **35**, 281–90.

BRIAN, M. V., HIBBLE, J. and STRADLING, D. J. (1965). Ant pattern and density in a southern English heath, *J. Anim. Ecol.*, **34**, 545–55.

CHERRETT, J. M. (1964). The distribution of spiders on the Moor House National Nature Reserve, Westmorland, *J. Anim. Ecol.*, **33**, 27–48.

DELANY, M. J. (1953). Studies on the microclimate of *Calluna* heathland, *J. Anim. Ecol.*, **22**, 227–39.

DUFFEY, E. (1968). An ecological analysis of the spider fauna of sand dunes, *J. Anim. Ecol.*, **37**, 641–74.

ELTON, C. S. (1966). "The Pattern of Animal Communities", Methuen, London.

GILBERT, O. W. (1956). The natural histories of four species of *Calathus* (Coleoptera, Carabidae) living on sand dunes in Anglesey, N. Wales, *Oikos*, **7**, 22–47.

GIMINGHAM, C. H. (1972). "Ecology of Heathlands", Chapman and Hall, London.

VAN HEERDT, P. F. and MÖRZER BRUIJNS, M. F. (1960). A biocenological investigation in the yellow dune region of Terschelling, *Tijdschr. Ent.*, **103**, 225–75.

HOBART, J. (1968). Insects of Anglesey. *In* "Natural History of Anglesey", 97–128 (Eifion Jones, W., ed.), Anglesey Antiquarian Soc., Llangefni.

KNÜLLE, W. (1957). Die Verteilung der Acari: Oribatei im Boden, *Z. Morph. Okol. Tiere*, **46**, 397–432.

KROGERUS, R. (1932). Uber die Ökologie und Verbreitung der Arthropoden der Triebsandgebeite an den Küsten Finnlands, *Acta zool. fenn.*, **12**, 1–308.

LUXTON, M. (1964). Some aspects of the biology of salt-marsh Acarina, *Acarologia*, **6** (fasc. h.s. C.R. Ier Congrès Int. d'Acarologie, Fort Collins, Col., U.S.A., 1963), 172–82.

LUXTON, M. (1967a). The zonation of saltmarsh Acarina, *Pedobiologia*, 7, 55–66.

LUXTON, M. (1967b). The ecology of saltmarsh Acarina, *J. Anim. Ecol.*, **36**, 257–77.

NIELSEN, C. O. (1949). Studies on the soil microfauna II. The soil-inhabiting nematodes, *Natura jutl.*, **2**, 1–131.

NIELSEN, C. O. (1955). Survey of a year's results obtained by a recent method for the extraction of soil-inhabiting enchytraeid worms. *In* "Soil Zoology", 202–14 (Kevan, D. K. McE., ed.), Butterworths, London.

NØRGAARD, E. (1951). On the ecology of two lycosid spiders (*Pirata piraticus* and *Lycosa pullata*) from a Danish *Sphagnum* bog, *Oikos*, **3**. 1–21.

NØRGAARD, E. (1956). Environment and behaviour of *Theridion saxatile*, *Oikos*, **7**, 159–92.

POINSOT, N. (1966). Étude écologique des Collemboles des dunes de Beynes (Haute Camargue), *Rev. Ecol. Biol. Sol*, **3**, 483–93.

RANWELL, D. S. (1972). "Ecology of Salt Marshes and Sand Dunes", Chapman and Hall, London.

REYNOLDSON, T. B. (1955). Observations on the earthworms of North Wales, *NWest. Nat.*, **3**, 291–304.

RICHARDS, O. W. (1926). Studies on the ecology of English heaths. III. Animal communities of the felling and burn successions at Oxshott Heath, Surrey, *J. Ecol.*, **14**, 244–81.

SCHJØTZ-CHRISTENSEN, B. (1957). The beetle fauna of the Corynephoretum in the ground of the Mols Laboratory with special reference to *Cardiophorus asellus* Er. (Elateridae), *Natura jutl.*, **6–7**, 1–120.

STRADLING, D. J. (1968). "Some Aspects of the Ecology of Ants at Newborough Warren". Ph.D. Thesis, University of Wales.

STRENZKE, K. (1950). *Oribatella arctica litoralis* n. subsp. eine neue Oribatide der Nord und Ostseeküste (Acarina: Oribatei), *Kiel Meeresforsch.*, **7**, 157–60.

STRENZKE, K. (1954). Zur Verbreitung und Okologie von *Passalozetes bidactylus*, *Faun. Mitt. Nord.*, **4**, 11.

STRENZKE, K. (1963). Die Arthropodensukzession im Strandwurf mariner Algen unter experimentell kontrollierten Bedingungen, *Pedobiologia*, **3**, 95–141.

SUDD, J. H. (1972). The distribution of spiders at Spurn Head (E. Yorkshire) in relation to flooding, *J. Anim. Ecol.*, **41**, 63–70.

TRAVÉ, J. (1958). Quelques remarques sur la microfaune des laisses d'étangs, 83e *Conr. Soc. sav.*, 611–18.

VLIJM, L. and KESSLER-GESCHIERE, A. M. (1967). The phenology and habitat of *Pardosa monticola*, *P. nigriceps* and *P. pullata* (Araneae, Lycosidae), *J. Anim. Ecol.*, **36**, 31–56.

WEYGOLDT, P. (1969). "The Biology of Pseudoscorpions", Harvard, Cambridge, Mass.

Chapter 8

The Forest Soil Fauna

The forest ecosystem is so complex and so varied that to do justice to all of its many aspects would require a book in itself. However, it is fortunate that we can recognize several separate, often geographically distinct, sub-divisions, such as peatland, coniferous forest, deciduous forest and tropical forest. Each of these has its own particular characteristics, its own range of microhabitats for soil animals. However, all have certain features which are common to forest formation-types, and which provide a framework within which the various ecological effects can be examined (Fig. 8.1).

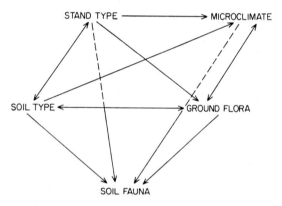

Fig. 8.1 The main environmental variables in forest soil ecosystems.

One of the most conspicuous features of forest is its vertical stratification. All communities show this stratification, but it is more pronounced, more complex and much more extensive in forest than, for example, in grassland, heath or moorland. Shading by the canopy influences the microclimate at the surface of the ground and also determines, to some extent, the type and distribution of the ground flora. In forests, the organic material entering the soil is derived from leaf fall and the decay of logs and branches, in the main. This contrasts with the situation in grasslands where much of the organic fraction of the soil originates from the *in situ* decomposition of rooting systems. Further, an important series of microhabitats is provided, in the forest, by decaying logs and stumps, and by debris which accumulates in tree holes. These microhabitats often provide suitable food and living space for members of the soil fauna which are thereby enabled to extend their vertical range beyond the confines of the soil. This topic is dealt with in Chapter 9.

Basically, there are two types of humus formation in forests, the mull and the mor (Müller, 1879, 1884). Mull humus is extensively mixed with mineral soil to produce a profile which is relatively homogeneous in character; it is neutral or slightly alkaline in pH, and often develops in sites where the parent material is rich in calcium carbonate. Mor humus is usually rather acidic and tends to form as a distinct layer below the litter and above the mineral soil. As a result, soil profiles containing mor humus are strongly structured, often with a sharp demarcation between the organic fraction and the mineral soil. Mor humus represents organic material which is less completely degraded than that in mull humus, and often has greater amounts of lignin than the latter. For this reason it is sometimes referred to as "raw humus". Intermediates between these two extreme forms of humus occur frequently.

The major environmental variables shown in Fig. 8.1 may act directly or indirectly to influence the composition and distribution of the soil fauna. Before going on to deal with these influences, however, something more must be said about the types of forests and their soils.

Forest types and forest soils

In the north temperate zone, the two main types are the coniferous and deciduous forest occupying broad, and geographically distinct, bands across North America and Eurasia. Generally speaking, coniferous species are associated with a mor humus formation, while deciduous species are mull-formers, although there are exceptions to this, as we will see. In wet sites, particularly in coniferous forests, leaf litter decomposition is retarded to such an extent that peat formation occurs, and a forest peatland develops. Rainfall is an important factor in determining the type of forest which develops in the tropics. In high rainfall areas is developed the typical tropical rain forest, but this gives way to a moist semi-deciduous association in rather drier areas. Each of these forest types is now considered in more detail.

Forest Peatland

In the account of fenland habitats given in Chapter 6, we noted that woody vegetation may become established on peaty soil. Although thicket vegetation of willow and alder may predominate in the fen carr, birch may also occur, together with spruce, and these two tree species sometimes become dominant to form a spruce–hardwood peatland forest. *Sphagnum* moss forms a conspicuous element of the ground vegetation, together with *Vaccinium*, *Carex* and *Equisetum*. This type differs from the majority of forest peatlands by the absence of pine, although on limestone sites intermediates between this and a true pine forest bog occur. Pine bogs may be classified into different types, depending mainly on the nature of the ground vegetation. The most characteristic type consists of stunted pine forest with a mixed ground vegetation including shrubs (*Andromeda*, *Oxycoccus*, *Vaccinium*, *Ledum* and *Calluna*) and herbaceous plants (*Rubus*, *Eriophorum* and *Carex*), together with *Sphagnum* which provides the main source of peat. In other types, one particular element of the ground vegetation, such as *Carex*, *Sphagnum* or *Eriophorum*, may be conspicuous. In most of these cases, the wet hummocky peat prevents the development of a firm forest type, and the pine trees are consequently stunted and unsuitable for harvesting. The drier soil conditions on drained forest peatland permit some decomposition of the peat, however, and a much healthier development of pine occurs, showing a marked trend towards a normal forest type. Only the presence of *Sphagnum* in the ground layer identifies this type with a forest peatland.

Coniferous Forest

The coniferous forest which extends as a broad vegetational belt across northern Eurasia and North America has, as its dominants, species of pine, fir, spruce, larch, poplar and birch (Fig. 8.2). To the north it forms an ecotone with the Arctic tundra, and to the south a broad ecotone with the deciduous forest. Within the conifer belt, various combinations of the dominant tree species occur from place to place, and with these are coupled variations in the type and extent of the ground flora. For example, in North America the main tree species in the coniferous forest proper are white spruce and balsam fir, the *Picea–Abies* association of the boreal forest. The dominant species in the mixed forest which links the boreal to the more southerly deciduous forest are eastern hemlock, white and red pine, white cedar and yellow birch. In northern Europe, on the other hand, the boreal forest is often dominated by Scots pine and Norway spruce which may occur in admixture with species of birch, notably *Betula verrucosa* and *B. odorata*.

These variations in species composition may cause differences in the degree of development and the character of the ground vegetation, the leaf litter and the humus layer. For example, on dry sandy sites supporting stands of Scots pine, the ground vegetation may be dominated by lichens

Fig. 8.2 Pine forest growing on Cretaceous sandstone in the Český ráj region of northern Czechoslovakia. (Photo. by author.)

of the genus *Cladonia*, whereas under mixed stands of Scots pine, Norway spruce and birch, the ground vegetation is more luxuriant, with the mosses *Hylocomium* and *Dicranum* and the shrub-like *Calluna* present in abundance. Under stands of Norway spruce, *Hylocomium* and *Pleurozium* mosses often form an extensive ground cover which decomposes slowly to form a relatively thick peat-like raw humus. Where birch occurs in any abundance, a richer herbaceous ground flora comes in, and a more completely decomposed, better drained, humus profile is developed. Table 8.1 shows the variation in the ground vegetation occurring within a Finnish boreal forest.

Most conifers have needle-shaped leaves which are shed intermittently so that the forest presents an evergreen aspect. Although this forest is predominant in the colder northern or alpine regions where the yearly range of temperature is often greater than in the more southerly deciduous forest, the persistence of foliage throughout the year often creates more moist microclimatic conditions and attenuates the seasonal variation in air temperature which is such a characteristic feature of deciduous forests. Equally significant, as far as the soil fauna is concerned, is the type of litter and humus produced by conifers. Conifer needles have a high carbon/nitrogen ratio and a high polyphenol content. These properties render them resistant to decomposition, and encourage the development of a well-defined organic layer of acid raw humus near the soil surface. Such a profile, which occurs

Table 8.1

The relationship between forest tree species and the type of ground vegetation in the boreal forest of Southern Finland. (Compiled after Karpinnen, 1955.)

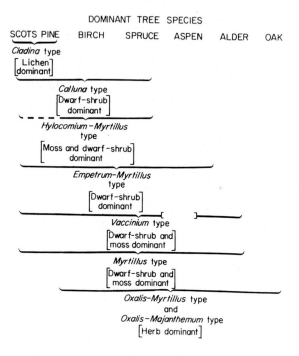

in podzols, commonly develops on well-drained sites where the underlying mineral soil is sandy. Even with good drainage, the high water-retaining capacity of raw humus ensures that these soils remain permanently moist.

The podzol profile (Fig. 8.3a) is characterized by strong leaching effects. Commonly formed from alluvial deposits or glacial drift with a high quartz and a low clay content, podzols are subject to percolating rainwater which removes some organic material and soluble minerals from the surface horizons, and re-deposits these lower in the profile in distinct bands. As a result, two organic zones develop, one immediately below the litter layer, comprising raw humus, and the other at a depth of approximately 35–85 cm in which the humus forms fine coatings around sand grains. Separating these two is a lighter-coloured zone consisting mainly of quartz from which soluble aluminium and iron oxides have been removed. These oxides are re-deposited lower down in the profile, sometimes as distinct layers, sometimes in admixture with the re-deposited organic material.

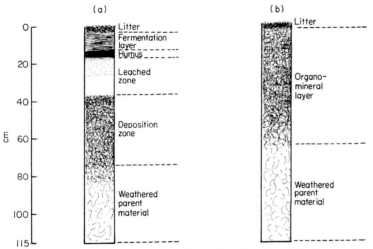

Fig. 8.3 Two contrasting types of forest soil profile. (a) Podzol, (b) a brown forest soil (altosol).

Deciduous Forest

Like the coniferous forest, the broad-leaved deciduous forest of the temperate zones is largely restricted to the northern hemisphere where it is extensive in the eastern part of the United States, in much of Europe, China and Japan. Among the dominant trees, beech, oak, maple, hornbeam and basswood are often prominent, with different species occurring in different geographical regions (Fig. 8.4). Under natural conditions, these forests are dense with closed canopies under which shrub vegetation gains little foothold, although where the canopy is more open, enough light may penetrate to allow the growth of bramble and bracken. The marked seasonal changes occurring in this forest type, particularly leaf-fall, permit a rich growth of herbaceous plants during the spring and early summer. To the north, as we have seen, this type forms an ecotone with the boreal forest. In more southerly regions the presence of pine is often conspicuous, maintained as a sub-climax by burning; examples of this are common in the south-eastern United States.

In general, the leaf litter produced by deciduous species has a lower carbon/nitrogen ratio than that produced by conifers. This leaf litter decomposes at a relatively rapid rate compared with conifer needles. In sites where the mineral soil is not too acid, for example where the parent material is chalk or limestone, lumbricid earthworms, millipedes and wood-lice occur in appreciable numbers, and the feeding activities of these animals may contribute to the decomposition process and the formation of a mull type of humus. In addition, these animals transport the organic material

Fig. 8.4 Typical beech woodland in the Chiltern Hills, southern England, as it appears just after leaf fall. (Photo. by author.)

from the soil surface to the deeper layers of the profile to produce a rendzina or brown earth soil type. The upper layers of such profiles may be affected by drought during a dry summer season.

Brown soils, sometimes known as altosols (Fig. 8.3b) are the commonest types found under natural deciduous woodland. They differ quite markedly from podzols in a number of respects. The surface litter decomposes rapidly and becomes intimately mixed with the mineral soil. There is not the distinct layering and raw humus formation which characterizes the podzol. Leaching effects are minimal and free iron oxides are fairly uniformly distributed down the profile. Parent materials from which brown soils may form are mainly silts and loams, basic or calcareous, originating from wind-borne deposits (loess), glacial drift, alluvial material and solifluction deposits. Intergrades between brown forest soils and podzols occur where leaching is intensive enough to remove basic cations in the drainage water, rendering the profile more acidic. This tendency is sometimes found in beech forest soils of the north temperate zone in which a mor-type humus is developed.

Tropical Forest

The most extensive developments of tropical forest occur in the equatorial regions of Africa, Indo–Malaysia and Central and South America. This vegetation type is much more heterogeneous in character than the temperate forests already discussed, and although certain tree species may be numerically dominant over a large area, for example rubber trees (*Ficus*, *Hevea* and *Castilloa*), bamboo (*Bambusa*), the cocoa tree (*Theobroma*) and papaya (*Carica*), diversity of species within a small area is considerable. This diversity often creates difficulties for the ecologist, for distribution patterns are often very complex indeed, and it may require an exhaustive search before two trees of the same species can be found. "Pure" stands, such as occur in temperate forests, are notably lacking in the tropical forest.

Despite its heterogeneous character, the tropical forest can be subdivided into three main types, namely the rain forest, cloud forest and semideciduous forest. Extensive tracts of the last-named type occur along the coastal region of West Africa. In Ghana, where two main associations have been identified, the *Celtis-Triplochiton* and *Antiaris-Chlorophora* associations, much of this type consists of secondary forest developed as a result of clearing the primary growth to allow cultivation of cassava and cocoa. Here, the undergrowth is thick, the atmosphere hot and humid. In contrast, the cloud forest which usually occurs sporadically in tropical Central and South America is cool and dank, with an almost continual shroud of mist or fog. Nevertheless, this forest type shares with the other two a characteristic diversity of plant species which is without parallel outside the tropics. The true rain forest undoubtedly represents the richest accumulation of living organisms in a terrestrial situation. The Amazon and Congo basins provide good examples of this vegetation type, which is also extensive in south-east Asia. It is very probable that the American, African and Asian rain forests have developed independently, for they differ quite markedly in their species composition, although all three have several general characteristics in common. Many of the tree species of the tropics have large, broad leaves which intercept much of the solar radiation. So effective is this barrier that the focal point of competition between individual plants largely shifts from the forest floor to the canopy. In the warm, moist tropical environment, this canopy is dense and complete, and the ground layer of vegetation is poorly developed except where clearing has opened up the forest. Selection favours tall growth, and the result is seen in the complicated, multi-storied canopy structure, characterized by several strata, or canopy levels, which allows maximum utilization of the available solar radiation. Another characteristic feature of the tropical forest is the growth of epiphytic plants, such as ferns, mosses, bromeliads, orchids and even cacti which grow independently of the ground on tree branches and trunks. Woody vines, or lianas, thread their way through the canopy from one tree to the next, and their foliage often forms an appreciable component of the total cover. The rich development of this canopy flora is another illustration of the need to compete in the canopy of the tropical forest, rather than on the ground.

It is almost as difficult to generalize about the character of the soil as it is to describe the character of the vegetation in a tropical forest. Many tree species are evergreen and shed their large leathery leaves throughout the year. However, this leaf-fall remains in a dry, shrivelled and largely unde-composed state during the dry season. As the wet season progresses, this surface litter is broken down rapidly and so completely that only the very surface of the mineral soil is lightly stained with humus. Occasionally, where the mineral soil is very sandy, a relatively thick layer of raw humus may be formed over a leached, podzolized profile. Where the underlying soil is clay, much of the fallen leaf material is carried away during heavy rains. Con-sequently, the soil under tropical forest often has a highly mineral character, and since chemical weathering occurs very rapidly through the profile, lateritic red earths, yellow-brown podzols, ochrosols and lithosols are fre-quently developed. Under these conditions, the most conspicuous members of the soil fauna belong to groups which flourish in rather acid mineral soils of low nutrient content, and which can excavate tunnels or burrow systems in these types of profile.

The general character of the forest soil fauna

Later in this chapter we will be considering some of the specific effects of the more important environmental factors on the composition and distri-bution patterns of the forest soil fauna. Before we do this, however, it is perhaps advisable to draw attention to the broad faunal distinctions that occur between coniferous, temperate deciduous and tropical forests.

Coniferous and Deciduous Forest

The acid conditions prevailing in coniferous forest soils are inimical to certain groups of animals, such as lumbricid earthworms, millipedes and woodlice. These prefer base-rich profiles, and their representation in coni-ferous forest soils is limited to a few acid-tolerant forms, such as the earth-worms *Dendrobaena octaedra* and *Lumbricus rubellus*, and the millipedes *Cylindroiulus punctatus* and *Iulus scandinavius*. The compacted mat of fine litter which develops under conifers provides little shelter for many of the larger non-burrowing arthropods, such as woodlice and the beetles belong-ing to the families Carabidae, Staphylinidae, Scarabaeidae and Elateridae. These groups are much more abundant in deciduous forest soils.

Soil fungi flourish better in acid raw humus than in the looser, base-rich soils of deciduous stands. These fungi form an important part of the diet of many cryptostigmatid mites and Collembola (Hale, 1967; Wallwork, 1967) and enchytraeid worms (O'Connor, 1967). These groups are not adversely affected by acid conditions, and even though they are non-burrowers for the most part, their small body size permits them to move easily through the small spaces in the litter mat and humus layer. In most raw humus soils, mites, Collembola and enchytraeids often occur in densities of tens or hundreds of thousands per square metre. Cryptostigmatid mites, in particular,

Fig. 8.5 The cryptostigmatid mite *Steganacarus magnus* (Nic.) ($\times$ 55), a microarthropod often encountered in coniferous leaf litter. (Photo. by B. Parry and reproduced through the courtesy of the British Museum, N.H.)

flourish in raw humus profiles under conifers, and "box" mites of the family Phthiracaridae (Fig. 8.5) have adapted to this environment to the extent that at least some of them burrow into conifer needles and twigs, and may complete their entire life cycle here (Jacot, 1939; Wallwork, 1957, 1967).

In order to gain some idea of population densities in coniferous forest soils, Table 8.2 has been compiled from data provided by Huhta *et al.* (1967). The range of density given in each case, in this table, reflects the fact that densities were higher, on average, in moist sites than in dry ones. Attention may also be drawn to the large nematode populations in coniferous soil, with densities of the order of a million per square metre. However, even at

Table 8.2

Population densities, expressed as average numbers per m², of the main animal groups in various coniferous forest soils in Finland (based on Huhta et al., 1967).

Group	DENSITY Average numbers/m²	
	Raw Humus Profiles	Moder Profile
Nematoda	911,000–1,223,000	1,521,000
Enchytraeidae	9,105–14,590	17,250
Lumbricidae	0·3–28·9	57
Collembola	14,700–34,300	25,400
Diptera larvae	10·7–26·9	69
Coleoptera adults	5·1–96·9	79·1
Coleoptera larvae	22·4–56·0	55·1
Acari	111,600–233,400	257,900

this high level, the nematode fauna can be compared unfavourably with that in grassland and deciduous forest soil, where densities of 20 million or more per square metre have been recorded (Nielsen, 1949; Volz, 1951).

Tropical Forest

In the tropical forest, as we have already noted, the main focal point of competition is located in the canopy layers. One result of this is that the horizontal stratification of the community is much more sharply defined in these upper levels than it is lower down. The limits of the soil/litter sub-system are difficult to define since the soil flora and fauna extends its vertical range upwards from the floor towards the canopy, in many cases. We find, here, groups of animals which are more closely associated with the soil in temperate localities, such as ants, termites and mites, living above the ground among the vegetation.

The astonishing diversity of the ant fauna in tropical forests is well known, and many species construct nests on the branches of trees, among the roots of epiphytic orchids and bromeliads, and in the foliage. The West African tree ant *Oecophylla* constructs a nest of leaves which are bound together by silken threads produced by the larvae. In the South American rain forest, ants of the genus *Azteca* are also truly arboreal and construct bag-like "carton" nests in the hollow stems of *Cecropia* trees. Similar habits are adopted by species belonging to the genera *Pseudomyrma*, *Camponotus* and *Iridomyrmex* (Wheeler, 1910). The termites, which resemble the ants in many biological respects, are also conspicuous members of the tree fauna in tropical forests. According to Harris (1961), carton nests of *Nasutitermes costalis* are common on the trunks of trees throughout Central America, and other wood-feeding termites of the genera *Amitermes*, *Syntermes* and *Copto-*

termes may also construct tree nests. However, it is doubtful if any of these forms can be considered as truly arboreal, for they can also produce mound nests or subterranean nests in the soil. Be this as it may, they do provide good examples of the way the soil fauna can expand its vertical range to include the arboreal habitat in tropical forests.

This arboreal component adds to the species diversity of a ground fauna which is far from impoverished. In one of the most detailed studies of this fauna published to date, Williams (1941) listed more than 200 species, representing five Phyla, 12 Classes and 37 Orders from leaf litter in the tropical rain forest on the island of Barro Colorado in the Panama Canal zone. This list included flatworms, nematodes, earthworms, leeches as well as a variety of crustaceans, myriapods, insects and arachnids. It was estimated that mites, Collembola and ants comprised more than 80% of all animals collected. A similar picture was reported by di Castri (1963) for the microfauna of the floor of broad-leaved forests in Chile, although here the ant fauna was of only minor importance, and mites and Collembola alone accounted for 80% of the fauna. The scarcity of ants in these forests suggests that the character of the soil fauna is approaching that found in temperate forest soils. Although few direct comparisons have been made between tropical and temperate forest soil systems, it seems that the density of the macrofauna is greater in the former than in the latter (Williams, 1941), whereas the reverse is the case as far as the microarthropods are concerned. Maldague and Hilger (1963)

Table 8.3

A comparison of mite and Collembola densities in temperate and tropical forest soils.

Habitat	Mean Density per cm²	Author
Temperate		
Beech mor	8·61	van der Drift (1951)
Spruce mor	14·6	Forsslund (1948)
Fen soil	3·19	Macfadyen (1952)
Hemlock mor	2·92	Wallwork (1957)
Tropical		
Rain forest (Brazil)	1·0–2·6	Beck (1967)
Sclerophyl forest (Chile)	1·0	di Castri (1963)

estimated mite populations within the range 27,000–50,000 per m², and Collembola between 12,000 and 22,000 per m² in leaf litter in the African equatorial forest. These densities are appreciably lower than those generally recorded for temperate forest soils (Table 8.3).

On the other hand, species diversity is probably greater in the tropical

8

forest. Beck (1962) found over 70 species of cryptostigmatid mites in a dozen samples of leaf litter from the Peruvian rain forest. This number exceeds that collected from several hundred samples taken from a hemlock/yellow birch forest floor in northern Michigan (Wallwork, 1957). This difference is probably due to the more varied nature of the leaf litter and to the more extensive pattern of vertical distribution in the tropical forest. As we have already noted, the soil fauna ranges upwards into the aerial vegetation much more readily than is the case in temperate regions. This vegetation provides a rich supply of microhabitats, and the opportunity for diversification of the fauna is enhanced. In temperate forests, on the other hand, extension of the vertical range of the soil fauna is often downwards, towards the mineral soil, where the number of suitable microhabitats is much more restricted. These effects, and others, which go to make up the forest soil ecosystem are now considered in more detail.

Effect of environmental variables

Forest ecosystems are very complex and very complicated. Environmental variables interact to such an extent that the separate effect of each is difficult to evaluate. However, a broad distinction can be made between biotic and abiotic factors, basically between the organic and physical factors in the environment. The organic milieu is of central importance since it is the repository of the food material and energy on which the soil community is built. The physical factors with which we will be concerned are principally those of microclimate. The effects of chemical factors on the soil fauna have already been discussed in the context of grasslands. In forests, chemical effects are exerted mainly through the organic environment, the litter and humus layers, and will be considered in this light.

Biotic Factors

Forest soil communities, whatever their particular characteristics, are all based essentially on detritus food chains. They contrast in this respect with many grassland systems where the flow of materials and energy is along grazing food chains. The main source of organic detritus in forest soils is, of course, the leaf litter which originates from the aerial vegetation. As we have seen, the character of this leaf litter varies with the type of forest, and may also be influenced by the presence of a ground flora. In pure stands, or monocultures, the homogeneous nature of the leaf litter provides a limited range of food resources for saprophagous soil animals; in mixed stands a greater diversity of food niches is available. These variations, which are basically horizontal ones, operate in conjunction with others which can be considered as vertical effects. Within each forest soil type, leaf litter decomposition proceeds along a series of successional stages, starting with the intact leaf and ending in finely particulate humus material. During this succession, organic particles become progressively smaller in size and more completely degraded, chemically. As the character of the organic material changes in this way, its availability as a substrate for micro-organisms and as food for soil animals also changes. In addition, this decomposing organic material becomes located at greater depths in the soil profile due to leaching and animal activity. Here, microclimatic features may be rather different from those at the surface, and this may have important repercussions on the character of the associated fauna. The patterns of faunal distribution in forest soils have, therefore, both horizontal and vertical components as we will see.

Horizontal effects: litter type. Within deciduous stands, there is some evidence that the character of the soil fauna may be influenced by the kind of tree species present, an effect which is exerted through the type of leaf litter produced. The leaves of deciduous species vary in the relative amounts of carbon and nitrogen they contain, and in their tannin content. There is a relationship between the C/N ratio and the rapidity with which leaves disappear from the litter. Thus, leaves with a low C/N ratio, such as alder, ash, hazel and elm, usually break down more or less completely during the first year after fall, whereas hornbeam, oak and birch, with larger C/N ratios, take rather longer, and beech, which also has a high tannin content, takes longer still (Lindquist, 1938; Wittich, 1942, 1943). The role of some of the larger-sized litter-feeding animals, such as earthworms, millipedes and woodlice, is evidently quite significant in this decomposition process. It has been shown, for example, that leaves with a relatively high content of available nitrogen are more palatable to, and therefore more readily accepted by, earthworms such as *Lumbricus terrestris* and *L. rubellus*, than leaves with a high C/N ratio (Satchell, 1967). The preference shown by various earthworm species for leaves of alder, elm, ash and birch, which disappear quickly from the litter under natural conditions (Lindquist, 1941), provides a convincing illustration of the importance of these animals in litter decomposition.

We must now consider what bearing these findings have on the distribution patterns of the soil fauna. Firstly, we can expect flourishing populations of earthworms and myriapods in soils under ash and elm, where there is an abundant supply of nitrogen-rich litter. Indeed, the vigorous feeding activity of earthworms in such sites often results in such a rapid and complete incorporation of leaf material into the soil that a distinct litter layer may not be formed or, if it is, this layer is thin and subject to desiccation during the summer. The organic profile formed under such conditions is of the mull type, and we will be looking in more detail at its faunal characteristics shortly. However, it may be noted at this point, that earthworms form a significant proportion of the fauna of the soil under these conditions, along with burrowing millipedes, such as *Glomeris marginata*, which can seek protection from desiccation by descending the profile. Various types of insect larvae, notably Diptera, which are present in the soil and litter for only a limited period of time during the cooler and wetter months of the year, also contribute to this fauna. Where there is no thick and permanent leaf litter in these mull sites, the fauna of mites and Collembola is relatively impoverished. This fauna reaches its best development in sites where leaf decomposition proceeds at a relatively slow rate. Secondly, if we compare the faunal characteristics of soils under ash and elm, where decomposition is rapid, with those under oak and beech, where litter accumulation occurs and, in some instances, may be converted slowly into an acidic raw humus, we can note a shift in emphasis from the earthworms to smaller-sized forms, such as the woodlice *Trichoniscus pusillus* and *Philoscia muscorum*, enchy-traeids, mites and Collembola, and to some of the larger, non-burrowing, moisture-loving millipedes, such as *Polydesmus* spp. To some extent, this shift probably reflects a change in feeding habit, with the fungi which grow in profusion in the layers of moist leaf litter becoming a more frequent component of the diet of many of the smaller soil animals. Higher moisture content and more equable temperature conditions are undoubtedly of great importance also. Beech litter with its high C/N ratio and high tannin content is not suitable food for many soil animals, at least in its freshly fallen state, and it is not until it has undergone some microbial decomposition that it can be utilized by the litter fauna. However, Elton (1966) points out that such "preparation" by microfloral decomposers can render beech litter richer in nitrogen, and as acceptable to the fauna as the fresh, nitrogen-rich litter of oak, ash and sycamore. Under these circumstances, the effect of the litter type manifests itself on the soil fauna as a change in numbers of individuals rather than of kind. Checklists drawn up for the fauna of oak, ash and beech litter may be strikingly similar, whereas there may be equally striking differences in the relative abundance of a given species in beech compared with ash. In parts of the Chiltern beechwoods, for example, the pill millipede *Glomeris marginata* occurs in beech litter, but it is not common and is frequently localized around nitrogen-rich sites, such as decaying logs containing large quantities of frass and faecal material produced by wood-boring insect larvae. van der Drift (1951) also noted that this species, which often develops large populations under ash and oak, occurred in a mixed beech/oak woodland, but only in places where oak leaves formed a part of the litter.

It is difficult in practice to isolate the effect of litter type on faunal distribution, since fallen leaf material provides only one component of the environment of soil animals. As this organic matter decomposes, it becomes associated in various ways with the mineral parent material to form an organic profile which may exert its own peculiar influence on the character and distribution patterns of the fauna. Even within pure stands of woodland, litter produced by patches of ground vegetation may provide local variations in the organic characteristics of the floor, and these may also influence microdistribution patterns.

Horizontal effects: humus type. The terms "mor" and "mull", to which we have already referred, were first coined by the Danish forester P. E. Müller to distinguish the strongly stratified profiles with a layer of acid, unincorporated raw humus, from the relatively homogeneous profile, often rich in calcium, in which organic and mineral material is intermixed to form a broad zone of mull humus which merges almost imperceptibly with the underlying parent material. In its extreme form, raw humus is represented by peat, and it is this humus type that we consider first.

Although there are few detailed investigations of the soil fauna of many of the various peatland forest types, the information available suggests that peaty soils do not support flourishing biotic communities. Low pH, low nutrient levels and waterlogged conditions are factors inimical to many soil organisms. The groups which predominate here are the Acari, Collembola, Nematoda and Enchytraeidae, and it may be no coincidence that these groups include a substantial amount of fungal material in their diet. Fungi tend to be associated with acidic soils supporting a variety of forest types, and it is not surprising that the development of a peat site in the direction of a drained forest soil produces no drastic change in the general composition of the soil fauna. However, in mature forest soils, other groups of arthropods often come in, such as millipedes, centipedes, woodlice, beetles and Diptera larvae. Some species are replaced by others during this transition from peatland to mature forest. As leaf litter accumulates on the surface of the peat, the effect of the latter becomes less direct, a wider range of microhabitats becomes available, and the fauna becomes richer and more varied, as a consequence. We have already noted this tendency in the fen habitats, and we find it again in a later stage in succession, the forest bog.

Forest bogs occur fairly commonly in the coniferous belt, and among the soil arthropods frequently encountered here are certain cryptostigmatid mites belonging to the families Nothridae and Camisiidae. Their distribution in a range of forest bogs has been investigated by Karppinen (1955) in a study which confirmed the faunal similarity between drained forest peatland and mature forest soil. However, specific differences in this fauna

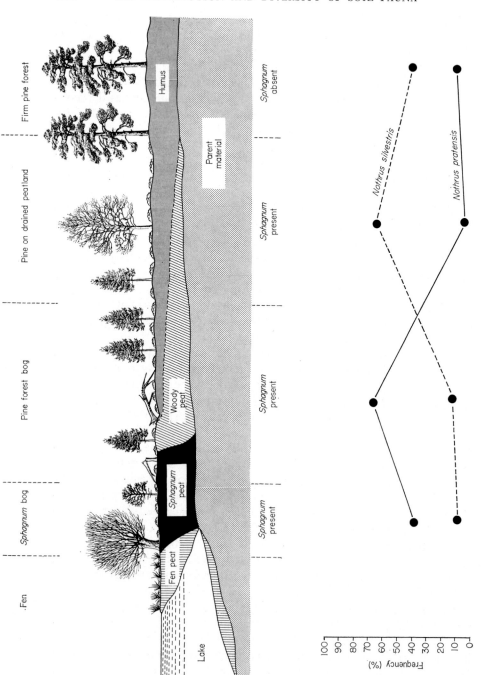

Fig. 8.6 Distribution patterns of two species of the oribatid mite sp. *Not-*

were apparent when comparisons were made between these two types and the forest bog habitat. For example, *Nothrus pratensis* was particularly common in the latter, but only occasionally encountered in the mature forest soil, and not recorded from drained peatland. In contrast, *Nothrus silvestris* was common in drained peatland and mature forest, but relatively scarce in the forest bogs (Fig. 8.6). Camisiid mites, such as *Heminothrus paolianus* and *Camisia lapponica* occurred commonly in mature forest soil, drained peatland and forest bog. In the wetter mature forests and forest bogs, *Neonothrus humicola* and *Platynothrus peltifer* made their appearance, although it is noteworthy that the latter species was more common in wet, treeless *Sphagnum* bogs and in the spruce–hardwood transition zone.

The extent to which these distribution patterns are influenced by food preferences cannot be estimated without more information on the feeding habits of the species concerned. It is possible that microclimatic effects are also important, and this topic is dealt with later in the chapter.

A less extreme type of raw humus, known as mor humus, commonly develops under conifers on well-drained sandy sites. Such profiles are usually characterized by strong leaching of bases (Fig. 8.3a). In contrast, mull humus occurs in brown forest soils (Fig. 8.3b), usually under deciduous vegetation growing on moderately drained sites where the parent material is clay or silt, and tends to have a mildly acidic, neutral or slightly alkaline pH. However, exceptions occur, for both mor and mull humus can be developed under spruce or under certain hardwoods, such as beech. Intermediate types can also occur, such as the moder or "acid mull" which has a lower pH than the true mull humus, yet resembles the latter in the extent to which it mixes with the mineral soil. This type provides conditions which are evidently less demanding on the soil fauna than those of raw humus profiles, as a comparison between the data presented in columns 2 and 3 of Table 8.2 indicate. The moder profile referred to here was developed under an old stand of spruce, and could be distinguished immediately from mor profiles by the absence of a discrete raw humus horizon below the litter. Faunal densities are at least as high as the maxima found in raw humus sites and, in the case of lumbricid earthworms, nematodes and Diptera larvae, appreciably higher than these. The distribution of these last two groups appears to be governed more by the moisture content of the soil than by the humus type, and probably the same holds true for the Enchytraeidae which have been recorded in densities of more than 134,000 per square metre from moder humus under Douglas fir in North Wales (O'Connor, 1957). However, it is not only differences in faunal densities that distinguish the biological characteristics of raw humus and moder profiles. Faunal diversity is greater in the latter, as a variety of insect larvae, woodlice, millipedes, centipedes and gastropod molluscs become established as part of the soil fauna.

In 1930 the Danish biologist C. F. Bornebusch published a study entitled: *The Fauna of Forest Soil*, which has since been regarded as a classic pioneer work in soil biology. Bornebusch's collecting techniques were less efficient than those in use today, and his absolute density estimates of numbers and biomass of various groups of the soil fauna were unrealistically low in some cases. However, it is a tribute to Bornebusch's work that his conclusions regarding the relative distribution of the commoner groups of soil animals in a range of forest sites have been confirmed repeatedly by recent workers. This author selected a number of sites in the forests of northern Denmark where the dominant species is oak or beech, under natural conditions. Clearing the forest, particularly where the soil is sandy, allowed the establishment of conifer stands (mainly spruce) within this deciduous formation. A variety of organic profiles are developed under both the beech and the spruce stands. These range from a good mull type occurring under healthy tree growth and a ground vegetation consisting of such herbaceous plants as *Mercurialis perennis, Asperula odorata, Oxalis acetosella* and *Anemone nemorosa*, to the acidic type of raw humus present under beech stands which are over-crowded and neglected, and under spruce growing on loose, sandy soil with a thick ground cover of *Hylocomium* moss. These two extreme conditions are linked by a series of intermediate types found mainly on higher, sandy ground under beech exposed to winds which remove leaf-fall. Under these conditions, where ground vegetation may consist only of a few species of grass (*Melica*) and herbs (*Oxalis*), organic decomposition is slower than in a good mull. Consequently, a heavy mull or a thin surface layer of raw humus, particularly under *Polytrichum* moss cover, is produced. Bornebusch compared the faunal characteristics of the various sites and his main findings are presented, diagrammatically, in Fig. 8.7. Several important points emerged from this work, and these are now considered briefly.

In the first instance, it is evident that the horizontal distribution pattern of a particular animal group may be influenced by more than one environmental factor. For example, the mites (Acari) are considerably more abundant in spruce mull than in oak or beech mull, suggesting that tree species, rather than humus type, may have a greater influence on the distribution pattern, in this case. On the other hand, these animals achieve their highest density in raw humus under beech, so that within the deciduous sites at least, the type of humus may have an important influence on the density of soil mites. Secondly, as far as the distribution of the total fauna is concerned, there is an inverse relationship between the number of animals and their biomass. This finding prompted Bornebusch to define the condition that: "The soil in which decomposition is most active contains the greatest weight of animals, but the lowest number; where decomposition is slow so that a heavy layer of raw humus is formed, we find the greatest number of animals, but, on average, these are very small, and their total weight is lower than that of the best soil. In other words, the good forest soil contains few and large animals; the inferior one, many and small animals."

However, these findings should not obscure the fact that the fauna of a

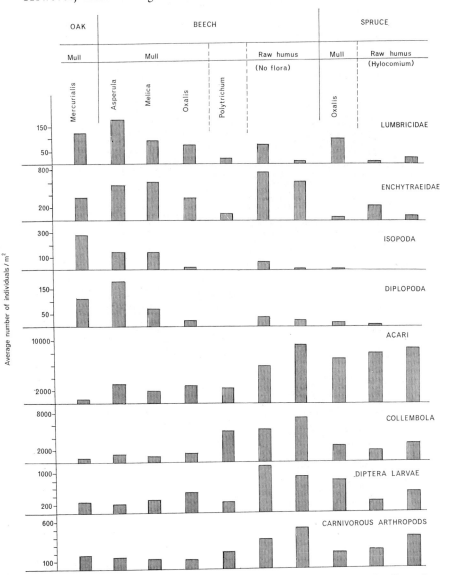

Fig. 8.7 Population densities of the fauna of various forest soils in Denmark. (From the data of Bornebusch, 1930.)

8*

Fig. 8.8 (a–l) Some typical representatives of the leaf litter fauna in a Chiltern beechwood. (a) The snail *Cepaea hortensis* (Müll.). (b) The geophilomorph centipede *Geophilus insculptus* Attems. This specimen is 28 mm long. (c) The carabid beetle *Nebria brevicollis* (Fab.). (d) The flat-back millipede *Polydesmus angustus* Latzel. (Photos. by A. Newman-Coburn.)

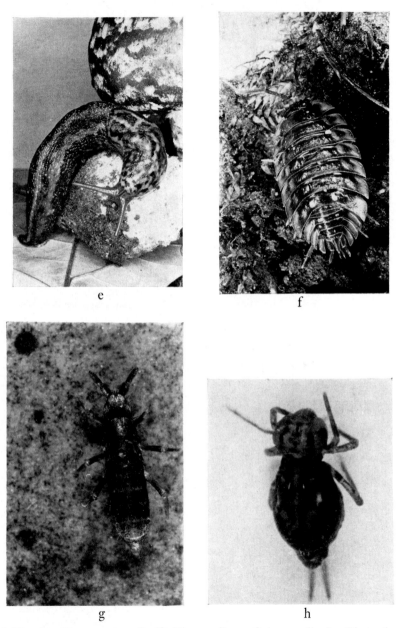

(e) The slug *Limax maximus* L. (f) The woodlouse *Oniscus asellus* L. (Photo. by A. Newman-Coburn.) (g) An arthropleone collembolan of the genus *Tomocerus* (3·5 mm long). (h) A symphypleone collembolan of the genus *Sminthurus*. (2 mm long). (Photo. by J. P. Harding.)

(i) An agelenid spider. This specimen is 4 mm long. (j) The pseudoscorpion *Neobisium muscorum* (Leach) (length: 2 mm). (Photo. by A. Newman-Coburn.) (k) A cryptostigmatid mite belonging to the genus *Hermanniella* (× 70). (l) A mesostigmatid mite belonging to the genus *Gütsia* (× 72).

good mull humus is at least as varied as that of a raw humus type, and often more so. Mull soils under beech in the English Chilterns harbour more species of Protozoa than do neighbouring raw humus sites (Stout, 1963). These mull soils also support a diverse fauna of annelids, molluscs and arthropods (Figs. 8.8a–1). This faunal diversity is rarely matched in raw humus soils except, perhaps, by the mites which may be represented by between 50 and 100 species within a small area of coniferous forest soil (Wallwork, 1957). Bornebusch showed that while mites were abundant in raw humus both under spruce and beech, Collembola and Diptera larvae were much more abundant in beech raw humus than that of spruce.

The maximum density of soil animals recorded by Bornebusch was 19,425 per square metre, the estimate obtained from beech raw humus. Of this number, 17,852 individuals were contributed collectively by the Acari, Collembola and Diptera larvae. To judge from more recent work in similar soil types, these densities may be under-estimates by a factor of 5 or 10. For example, Evans (1951) recorded a figure of 154,600 per square metre for soil arthropod densities in an English Sitka spruce forest soil, and Wallwork (1959, 1967) gave an estimate of the same order of magnitude for mites and Collembola in a hemlock mor in northern Michigan. Lebrun (1965) reported densities of cryptostigmatid mites ranging between 70,000 and 180,000 per square metre during a yearly cycle in a moder-type soil under a *Quercus/Carpinus* association in Belgium.

Horizontal effects: ground vegetation. The relationship between tree species, humus type and the species composition of the soil fauna is not as direct as the account given above may indicate. As we have already noted, mor humus is not confined to coniferous stands, nor is the mull type completely restricted to profiles under deciduous species. The reasons for this are not immediately clear, although the possibility cannot be ignored that the ground vegetation may influence the type of humus produced. Certainly the character of this vegetation appears to influence the relative abundance of the soil fauna populations. This is suggested by the comparisons drawn in Fig. 8.9 which considers the horizontal distribution of some common species of cryptostigmatid mites in relation to the character of the ground flora. Clearly, *Oppia neerlandica*, *Tectocepheus sarekensis* and *Suctobelba subcornigera* are widely distributed under various types of ground vegetation, evidently a reflection of the ecological tolerance of these species. On the other hand, *Nanhermannia nana*, *Ceratozetes gracilis* and *Chamobates schützi* apparently show some preference, the first for the *Hylocomium-Myrtillus* type, the second for *Calluna*, and the third for *Vaccinium* and the *Hylocomium-Myrtillus* vegetation under birch stands. The relative abundances of *Oppia translamellata* in the *Hylocomium-Myrtillus* under birch and under

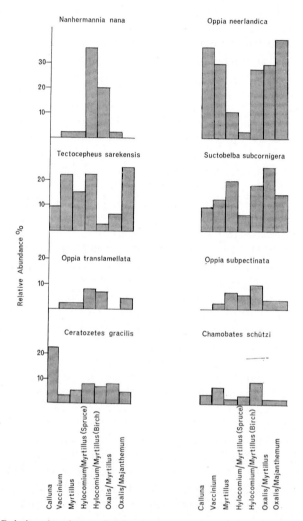

Fig. 8.9 Relative abundance of eight common species of cryptostigmatid mites in seven different forest sites in Finland. (From Karppinen, 1958b.)

spruce are very similar, which suggests that tree species have less influence than ground vegetation in this particular case. Similarly, in forests dominated by Scots pine, *C. gracilis* is relatively more abundant under *Calluna* than under *Vaccinium*.

These variations in the horizontal distribution pattern are rather subtle, and are expressed as differences in numbers of individuals, rather than in numbers of species. This point is made clearly by Forsslund (1945) who could not define species associations of mites and Collembola which were characteristic of the different types of ground vegetation occurring in a spruce–pine–birch raw humus soil. However, these microarthropods were present in higher densities under a ground layer of *Vaccinium* than under *Dryopteris* or *Geranium* in the same forest type, whereas larger soil animals, such as lumbricids, molluscs, spiders and beetles, were more abundant under *Dryopteris* than under the other two types. Similarly, Elton (1966) provides observations on the fauna of decaying bracken, a type of ground vegetation which often occurs in forest glades where the canopy is open, but which also forms an important component of the field layer at the edge of woodlands. When bracken dies back it forms a very thick layer of litter which often remains moist, even during prolonged dry spells, and which decomposes slowly to form a peaty humus. This litter often contains a rich fauna of larger arthropods and, on limestone sites in Wytham Woods, Elton found higher numbers of the woodlouse *Trichoniscus pusillus* in this habitat than in neighbouring oak litter. Earlier studies in this same locality had recorded the pill millipede, *Glomeris marginata*, in abundance, along with woodlice and staphylinid beetles. As far as the microfauna was concerned, notably enchytraeids, mites, Collembola, pseudoscorpions and small spiders, Elton noted these "in small numbers", and drew the general conclusion from these observations that "bracken litter in the spaces of woodlands may have a flourishing litter fauna, and that many species are similar to, and probably most are actually the same as under the woodland canopy". Clearly, the effect of this ground vegetation is reflected in quantitative, rather than qualitative, variations in the horizontal distribution of the soil and litter fauna.

The influence of ground vegetation on the species diversity of the soil fauna becomes more apparent when comparisons are made between the fauna under virgin stands, where ground vegetation is sparse, and that under secondary, or partly cleared, stands where opening the canopy has permitted a ground layer to become established. Karppinen (1958a) noted a much more varied fauna of soil mites under forest stands having rich ground vegetation, such as the *Calluna-Myrtillus* type under birch and spruce, the *Myrtillus* type under pine and the *Oxalis-Majanthemum* type under alder, compared with that under virgin aspen and virgin mixed woodland of the *Oxalis-Myrtillus* type. Figure 8.10 makes this comparison. The figure shows that the number of species (black columns) is relatively low under the two virgin stands. In these, the "dominant" species (i.e. those with a relative abundance greater than 5%) contributed between 80 and 90% of the total number of individuals present. In other habitats, where ground vegetation is more strongly developed, there are larger numbers of species, and here the dominant species make a smaller total contribution (between 55 and

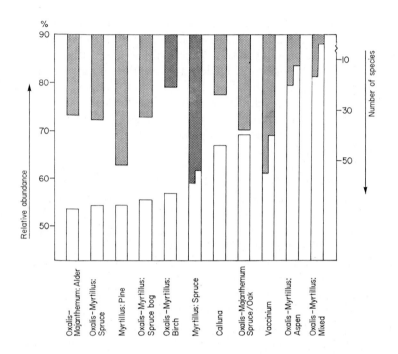

Fig. 8.10 The relationship between the number of species of cryptostigmatid mites (shaded columns) and the numerical contribution of the "dominant" species to the total faunal numbers, expressed as a percentage (open columns) in various forest sites in Finland. (From Karppinen, 1958a.) See text for further explanation.

70%) to the total number of individuals. In other words, the total number of individuals is more evenly distributed among the total number of species present, and the equitability component of species diversity (see Chapter 2) increases. The presence of ground vegetation increases the diversity of microhabitats available to soil animals and, for example, encourages the establishment of populations of such mites as *Carabodes labyrinthicus*, *Scheloribates laevigatus*, *Achipteria coleoptrata* and members of the Galumnidae which often move between the litter and the vegetation. These mites are rare or absent in localities where ground vegetation is poorly developed.

In the study discussed above, the density of cryptostigmatids in the virgin aspen and mixed woodland sites is approximately the same as that under the richer *Myrtillus* type under pine and spruce, rather higher than that under *Calluna* and *Vaccinium*, and considerably higher than that of the *Oxalis-*

Myrtillus type under birch. However, the total density is considerably lower than that under the *Oxalis-Majanthemum* types under alder, spruce and oak, and the *Oxalis-Myrtillus* type under spruce. This suggests that there is no simple, direct relationship between the number of species present in a site and the number of individuals per species. It might be expected that in species-poor sites, the number of individuals per species would be high, but

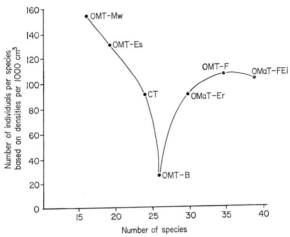

Fig. 8.11 The relationship between the number of species and the number of individuals per species of Cryptostigmata in seven boreal forest habitats in Finland. (Compiled from data provided by Karppinen, 1958a.) Key to habitats—OMT-Mw: Oxalis-Myrtillus under mixed woodland; OMT-Es: Oxalis-Myrtillus under aspen; CT: Calluna under pine, birch and spruce; OMT-B: Oxalis-Myrtillus under birch; OMaT-Er: Oxalis-Majanthemum under alder; OMT-F: Oxalis-Myrtillus under spruce; OMaT-FEi: Oxalis-Majanthemum under spruce and oak.

that this number would decrease as more species are added. The data plotted in Fig. 8.11 show that up to a certain point the proposition holds true that in the more restricted habitats, the number of species is low but the number of individuals per species is relatively high. Beyond this point, the relationship becomes less direct – some species associations are more flourishing, being richer in species and in numbers of individuals per species while others are poorer in these respects, than we might expect. There is probably no simple explanation for this, although the amount of available living space may be important. A comparison of the faunal composition under the *Oxalis-Myrtillus* vegetation under birch (OMT-B, Fig. 8.11) with that under *Oxalis-Myrtillus* under spruce and with *Oxalis-Majanthemum* under alder, spruce and oak (OMT-F, OMaT-Er, OMaT-FEi, respectively) indicates as much. The fauna under OMT-B is less varied than that under the other types, and also the number of individuals per species is considerably lower. Now it could be expected, if the initial proposition held, that the number of individuals per species would be higher under OMT-B than under the other types, which obviously is not the case. One possible explanation for this is that there is less available space for populations to expand into, in the OMT-B profile. The ground vegetation above this profile is mainly composed of grass species, the soil is somewhat drier than that of the other sites considered, and these two features suggest that the organic profile is poorly developed, with a shallow surface litter layer lying on top of mineral soil. Cryptostigmatid mites do not occur in abundance in mineral soil and there exists, therefore, only a narrow surface litter layer for them to colonize, under OMT-B ground vegetation. The depth distribution of these mites in OMT-B indicates as much, for more than 80% of the total number of individuals in the profile is located in the top centimetre of litter. In other habitats, as much as 50% of the total may occur below the surface in the deeper organic horizon, and in these cases, larger total densities are recorded. This kind of analysis illustrates the fact that faunal distribution patterns have a vertical, as well as a horizontal, component. This is now considered in more detail.

Vertical effects: the leaf litter microsere. In many forest soils, particularly those of the podzol type, there is a vertical sequence of organic horizons (Fig. 8.3a) which provides a series of distinct microhabitats for soil animals. These horizons correspond to different stages in the decay of leaf litter. The initial stage in this sequence is represented by freshly-fallen leaves which accumulate at the surface of the profile. As decomposition proceeds this organic material becomes fragmented and chemically degraded. The second stage is marked by the accumulation of partly decomposed leaf litter in a zone, the fermentation layer, which lies directly beneath the litter layer. As chemical degradation proceeds, particle sizes are reduced even further, and the organic complex known as humus develops. In this decomposition process there are, essentially, three phases, namely the leaf litter phase (L), the fermentation phase (F) and the humus phase (H), which form a vertical sequence from the ground surface downwards.

The initial phase in this sequence, fresh litter, represents a new habitat for the soil fauna, and it is rapidly colonized by fungal-feeding species of mites, enchytraeids and Diptera larvae. During this phase, which lasts for about 6 months in the case of beech and chestnut, the leaf litter is not attacked directly by soil animals (Anderson, 1973a) due to the tough texture of the leaves and high contents of distasteful polyphenols and tannins (Anderson, 1973b). However, species diversity increases during this phase as feeding niches become filled.

As decomposition proceeds into the second and third phases, the fermentation and humus phases, a variety of breakdown products accumulate in the organic horizons. The concentration of polyphenols is reduced, and this allows soil animals to feed directly on the leaf litter. Thus, there is an increase in the variety of food material, and this is paralleled by an increase in the species diversity of the soil fauna. At the same time, the equitability component of this diversity tends to become constant as the common species become less common and the rarer species less rare. This is a reflection of the ecological shift from species which are predominantly fungal-feeders to the macrophytes which feed on leaf litter. As a result of the feeding activities of this fauna a variety of metabolic products is released into the organic environment, biochemical diversity increases as does the microhabitat complexity. At this stage in the microsere, niche diversification is at a maximum, and rare species continue to invade the habitat until it is saturated. From this point on, biochemical diversity decreases as inorganic nutrients are taken up by plants, removed by leaching, or locked up in humic compounds. The soil community is similarly degraded to a few abundant species which are adapted to the unfavourable environmental conditions of the humus layer.

Each organic horizon in the forest soil profile presents its own distinct complex of environmental features, and has its own particular association of animal species (Table 8.4). Although seasonal movements up and down the profile do occur in some cases, the centres of population density of many species remain in particular horizons throughout the year. Recent developments in the technique of soil sectioning have confirmed this (Anderson, 1971). Thus, the heterogeneity of the organic profile and the diversity of microhabitats occurring here, encourages a spatial separation of species populations, reduces inter-specific competition and increases species diversity. This diversity is enhanced by two factors. Firstly, the various organic horizons provide different ranges of substrates on which soil animals can feed. Secondly, there is a decrease in the particle size of the organic material, with depth, from the litter to the humus. This means that there is a progressive reduction in the size of the soil spaces, the living space available to non-burrowing soil animals. As a consequence of this, the larger-sized species of mites and Collembola tend to be confined to the surface layers of the profile, while smaller forms predominate at lower levels.

Effect of Microclimate

Earlier in this chapter, attention was drawn to the fact that populations of various soil animals in coniferous sites tend to be larger in moist areas than in dry ones. This appears to a fairly general phenomenon, and it applies equally to the fauna of deciduous soils where, in some instances at least,

Table 8.4

The dominant cryptostigmatid mites of the main organic layers in various forest sites in South-East England (after Evans et al, 1961).*

Organic Layer	Oak	Oak/Beech	Scots Pine	Sitka Spruce
			FOREST	
Litter	Brachychthonius spp. (24·7)	Oppia ornata (10·0)	Tectocepheus velatus (16·0)	Oppia ornata (32·7)
	Tectocepheus velatus (13·1)	Tectocepheus velatus (9·6)	Suctobelba subtrigona (6·2)	Steganacarus striculus (7·3)
	Oppia ornata (9·6)	Eniochthonius pallidulus (6·2)	Oppia nova (5·2)	Suctobelba sarekensis (7·0)
			Brachychthonius spp. (5·3)	
Fermentation	Oppia ornata (27·0)	Oppia nova (38·9)	Brachychthonius spp. (19·6)	Suctobelba sarekensis (28·8)
	Suctobelba sarekensis (18·2)	Suctobelba sarekensis (4·4)	Pseudotritia minima (18·7)	Oppia nova (27·4)
	Brachychthonius spp. (13·1)	Tectocepheus velatus (3·7)	Suctobelba subtrigona (13·5)	Brachychthonius spp. (8·9)
Humus	Carabodes femoralis (23·9)	Suctobelba sarekensis (26·3)	Suctobelba subtrigona (34·5)	Oppia quadricarinata (27·2)
	Oppia ornata (19·7)	Pseudotritia minima (21·0)	Oppia minus (20·0)	Oppia fallax obsoleta (18·1)
	Brachychthonius spp. (14·5)	Oppia nova (19·5)	Brachychthonius spp. (19·3)	Suctobelba sarekensis (18·1)

* Relative abundances given in parentheses.

differences in the species composition of the fauna from place to place can be interpreted in terms of microclimatic effects. A good illustration of this is provided by Thiele and Kolbe (1962) who compared the beetle fauna of the soil under two deciduous stands, an oak–hornbeam association and a beech–holmoak type. These two habitats differed not only in their vegetational characteristics but also in climatic features, such that evaporation rate, temperature and humidity fluctuations were greater in the beech–holmoak than in the oak–hornbeam stand. Consequently, the former represented a drier and warmer habitat than the latter, at least during the day.

The beetle fauna of these two sites was markedly distinct in a number of respects and, of the total of 68 species encountered, only 22 were common to both types of forest, 33 species were restricted to the oak–hornbeam, compared with only 13 to the beech–holmoak. The richer development of the fauna in the oak–hornbeam was attributed to the cooler, wetter conditions prevailing there. These conditions were more suitable than those of the drier beech–holmoak, particularly for the Carabidae. Forest-dwelling carabids breed in spring, in most cases, with larval stages appearing during the summer. The susceptibility of these larvae to drought conditions precludes their establishment in dry sites. The fauna of the oak–hornbeam also included a richer representation of species commonly associated with hilly or mountainous regions, such as *Abax ovalis*, *Molops* spp., *Domene sabricollis* and *Epipolaeus caliginosus*. Again, this apparently was a reflection of the preferences of these species for cool, moist habitats. This was also borne out by the observation that the few montane species inhabiting the beech–holmoak type (*Feronia cristatus*, *Trichotichnus laevicollis* and *Philonthus decorus*) were much more abundant in the wetter parts of this site than in the drier areas.

The relationship between the species composition of the forest-dwelling carabid beetles and the environmental moisture conditions is frequently well marked. In general, the fauna of wet sites consists mainly of species which overwinter as adults and breed in the spring, such as members of the genera *Agonum*, *Bembidion* and certain *Feronia* species. The fauna of dry sites includes many of the species which breed in late summer and autumn, and overwinter as larvae, such as *Abax ater* and certain species of the genus *Carabus*. These autumn breeders tend to avoid wet sites where their developing larvae may be exposed to flooding (Murdoch, 1967).

Beetles are richly represented in the forest floor fauna of the tropics. Schubart and Beck (1968) recovered representatives of 35 families from leaf litter and mineral soil in the Brazilian rain forest, although more than 70% of the total catch belonged to five families, notably the Staphylinidae, Pselaphidae, Ptiliidae, Carabidae and Scydmaenidae. For the most part, this fauna consisted of relatively small-bodied forms, which were much more abundant in the leaf litter layer than in the underlying mineral soil. In the latter, small, elongate species belonging to the Pselaphidae, Ptiliidae and Scydmaenidae predominated, along with some burrowing carabids and staphylinids. Horizontal distribution patterns showed some striking

Table 8.5

Frequency and abundance of the five most common families of beetles in rain forest litter of wet and dry biotopes in Brazil (based on samples 1 m²) (from Schubart and Beck, 1968).

| | Dry Sites | | Wet Sites | |
	Frequency %	Mean Density	Frequency %	Mean Density
Staphylinidae	63	4·6	88	15·3
Pselaphidae	63	5·0	38	14·0
Ptiliidae	25	1·0	13	7·0
Carabidae	25	1·0	100	16·4
Scydmaenidae	63	2·2	63	3·8

similarities to those in temperate forests with differences in the composition and abundance of the beetle fauna in wet and dry sites. In wet sites flooded periodically by river overflow, the larger Carabidae were relatively more abundant than in dry sites, whereas the Scydmaenidae showed no strong trend (Table 8.5). Taking the larger beetles as a whole, densities were higher in wet sites than in dry ones by a factor of 4. This difference is, to some extent, a reflection of the great mobility of these beetles which allows them to move in advance of the floodwater to the marginal zone between wet and dry areas, where they often congregate in large numbers.

The variation in microclimate within a particular site is the result of a number of other effects, notably aspect, topography, drainage, tree cover and seasonal variation in rainfall. In the Chiltern beechwoods, for example, one of the commoner microarthropods of the litter is the large cryptostig-matid mite *Damaeus onustus*, a species which apparently prefers conditions that are not too wet or too dry. This mite can be found in relatively large numbers where beech litter has accumulated in sheltered depressions (Fig. 8.4), but it is less common on rather steep slopes. During the drier parts of the year, in spring and summer, the deeper parts of this litter remain moist while the upper parts, which harbour the bulk of the *Damaeus onustus* population, are rather drier. There is, however, a considerable variation in the abundance of this species, even from day to day at the same site, with much lower numbers being encountered when the upper layers of the litter have been soaked with rain. This may be due to a general movement of the population to rather drier sites, for example around the bases of trees or in the shelter of fallen logs. Again, there are sites in this beechwood, usually fairly exposed situations, where the leaf-fall produces only a narrow zone of surface litter. During the winter, when this organic layer is thoroughly moistened for long periods of time, it supports rich populations of enchy-traeids and woodlice, such as *Trichoniscus pusillus*, but is very sparsely colonized when it dries out more or less completely in the summer. One possible explanation for this change is that a large proportion of the litter fauna may move down the profile into more moist humus soil during the summer. There is abundant evidence that seasonal movements of this kind occur in many groups of temperate forest soil animals, although the picture appears to be rather different in the tropics.

In tropical forest soils, the density of the litter fauna is severely reduced during the dry season (Dammerman, 1925; Williams, 1941; Beck, 1964).

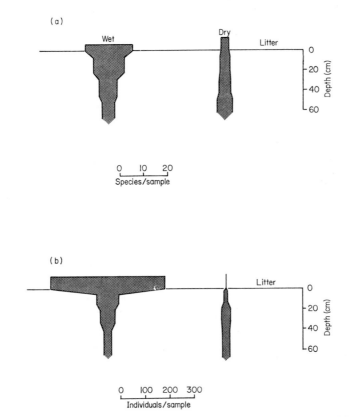

Fig. 8.12 A comparison of the vertical distribution patterns of cryptostigmatid mites in wet and dry seasons in an El Salvador rain forest soil. (From Beck, 1964.) (a) Number of species per sample, (b) number of individuals per sample. Sample size: 500 cm³.

This reduction does not appear to be correlated with any corresponding increase in the faunal density lower down the profile (Fig. 8.12), and we may conclude that the major portion of the litter fauna is unable to withdraw to deeper layers and, consequently suffers heavy mortality from exposure to seasonal drought. It is interesting to note, in passing, that there appears to be a difference here between tropical forest and tropical grassland, in the response of the soil fauna to seasonal changes in climate. In the grassland (see Chapter 4), centres of population density apparently shift downwards in the profile during the dry season. Climatic factors undoubtedly have an important effect on the distribution patterns and population sizes of the forest soil fauna in the tropics, and may overshadow biotic effects. For example, Maldague and Hilger (1963) could find no correlation between the numbers of Collembola and mites and their natural food organisms, bacteria and fungi, in African tropical forest. Predation by ants may control population sizes of beetles, mites and Collembola, although this effect has not received enough attention as yet.

Effect of Management Practices

Much of the forest in Europe and North America is under some kind of management, designed to increase timber yield. The effect of Man's intervention in natural systems has already been discussed at some length in Chapter 5, although the effect of forestry practice was not considered in any detail.

As a starting point, we can return to the work of Bornebusch mentioned earlier which included some findings on the effect of silvicultural practice on the soil fauna. The most unsuitable sites of the ones Bornebusch studied, as far as the soil fauna is concerned, were the heavy *Oxalis* mull and the thin raw humus developed under *Polytrichum* in beech stands. These sites had developed in localities exposed to wind and sun by the thinning of timber, or on a sandy mineral soil which was relatively infertile. Forest clearing often changes the character of the soil environment in several ways, one of the most obvious of which is the development of moss vegetation when the forest canopy is opened. The growth of moss apparently inhibits the development of mull humus, particularly in sites exposed to wind, for leaves of tree species are blown away before they have a chance to decompose *in situ*. On the other hand, the moss plants may provide suitable habitats for some species of soil animals, especially rotifers, tardigrades and nematodes. Bornebusch found that such moss cover had a richer fauna of the earthworm *Dendrobaena octaedra* and the large cryptostigmatid mites of the family Camisiidae, than was present in raw humus where moss cover was lacking. Huhta *et al.* (1967) also paid attention to the effects of forest practice on the fauna of coniferous soil in Finland, with particular emphasis on clear-cutting, thinning, burning, the addition of fertilizers and insecticides, and restocking. Burning over and insecticide treatments produced decreases in population sizes of almost all of the annelid, nematode and arthropod groups present, although in sites where the organic layer was well developed some measure of protection from these effects could be found deeper in the profile. After burning, in particular, initial population decreases were followed by a resurgence in numbers, notably of lumbricid and enchytraeid worms and Collembola, and this suggests that these populations may find deeper refuges in the profile.

In addition to the immediate mortality caused by burning, this treatment also destroys a considerable amount of the organic matter, vegetation, litter and humus on which many of the saprophages feed. The absence of any signs of recovery of populations of fungus-feeding mites and nematodes could be attributed to the destruction of the main item of their diet. An alteration of microclimatic conditions also results from burning, with notable increases in temperature and moisture fluctuations in the upper parts of the profile. These changes were evidently severe enough to prevent any marked recovery of populations of beetles and spiders for up to four years after the initial effect.

Clear-cutting of timber produces some effects that are similar to those of burning, notably an opening up of the canopy to allow more light to reach the soil surface, greater fluctuations in temperature, and a change in soil moisture conditions such that the surface layers become liable to desiccation. However, unlike burning, clear-cutting adds an appreciable quantity of organic material to the surface of the soil, in the form of leaves, branches, bark and other felling residues, in a short period of time. This material, and the disturbance caused by such operations, may destroy much of the ground flora and reduce the species diversity of the soil fauna. On the other hand, the additional organic material added to the soil surface may promote the growth of fungi and bacteria on which many soil animals feed. This is likely to be a short-term effect for once the clearing has been completed, the leaf litter is no longer replenished periodically, and when the initial supply of organic material is exhausted the soil will be unable to support a flourishing biotic community. Huhta et al. (1967) point out that the effects of clear-cutting on the soil fauna appear to be rather complex and difficult to analyse, for various populations reacted in different ways. Some, such as Diptera larvae and adult beetles, showed a sharp and immediate increase in numbers, while a more gradual increase was shown by nematodes, enchytraeids and Collembola. Populations of lumbricid earthworms and spiders apparently could not tolerate the changing environment, and their numbers decreased. The mites and larval beetles underwent an initial numerical increase, but subsequently their numbers declined, possibly as their food supplies became exhausted. It is also possible that these populations could not tolerate the increased fluctuation in moisture and temperature conditions. The effects of thinning are much the same as those following clear-cutting, but less marked, as would be expected.

The long-term effects on the soil fauna of re-stocking after clear-cutting are poorly understood, mainly because the necessary information has to be collected over a period of 10, 20 or 30 years. It could be predicted that the effects of the clear-cutting would gradually fade as the new timber becomes established. As the growing vegetation provides more protection for the soil surface, temperature fluctuations will be less violent, and the leaf litter layer less susceptible to drought. Movement of water through the profile will increase as plant roots tap their moisture supplies. The restoration of the soil fauna to its "normal" undisturbed character evidently is a slow process, for Huhta et al. (1967) could detect no real progress in this direction during a period of 15 years after re-stocking. It is also common forestry practice to re-stock with rapidly-growing conifers in sites which formerly supported deciduous stands. This artificial succession undoubtedly produces permanent changes in the character of the soil humus layers and, in turn, on the character of the soil fauna, although little attention has been given to these changes up to now.

Finally, we turn to the effects of applying fertilizers to forest soil. Chardez et al. (1972) noted that the addition of fertilizers to an acid brown forest soil under beech in the Ardennes region of Belgium had marked effects on the abundance and species composition of the rhizopod protozoan fauna. As a result of treatment with urea, rhizopod biomass was reduced by about two-thirds of that on untreated plots, whereas biomass doubled on plots receiving a mixture of urea and potash. The application of potash, alone or in combination with urea, promoted a shift in the species composition as moss-dwelling forms were replaced by soil-dwellers. Huhta et al. (1967) studied the response of the soil fauna under pine and spruce to the application of a fertilizer rich in lime. The purposes of this treatment were to increase the amount of available nitrogen, phosphorus and potassium and, through the presence of lime, to neutralize the rather acid pH of the sites. As far as most groups of the soil fauna were concerned (nematodes, cryptostigmatid mites, beetles, Diptera larvae, centipedes and spiders), the effect was negligible or inconclusive. In the case of Collembola, predatory mites, enchytraeids and lumbricids, an initial decrease in population size was followed, after a period of 1–3 years depending on the group, by a marked increase in numbers. This delayed response may be a reflection of adjustments made by the populations concerned to the changed pH conditions or, more probably, to changes in the physical character of the humus.

Summary

The diversity of the forest soil fauna precludes any simple treatment of its ecology. We are still in the process of discovering and describing the range of microhabitats which are available to soil animals in these ecosystems. However, our understanding has reached a point at which a tentative synthesis may be attempted, at least as far as broad outlines are concerned.

The four main forest types considered here, peatland, temperate deciduous and coniferous forest, and tropical forest, vary quite distinctly in the range of microhabitats suitable for soil animals. This can be illustrated by examining the relationship between the composition and diversity of the soil fauna and the degree of stratification of the forest. This stratification can provide a measure of the amount of living space, in the vertical dimension, available for this fauna. (Fig. 8.13.)

Forest peatland undoubtedly provides the most restricted range of microhabitats, and the tropical forest the most extensive. Within these two extremes are located the deciduous and coniferous formations of the temperate regions. The limitations imposed in peatland can probably be attributed to two main factors, namely the presence of the water table at or near the surface of the soil, and the strongly acidic humus layer. Under such conditions, the fauna is largely composed of groups which can carry on an aquatic or semi-aquatic mode of life in acid conditions, such as the Protozoa, nematodes, enchytraeids, mites and Collembola. The macrofauna, for the most part, finds living space only at the ground surface, and even here it may be exposed to the dangers of periodic flooding.

In coniferous forest growing on well-drained soil the diversity of microhabitats is greater than that of peatland, mainly because the water table is much deeper and because the organic profile is sub-divided into distinct horizons, namely the litter layer, the fermentation zone and the humus layer. The closed canopy provides a shade effect which inhibits the growth of ground vegetation and so suitable living space is found in the microhabitats which extend from the soil surface downwards through the organic profile. Conifer leaf litter has a high carbon/nitrogen ratio and large amounts of tannin and polyphenols which render it unpalatable to many of the larger saprophages, such as the earthworms, woodlice and millipedes. Consequently, these groups are poorly represented, and the bulk of the fauna consists of small-sized fungivores and detritus-feeders, such as mites, Collembola and certain nematodes. These animals do not burrow but utilize existing soil spaces, and their depth distribution is governed by the cavity structure of the soil. The diameter of soil spaces decreases progressively with depth. In the case of the cryptostigmatid mites and Collembola in particular, this is paralleled by a succession of species in which the largest forms ($\geqslant 0.75$ mm in length) are virtually confined to the surface litter layer, medium-sized forms (0·5–0·75 mm) predominate in the fermentation layer, while the smallest species (0·2–0·5 mm) form the bulk of the humus fauna. The physical separation of species populations in this way reduces inter-specific competition for living space and food resources. Some degree of feeding specificity may also be associated with this vertical pattern, with fungivores predominating in the litter layer, macrophytes in the fermentation layer, and suspension and liquid feeders in the humus layer.

The soils under deciduous forest in temperate regions, with certain exceptions, are characterized by an organic profile of the mull type. Here, there are no strongly structured horizons but a rather extensive zone of organo-mineral complexes, which ranges downwards from the litter layer to the zone of weathered parent material. The leaf litter produced by deciduous species has a lower carbon/nitrogen ratio than that of conifers and is more readily attacked by large saprophages, such as earthworms, woodlice, molluscs, millipedes and insect larvae. The burrowing activities of earthworms in particular serve to break down the leaf material into smaller particles and to mix these intimately with the mineral soil, which is often of a calcareous nature. These organo-mineral complexes provide suitable substrates for bacterial colonies which promote the decomposition of humus and, thereby, increase the biochemical diversity in the profile. This diversity presents a wide range of food materials for the microfaunal Protozoa, nematodes, enchytraeids, mites, Collembola and insect larvae. Here, in contrast to coniferous soils in which the microfauna predominates, we have a balanced system in which both macrofaunal and microfaunal elements are present. This diversity is further enhanced by the fact that microhabitats above the soil surface are available in the temperate deciduous forest. The presence of ground vegetation, at least during the spring and early summer, provides additional living space for those populations which can tolerate some degree of exposure to dry conditions. Population movements between the leaf litter and the aerial microhabitats provided by herbaceous vegetation and decaying logs occur, often in response to changing microclimatic conditions. Such movements may also be associated with a distinct phase in the life history of species in which juveniles have different microhabitat requirements from adults.

The situation in tropical forests contrasts sharply with that in temperate forests for, here, the range of microhabitats extends upwards away from the leaf litter into the aerial vegetation. The absence of a well-developed organic horizon in the soil restricts the microhabitat diversity of this stratum, and groups of animals which are usually associated with the soil in temperate regions are encouraged to seek microhabitats above the ground. The warm, moist microclimate reduces the danger of desiccation, and abundant food

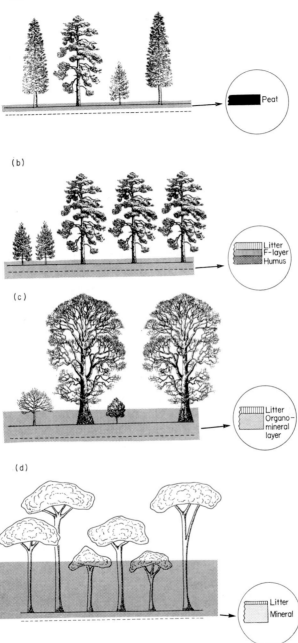

Fig. 8.13 The vertical range of habitats available for soil animals (shaded areas) in four contrasting forest types. (a) Forest peatland, (b) coniferous forest, (c) deciduous forest, (d) tropical rain forest.

material is provided in the form of decaying organic matter which accumulates in tree holes, the axils of branches and between the leaves of plants such as the bromeliads. This upward shift in the microhabitat range, compared with temperate forests, is a reflection of the need to compete for survival in the canopy layers of the tropical forest, rather than on the ground.

The forest soil fauna, especially in temperate regions, is essentially mesophilous in character. The majority of its members prefer conditions that are neither too wet nor too dry. Permanently wet soils, such as peatlands, have an impoverished fauna in which mesophiles are poorly represented. This element of the fauna also suffers long-term reduction in diversity as a result of management practices. Burning and clear-cutting in particular open up the canopy and expose the surface layers of the soil to marked fluctuations in temperature and moisture. Under natural conditions, microhabitat diversity is influenced not only by the variety of organic substrates but also by patterns of microclimate. Mosaics of wet and dry areas are formed where the topography is irregular, where clumps of ground vegetation produce shade effects, or where drainage patterns are variable. In addition to these horizontal mosaics, there are vertical ones particularly in strongly structured profiles where the various organic horizons differ in their moisture contents. The presence of these mosaics provides yet another opportunity for habitat separation to occur, with consequent increases in species diversity of the mesophilous fauna.

There is still much to be learned about forest ecosystems and their soil fauna. Some idea of the complexity of these systems is now apparent and it is clear that a wide range of micro-environmental situations can be represented within a single forest type. It is at this level, the microhabitat level, that further research promises to yield the most information.

REFERENCES

ANDERSON, J. M. (1971). Observations on the vertical distribution of Oribatei (Acarina) in two woodland soils, IV. Colloquium Pedobiologiae, C.R. 4ème Coll. Int. Zool. Sol., 257–72.

ANDERSON, J. M. (1973a). The breakdown and decomposition of Sweet Chestnut (Castanea sativa Mill.) and Beech (Fagus sylvatica L.) leaf litter in two deciduous woodland soils. I. Breakdown, leaching and decomposition, Oecologia (Berl.), 12, 251–74.

ANDERSON, J. M. (1973b). The breakdown and decomposition of Sweet Chestnut (Castanea sativa Mill.) and Beech (Fagus sylvatica L.) leaf litter in two deciduous woodland soils. II. Changes in the carbon, hydrogen, nitrogen and polyphenol content, Oecologia (Berl.), 12, 275–88.

BECK, L. (1962). Zur Ökologie und Taxionomie der neotropischen Bodentiere I. Zur Oribatiden-Fauna Perus. Thesis, Braunschweig.

BECK, L. (1964). Tropische Bodenfauna im Wechsel von Regene-und Trockenzeit, *Natur und Museum*, **94**, 63–71.

BECK, L. (1967). Die Bodenfauna des neotropischen Regenwaldes, *Atas do Simposio sobre a Biota Amazonica*, **5** (Zool.), 97–101.

BORNEBUSCH, C. H. (1930). The fauna of forest soil, *Forst. ForsVaes. Danm.*, **11**, 1–224.

CHARDEZ, D., DELECOUR, F. and WEISSEN, F. (1972). Évolution des populations thécamoebiennes de sols forestiers sous l'influence de fumures artificielles, *Rev. Écol. Biol. Sol*, **9**, 185–96.

DAMMERMAN, K. (1925). First contribution to a study of the tropical soil and surface fauna, *Treubia*, **6**, 107–39.

DI CASTRI, F. (1963). Etat de nos connaissances sur les biocenoses édaphiques du Chili. *In* "Soil Organisms", 375–85 (Doeksen, J. and Drift, J. van der, eds.), North Holland, Amsterdam.

DRIFT, J. VAN DER (1951). Analysis of the animal community in a beech forest floor, *Tijdschr. Ent.*, **94**, 1–168.

ELTON, C. S. (1966). "The Pattern of Animal Communities", Methuen, London.

EVANS, G. O. (1951). Investigation on the fauna of forest humus layers, in Forestry Commission, *Rep. Forest. Res. Lond.* (1950), 110–13.

EVANS, G. O., SHEALS, J. G. and MACFARLANE, D. (1961). "Terrestrial Acari of the British Isles. I. Introduction and Biology", *Brit. Mus. Nat. Hist.*, London.

FORSSLUND, K-H. (1945). Studier över det lägre djurlivet i nordsvensk skogsmark, *Meddn. St. SkogsförsAnst.*, **34**, 1–283.

FORSSLUND, K-H. (1948). Nagot om insamlingsmetodiken vid markfaunaundersökningar, *Meddn. St. SkogsförsAnst.*, **37**, 1–22.

HALE, W. G. (1967). Collembola. *In* "Soil Biology", 397–411 (Burges, N. A. and Raw, F., eds.), Academic Press, London and New York.

HARRIS, W. V. (1961). "Termites. Their Recognition and Control", Longmans, London.

HUHTA, V., KARPPINEN, E., NURMINEN, M., and VALPAS, A. (1967). Effect of silvicultural practices upon arthropod, annelid and nematode populations in coniferous forest soil, *Ann. Zool. Fenn.*, **4**, 87–143.

JACOT, A. P. (1939). Reduction of spruce and fir litter by minute animals, *J. For.*, **37**, 858–60.

KARPPINEN, E. (1955). Ecological and transect survey studies on Finnish Camisiids, *Ann. Zool. Soc. 'Vanamo'*, **17**, 1–80.

KARPPINEN, E. (1958a). Uber die Oribatiden (Acar.) der Finnischen Waldböden, *Ann. Zool. Soc. 'Vanamo'*, **19**, 1–43.

KARPPINEN, E. (1958b). Untersuchungen über die Oribatiden (Acar.) der Waldböden von Hylocomium-Myrtillus-Typ in Nordfinnland, *Ann. Ent. Fenn.*, **24**, 149–68.

LEBRUN, P. (1965). Quelques caractéristiques des communautés d'oribatides dans trois biocénoses de Moyenne-Belgique, *Oikos*, **16**, 100–8.

LINDQUIST, B. (1938). Dalby Söderskog: en skansk lövskog i forntid och nutid, *Acta Phytogeogr. Suec.*, **10**, 273 pp.

LINDQUIST, B. (1941). Undersökningar över nagra Skandinaviska daggmaskarters betydelse för lövförnans omvandling och för mulljordens struktur i svensk skogsmark, *Svenska SkogsvFören. Tidskr.*, **39**, 179–242.

MACFADYEN, A. (1952). The small arthropods of a *Molinia* fenatt Cothill, *J. Anim. Ecol.*, **21** 87–117.

MALDAGUE, M. E. and HILGER, F. (1963). Observations faunistiques et microbiologiques dans quelques biotopes forestiers équatoriaux. *In* "Soil Organisms", 368–74 (Doeksen, J. and Drift, J. van der, eds.), North Holland, Amsterdam.

MÜLLER, P. E. (1879). Studier over Skovjord, som Bidrag til Skovdyrkningens Theori. I. Om Bøgemuld og Bøgemor paa Sand og Ler, *Tidsskr. Skovbr.*, **3**, 1–124.

MÜLLER, P. E. (1884). Studier over Skovjord, som Bidrag til Skovdyrkningens Theori. II. Om Muld og Mor i Egeskove og paa Heder, *Tidsskr. Skovbr.*, **7**, 1–232.

MURDOCH, W. W. (1967). Life history patterns of some British Carabidae (Coleoptera) and their ecological significance, *Oikos*, **18**, 25–32.

NIELSEN, C. O. (1949). Studies on the soil microfauna. II. The soil-inhabiting nematodes, *Natura jutl.*, **2**, 1–131.

O'CONNOR, F. B. (1957). An ecological study of the enchytraeid worm population of a coniferous forest soil, *Oikos*, **8**, 161–99.

O'CONNOR, F. B. (1967). The Enchytraeidae. *In* "Soil Biology", 213–57 (Burges, N. A. and Raw, F., eds.), Academic Press, London and New York.

SATCHELL, J. E. (1967). Lumbricidae. *In* "Soil Biology", 259–322 (Burges, N. A. and Raw, F., eds.), Academic Press, London and New York.

SCHUBART, H. and BECK, L. (1968). Zur Coleopterenfauna amazonischer Böden, *Amazoniana*, **1**, 311–22.

STOUT, J. D. (1963). Some observations on the Protozoa of some beech wood soils on the Chiltern Hills, *J. Anim. Ecol.*, **32**, 281–87.

THIELE, H-U. and KOLBE, W. (1962). Beziehungen zwischen bodenbewohnenden Käfern und Pflanzengesellschaften im Waldern, *Pedobiologia*, **1**, 157–73.

VOLZ, P. (1951). Untersuchungen über die Microfauna des Waldboden, *Zool. Jb. (Syst.)*, **79**, 514–66.

WALLWORK, J. A. (1957). "The Acarina of a Hemlock-Yellow Birch Forest Floor". Thesis, University of Michigan.

WALLWORK, J. A. (1959). The distribution and dynamics of some forest soil mites, *Ecology*, **40**, 557–63.

WALLWORK, J. A. (1967). Acari. *In* "Soil Biology", 363–95 (Burges, N. A. and Raw, F., eds.), Academic Press, London and New York.

WHEELER, M. W. (1910). "Ants, Their Structure, Development and Behaviour", Columbia U.P., New York.

WILLIAMS, E. C. (1941). An ecological study of the floor fauna of the Panama rain forest, *Bull. Chicago Acad. Sc.*, **6**, 63–124.

WITTICH, W. (1942). Uber die Aktivierung von Auflagehumus extrem ungünstiger Beschaffenheit, *Z. Forst-u Jagd.*, 718.

WITTICH, W. (1943). Untersuchungen über den Verlauf der Streuzersetzung auf einem Boden mit Mullzustand II, *Forstarchiv*, **19**, 1–18.

Chapter 9

Fauna of Decaying Wood, Rocks and Trees

The forest type provides a whole series of ecosystems, some more sharply delimited than others, in which rather specialized, complex communities develop. Soil ecologists, not unnaturally, usually focus their attention on the ground, and give little more than a passing glance to communities above the soil surface. However, rocky outcrops, growths of fungi, moss and lichens on trees, "suspended soils" and decaying wood all have their own peculiar faunal characteristics, and it is of some interest to determine the extent to which these characteristics are derived from those of the soil fauna. In the previous chapter, attention was drawn to the way in which the soil fauna extends its vertical range above the ground in tropical forests. This phenomenon can now be considered in more detail, particularly as it applies in temperate regions. The fauna of decaying wood is taken as a starting point.

Decaying wood can be classified, ecologically, into three major types, namely dead branches remaining *in situ* on the tree, decaying stumps and, finally, rotting logs and branches lying on, or partly buried in, the litter layer of the soil. The first of these three categories need not concern us here, for dead branches which remain attached to the tree usually undergo dry decomposition, and the faunal succession they support shows little or no affinity with that of the soil. Decaying tree stumps and fallen branches are generally invaded by members of the soil fauna which become intimately associated with the decay process, and it is with these habitats that we are particularly concerned.

Decaying stumps

The rate at which a tree stump decomposes and the particular path which this decomposition takes vary considerably from place to place, and are influenced not only by the type of wood, but the time of year at which felling occurs, the exposure, topography and prevailing microclimatic conditions, In certain parts of the Chiltern beechwoods of southern England, for example, the distribution of stumps in varying stages of decay suggests that decomposition proceeds much more rapidly on south-facing slopes, compared with those facing north. Furthermore, in sites where decomposition is apparently proceeding more rapidly, there is developed a much more flourishing fauna. This consists of specialized forms which are more or less restricted in their distribution to decaying wood, together with elements of the soil fauna which have invaded this habitat from the surrounding substratum. These elements, together with the growths of moss, lichens, algae, fungi and bacteria, form a complex community whose characteristics are influenced, to a large extent, by the moisture content of the wood.

In selecting this one important property of decaying wood, namely its moisture content, three different pathways can be recognized along which the decomposition process develops. These are dry decomposition, wet decomposition and an intermediate type which we can call moist decomposition. The last-named of these is probably the most interesting from the biological point of view.

Dry Decomposition

The participation of the soil fauna in the decomposition of dry stumps is minimal, since few of these animals have sufficient tolerance of low moisture content to become established permanently in such sites. In the main, the disintegration of the wood occurs through the activities of certain insects, notably Coleoptera, Hymenoptera and Isoptera, which feed on the wood and often produce burrows or gallery systems in it. Among the Coleoptera, beetles belonging to the families Scolytidae, Buprestidae, Anobiidae and Cerambycidae may be singled out as capable of colonizing dry wood. These, along with the hymenopteran wood wasps (Siricidae) and carpenter bees (*Xylocopa* spp.), reduce the wood to a fine, dry powder which appears to be quite resistant to further chemical decay. Dry stumps also provide nesting sites for certain species of ants, such as members of the genus *Camponotus* and the red ant, *Formica rufa*. These animals may chew the wood to a papery pulp, or "carton", which is used in the construction of the nest. A similar habit is shown by many of the termites. This group is mainly tropical in distribution, as we have already seen, and its members often build up large colonies in moist wood. However, the southern European *Calotermes flavicollis* is a dry wood termite, and its feeding activities produce the same kind of effects as those of the beetles, wasps, bees and ants already mentioned. A common feature of dry wood decomposition is that the woody frass which accumulates through the activities of the fauna does not become thoroughly mixed with the surrounding soil. This suggests that the dry wood fauna is a rather isolated one, and shows little overlap with that of the soil.

Wet Decomposition

Fig. 9.1 Bracket fungus (Polyporaceae) infecting a beech tree. (Photo. by author.)

Tree stumps undergoing wet decomposition are usually characterized by extensive growths of microfloral decomposers. Where fungi predominate over bacteria a condition known as "white rot" develops (Fig. 9.1). Where the reverse is the case putrefaction, or "brown rot", occurs. During the latter process, cellulose is decomposed more than lignin, while the white rot type of decomposition is characterized by the selective breakdown of lignin. As a general rule, the conditions associated with white rot do not favour many groups of soil animals (although various Diptera larvae, such as the Tipulidae, Mycetophilidae and Sciaridae are exceptions) for the branching meshwork of hyphae is avoided by the more active Collembola and mites (Kühnelt, 1961). In contrast, brown rot favours the development of large populations of rhabditoid nematodes, certain Collembola, woodlice, centipedes and a varied assortment of insects and spiders. However, we should perhaps be wary of drawing too rigid a distinction between the faunal characteristics of these two types of wet decomposition, since moisture conditions may vary, even within a given stump. White rot may be developed extensively in the heartwood, while rather drier conditions may occur beneath the bark, and here mites and Collembola may be abundant. Furthermore, the extensive development of fungal fruiting bodies, which is associated with white rot, often provides niches for various animals, as we will see later in this chapter.

Moist Decomposition

It will be evident from the above remarks, that although extreme dry and extreme wet decomposition of wood is characterized by a rather impoverished fauna, this fauna is more closely related to that of the surrounding soil and litter in wet wood compared with that undergoing dry decomposition. A condition intermediate between these two extremes, namely moist decomposition, provides very suitable moisture conditions for soil animals. Here, the decomposition of a tree stump proceeds along three, fairly well-

Fig. 9.2 Gallery systems of engraver beetles in birch bark. Each of the two central chambers shown is about 3 cm long. (Photo. by A. Newman-Coburn.)

defined avenues. Firstly, as the bark becomes loose and starts to peel away from the heart wood, various groups of animals colonize this outer layer and the crevices beneath it. Such groups often include bark beetles of the families Scolytidae and Bostrichidae, which may form characteristic burrow systems in the wood (Fig. 9.2). Other wood-feeders are found among the stag beetle family Lucanidae, a group of considerable importance in moist wood decomposition, which includes such forms as *Dorcus parallelepipedus* and *Sinodendron cylindricum* in beech, and *Lucanus cervus* in oak. The larvae of certain click beetles (Elateridae), notably *Melanotus*, are generally associated with moist decaying wood. Some members of this family are true wood-feeders, while others occur here as predators of other wood-inhabiting insects. Few of these Coleoptera can be considered as representative of the surrounding soil fauna, but they are joined in and beneath the bark by many other animals that are. This rich fauna includes lumbricid earthworms, enchytraeids, centipedes, woodlice, millipedes, molluscs, Diptera larvae, Collembola and mites. Some of these, for example the earthworms *Bimastos eiseni*, *Dendrobaena* spp. and *Lumbricus rubellus*, the centipede *Lithobius variegatus* (Fig. 9.3) and the slug *Arion hortensis*, are seasonal visitors seeking the shelter of crevices beneath the bark. Others, such as the "box mites" of the family Phthiracaridae (Fig. 9.4), which can globulate completely when disturbed, feed actively on the woody tissue and may be permanent inhabitants. The ecology of all these groups will be considered in more detail when we deal with the fauna of decaying logs later in this chapter.

Direct invasion of the heartwood also occurs via the cut or broken face of the stump which is soon overgrown with algae, fungi, moss and lichens. Lichens frequently have a rather specialized fauna of mites and Collembola, whereas algal growths attract various snails and slugs; fungal fruiting bodies also have their own peculiar fauna (see later). The faeces produced by these animals accumulate in cracks and fissures in the wood, and not only are colonized by nematodes, enchytraeids and Diptera larvae, but also promote the growth of moss which may help to prevent the wood from drying out. A faunal succession then develops in this moist substratum, in which many of the groups also found under bark are represented. Lumbricid earthworms, nematodes and Diptera larvae of the family Tipulidae are common under

Fig. 9.3 The centipede *Lithobius variegatus* Leach commonly found beneath the bark of decaying tree stumps. (Photo. by A. Newman-Coburn.)

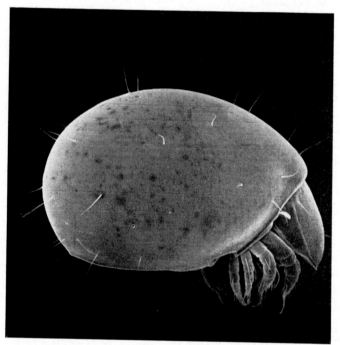

Fig. 9.4 The "box mite" *Phthiracarus anonymum* Grndj. ($\times$ 168), a typical representative of a group of cryptostigmatid mites which has developed a wood-boring habit. Photo. by B. Parry and reproduced through the courtesy of the British Museum, N.H.)

moss growths in such sites. As the decomposition of the underlying wood proceeds, wood-feeding mites, Collembola, lucanid beetles and larval Elateridae continue and, indeed, may accelerate this process. The result of all this activity is an accumulation of finely particulate, friable woody material in the interior of the hollowed stump.

The breakdown of the central heartwood often proceeds more rapidly than that of the peripheral tissue, which eventually forms the wall of the hollow stump, for the outer layers of the stump are often drier than the inner zone and, consequently, decompose more slowly. Furthermore, decomposition of the inner wood occurs not only from above, i.e. from the upper end surface of the stump, but also from the root zone buried in the soil. (Fig. 9.5). The attack from below is often carried out by root-feeding nematodes and the larvae of the chafer beetles belonging to the family Scarabaeidae, together with various detritus-feeding mites and iuloid millipedes. The activities of these animals provide a good illustration of the direct participation of the subterranean fauna in woody decomposition.

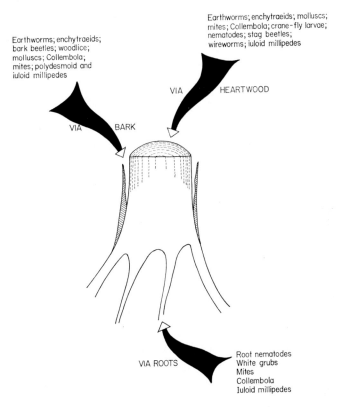

Earthworms; enchytraeids; molluscs;
mites; Collembola; crane-fly larvae;
nematodes; stag beetles;
wireworms; iuloid millipedes

Earthworms; enchytraeids;
bark beetles; woodlice;
molluscs; Collembola;
mites; polydesmoid and
iuloid millipedes

VIA   HEARTWOOD

VIA   BARK

VIA ROOTS

Root nematodes
White grubs
Mites
Collembola
Iuloid millipedes

Fig. **9.5** Major pathways of faunal invasion of a decaying tree stump.

The end result of all this plant and animal activity is the reduction of the woody tissue to small, partly decomposed particles which become intimately mixed with the surrounding soil after their passage through the gut of earthworms and molluscs. Many of the mites and Collembola, which feed on the wood, subsequently distribute it as faecal pellets deeper in the surrounding soil, where it may be degraded further by micro-organisms. Sometimes, the tree stump persists merely as a dry, hollowed shell; in other cases, where moisture conditions are adequate, the entire structure breaks down and eventually becomes covered with soil and ground vegetation. The changing pattern of faunal distribution which accompanies this decay process has been shown by Wallace (1953) to be characterized by an increase in species diversity and an ecological shift from wood-feeding forms, which predominate in the early stages of decay, to fungus- and detritus-feeding species in the later stages.

This general outline of the course of decomposition of tree stumps must not obscure the fact that the species composition of the associated fauna may vary from site to site, and there is sometimes a sharp distinction between hardwoods and conifers in this regard. For example, earthworms may be less abundant in conifer stumps than in beech or oak, and this may be a reflection of the different ecological status of these animals in the soil of these two major vegetation types. Such effects have already been considered in the previous chapter and need not be recalled here. However, we can go on to look at a few detailed studies of the faunal characteristics of decaying wood, based on observations made on logs and branches lying in the litter. To some extent, the course of the decomposition process occurring in these sites differs from that of decaying stumps, for smaller entities are involved and one major pathway of faunal invasion, namely that via the roots, is not available. Furthermore, small logs can be manipulated for experimental purposes more precisely than can fixed stumps. However, there is no reason to believe that many of the ecological events occurring in decaying wood are not common to stumps and smaller logs, so that the following discussion is relevant to what has been said in the foregoing pages.

Decaying logs

Decaying logs lying on the surface of the leaf litter form an extremely important component of the forest soil ecosystem, for they support an astonishing variety of animal life. Elton (1966) had this to say about this fauna: "It will be evident, even if we did not have the wealth of previous information from forest and other entomologists to confirm it . . . that we have here one of the richest series of animal communities in the country. It is on a level of scale in variety, interest and ecological power with the communities of the canopy, of field layers, of the ground zone and the topsoil, and far beyond any section of any aquatic system in these islands." The activities of this community, as with those in decaying stumps, are often centred around, and dominated by, various wood-boring insects, notably bark beetles of the family Scolytidae, stag beetles (Lucanidae), larval click beetles (Elateridae), weevils (Curculionidae) and crane-fly larvae of the genus *Tipula*, among others. Many of these forms are adapted morphologically, physiologically and behaviourally to this mode of life, being equipped with powerful chewing mouthparts and digestive enzymes which can cope with some, at least, of the cell contents and structural carbohydrates in their woody food (Parkin, 1940). A detailed account of the ecology of various wood-feeding insects has been given by Savely (1939) and this study is of particular interest because it compares the faunal succession in pine and oak logs. Savely found that the general course of this succession was similar in both kinds of wood, with an initial invasion of cerambycid, buprestid and scolytid beetles, which are primarily phloem-feeders, promoting the growth of fungi which render the sapwood soft and moist. At this stage, species which prefer soft woody food, together with fungivores and predators, come into the succession. At a later stage in decomposition, the fauna of the sapwood becomes depleted, and as the decomposing woody tissue becomes incorporated in the soil, a new succession of animal species, members of the true soil fauna, occurs. Despite this general similarity of ecological events in the two types of logs, there were marked differences between the species composition of the two faunas. In a total species list of over 250, only 25 were common to both types of log. In comparing the fauna of pine and oak at roughly the same stage in decay, Savely noted a much richer fauna in the latter than in the former. He attributed this to the fact that oak decomposes much more slowly than pine, and provides a more stable and persistent habitat. These faunal distinctions between the two kinds of logs were maintained until decomposition had become far advanced, although in all probability these differences would be less marked when the woody tissue became part of the organic layer of the soil.

In addition to the rather specialized forms, there is a part of the fauna of decaying logs that is derived from the true soil and litter fauna. Here, we can define two main groupings, although there are others, as we will see later. There are the animals which move into decaying logs to seek shelter during certain times of the year, and there are those which have been able to colonize this habitat more or less permanently. It is this latter group, consisting of animals ecologically plastic enough to extend their habitat range, which is of the prime importance from the functional point of view; they become intimately involved in the ecological events that occur in decomposing wood. The former group can be regarded more as transient members of this community, although they make some contribution by depositing faecal material and by being available to predators. Consequently, it is perhaps wise not to attempt to make too sharp a distinction between these two ecological groupings.

Logs as Seasonal Refuges

Lloyd (1963) has drawn attention to the considerable amount of faunal movement that occurs between decaying branches and the surrounding litter in a beech woodland in southern England. Working with fallen branches averaging 5 cm in diameter, he devised an experiment to test the hypothesis that animals tend to move from such branches into the surrounding litter during periods of cold weather, and return to these branches when the weather conditions improve. Eleven isolated groups of branches were selected, and these were sawed into paired lengths. One of each pair was left in the litter, and the other removed to the laboratory for dissection and the collection of the animals it contained. This initial operation was carried out as the snow cover was beginning to disappear at the end of February, on a day when ground temperatures approximated 0°C at the time of sampling. Four days later most of the snow had disappeared and the ground temperature had risen to between 7°C and 11°C, when the second series of branches was collected. These too were analysed for their faunal inhabitants. If the original hypothesis is correct, larger populations should be

Table 9.1

Total numbers of animals occurring in fallen beech branches collected in snow (C) and in warmer weather 4 days later (W), based on samples of 11 units (from Lloyd, 1963).

	Numbers of individuals		Probability
	C	W	
Lithobius crassipes	9	23	<0.01
Lithobius variegatus	2	14	<0.01
Enchytraeidae	67	140	>0.10
Lumbricidae	30	47	<0.01
Slugs	0	17	<0.01
Snails	17	62	<0.004
Oniscus asellus	48	184	<0.003
Trichoniscus pusillus	8	90	<0.02
Diptera larvae	256	232	>0.67

found in the second series of logs, which had been exposed to a period of warmer weather. A summary of most of the main findings is given in Table 9.1. These data show clearly that a wide variety of soil and litter animals, such as lithobiomorph centipedes, woodlice, snails, slugs and lumbricid earthworms, move into fallen branches as the ground temperature increases. The two notable exceptions to this pattern are the enchytraeids and Diptera larvae, and the author attributes this to the lack of mobility on the part of these groups. Lloyd also considered the possibility that movements of animals between fallen branches and the surrounding litter might follow a diurnal pattern, with a movement out of the branches into the litter at night, and a reversal of this during the day. Such diurnal activity could not be detected, however.

It is pertinent to enquire, at this stage, why movements such as those described above should occur at all. What special features of the decaying wood habitat attract soil and litter animals? There is, perhaps, no one simple answer to this question, but we can consider several possibilities. The idea that animals seek to avoid unfavourable microclimatic conditions by moving into the logs does not seem feasible on the basis of the evidence presented above, for the movement is the other way, namely into the surrounding litter during the colder weather. However, studies carried out by Fager (1968) in this same general locality revealed that greater abundances of earthworms and centipedes occurred in oak logs during winter (October to April) than during summer (May to October). This was interpreted as a movement from litter into a more sheltered habitat during the colder part of the year. The suitability of a log as a refuge or overwintering quarters will depend, among other things, on its size. The larger the log, the greater the protection it can afford. Fager's logs consisted of oak boxes, and his study is not then strictly comparable with that of Lloyd. However, Fager did find that twenty species of arthropod, belonging to the Collembola, Acari, Diptera larvae and Coleoptera larvae, were significantly more abundant in logs in summer than in winter, compared with only nine species which showed the opposite pattern. The prevalence of insect larvae in the summer logs was attributed to seasonal reproductive patterns, and this explanation may also apply to some species of mites and Collembola which showed seasonal replacements of one species with another. For example, the collembolan *Isotoma westerlundi* was present in all winter logs examined, often as the most abundant animal, but was absent from all summer logs, whereas *Tomocerus minor* occurred in almost all of the summer logs in great abundance, but in much reduced densities in winter logs. On the other hand, some litter animals evidently show a bimodal peak of occurrence in logs, and an example of this is furnished by a population of the carabid beetle *Feronia oblongopunctata* inhabiting an acid mull soil in mixed deciduous woodland in Scotland (Penney, 1967). This species is highly dependent on decaying logs, aestivating during the summer and hibernating in winter in cells, neatly constructed, beneath the bark. At other times of the year, adult beetles were active in the surrounding litter. So complete was this transition from activity in the litter during spring and autumn to quiescence beneath the bark in mid-summer and winter, that the availability of such hibernacula may be an important factor in limiting the distribution of *F. oblongopunctata* to woodland in Britain.

Logs and Life Histories

The possibility that decaying logs are selected as suitable habitats for certain stages in the life cycles of soil and litter animals deserves further attention. Banerjee (1967) has found definite evidence that the millipede *Cylindroiulus punctatus* (Fig. 9.6) moves upwards from the mineral soil and humus layer, where the adults overwinter, to decaying logs during the spring and early summer. Here mating occurs and the eggs develop to the end of the third juvenile instar, after which there is a movement out of the logs.

Fig. 9.6 The millipede *Cylindroiulus punctatus* (Leach). (Top) Active, (bottom) quiescent. (Photos by A. Newman-Coburn.)

There are three possible benefits to be derived from this pattern although, as far as one can be aware, none of these has been confirmed. Firstly, the concentration of the adult population in spring in a clearly delimited area may facilitate breeding. Alternatively, the development of the eggs and the early juvenile stages may be stimulated by their location in a habitat where higher temperatures occur, at least during the day. Finally, the decaying log habitat may provide some specific food requirement for the early juvenile stages. This third alternative apparently is used by certain crypto-stigmatid mites, and an example can be taken from a study of microarthro-pod distribution patterns in decaying yellow birch and hemlock branches, carried out in the Ottawa National Forest of northern Michigan. The nymphal mite illustrated in Fig. 9.7 was found to be almost entirely

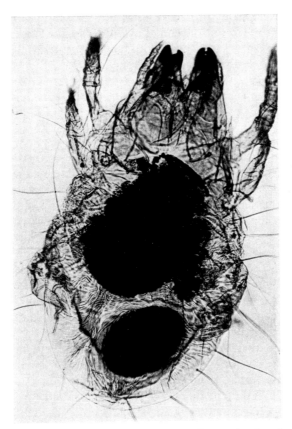

Fig. 9.7 A cryptostigmatid mite nymph of the genus *Hermannia* taken from a lenticel in yellow birch bark. The dark-coloured gut contents consist of corky material. (Photo. by author.)

restricted to the bark layer of hemlock and yellow birch. It was particularly common in the lenticels of the birch bark, about 30% of which were found to be attacked. Adults were not encountered in these situations, and although the nymphs also occurred in the surrounding litter, they were rare there. The presence of numerous faecal pellets consisting of the corky material of the lenticel, together with the appearance of the gut contents of the mites, clearly suggest that these animals excavate a chamber in the lenticel by their feeding activities. As a general rule, only one individual occurred in each lenticel, and since the chamber in which this animal lived was closed to the extent that each lenticel had to be dissected to reveal its occupant, it may be presumed that the infestation originated either by the direct deposition of an egg into the lenticel or on its surface. In the latter case, the invasion would then be made by one of the early developmental stages. As post-embryonic development proceeds, the cavity in the lenticel is gradually enlarged until, probably at the end of nymphal life, the animal emerges and moves away from the site, as an adult. This is an interesting example of a very specialized feeding habit in a group of mites, the Cryptostigmata, where such specialization is quite rare. It also serves as an example of an ecological link between the temporary and permanent fauna of decaying wood.

Logs as Food Sources

The question of why certain animals should be attracted to woody tissue as a source of food is something of a mystery, for this tissue is not, in itself, particularly nutritious. Its nitrogen content is low, and in order to digest the cellulose component, most of the animals concerned require gut symbionts such as bacteria and flagellates. On the other hand, decaying logs are probably enriched by small particles of organic debris, blown in by wind, washed in by water, or deposited as faeces by temporary invaders from the litter and soil fauna. The extensive fungal growths that often occur in decomposing wood may also provide food for various mites, Collembola, enchytraeids, nematodes and insect larvae. To what extent the fauna of decaying logs relies on these food sources is not known but it would seem that, potentially, they are of considerable significance, particularly for the animals living in spaces in the wood and in the crevices under the bark. In addition to these forms, there are others which feed directly on the wood and in doing so excavate tunnels in this tissue, promoting physical decomposition by comminuting the food particles, and contributing to the spread of fungal infections. To this ecological group belong a number of specialists, such as the various beetles which as adults or larvae, or both, occur only as wood-borers. As already mentioned, we are not concerned particularly with this group, since it is not derived from the soil fauna. However, it must be pointed out that finely shredded frass and faeces produced by these insects often have an associated fauna, consisting of mites, Collembola, nematodes and enchytraeids, which has probably extended its habitat range from the soil and litter into the tunnels of the wood-borers. In addition, there are certain species of mites which have developed the wood-boring habit, and these produce burrow systems in the heartwood, in which a very great proportion of the life cycle is spent. These mites belong to the order Cryptostigmata and are closely related to, if not identical with, species occurring in the surrounding soil and litter.

Microdistribution Patterns

In order to document some aspects of faunal distribution in decaying twigs and branches, we can now refer to the results of a study, mentioned earlier, of the mite fauna of yellow birch and hemlock wood at a site in northern Michigan (Wallwork, 1957). Although this was not an exhaustive investigation, the results obtained illustrate various patterns, and a summary

Table 9.2

A comparison of the depth distribution of various ecological groupings of mites in yellow birch and hemlock branches.[1,2]

	Yellow Birch Zones				Hemlock Zones			
	I	II	III	IV	I	II	III	IV
Wood-borers:								
Steganacarus magnus	0	3	6	23	0	247	3	0
Cepheus sp.	0	0	1	4	0	0	0	0
Hermannia sp. (nymph)	0	27	1	0	0	34	4	0
Rhysotritia ardua	0	1	0	5	0	26	0	0
Associates:								
Oppia spp.	7	46	14	34	0	0	0	0
Schwiebea talpa	0	5	17	9	1	2	1	0
Galumna formicarius	2	0	0	5	0	0	0	0
Scheloribates laevigatus	1	7	0	6	0	0	0	0
Heminothrus sp.	3	0	0	0	0	0	0	0
Brachychthonius jugatus	2	0	0	0	0	0	0	0
Predators:								
Gamasolaelaps sp.	7	62	23	32	1	0	1	0
Parasitus sp.	0	0	0	0	1	0	0	0
Bdella sp.	1	1	0	0	0	0	0	0
Total numbers of individuals	23	152	62	118	3	309	9	0
Average numbers per branch	4	25	10	20	1	103	3	0

[1] Data represent total counts from 6 yellow birch and 3 hemlock branches.

[2] The zones indicated are: I—outer surface of bark; II—in bark; III—beneath bark; IV—in heartwood.

of the data is presented in Table 9.2. The branches which were dissected and examined in detail were roughly comparable in size, being 15–20 cm long and approximately 1 cm in diameter, and the sample comprised six such units of yellow birch and three of hemlock. The objectives of this work were: (1) to define the trophic structure of the decaying wood community, (2) to determine the pattern of faunal distribution through the branches, from the surface of the bark to the innermost part of the heartwood, (3) to identify any differences which occur between the two different types of branches with regard to the trophic structure of the community and the distribution patterns of the constituent species.

Dealing first with the trophic structure and distribution of the mite component of the community, we can consider three main categories, namely wood-borers, associates and predators. The four species of wood-borers are all members of the Cryptostigmata and, with the exception of *Cepheus* sp. which was found only in yellow birch, they were common to both types of branch. *Steganacarus magnus* was found mainly in the heartwood in yellow birch and the bark of hemlock, often producing a complex network of burrows in which there was a preponderance of nymphal forms. Each burrow was packed with faecal pellets consisting of finely comminuted wood. All the indications suggested that, in this site at least, *S. magnus* can spend the whole of its life cycle in the wood, for it was rarely encountered in the surrounding litter. Elsewhere, in beech woodland for example, this species is among the commonest mites in the litter, so we have here a clear change of habit, possibly a reflection of the fact that the surrounding litter consisted for the most part of hemlock needles which may be unsuitable as food. Eggs, which presumably develop parthenogenetically, and the cast skins of larval and nymphal stages were found among the faecal material in the burrow systems of *S. magnus*. These skins could often be located by noting where there occurred a marked widening of the burrow and an increase in the diameter of the faecal pellets. Side branches from the main burrow harboured early developmental stages. From all these observations, it is possible to build up a picture of the life history and activity of this mite which, in view of its abundance, is probably one of the more important agents in opening up the interior of these decaying branches to fungal and bacterial infection. Of the other wood-boring species, only the nymphal *Hermannia* sp. (Fig. 9.7) would appear to be of much ecological importance, and some aspects of its biology have been considered already. *Rhysotritia ardua* has a distribution pattern which is very similar to that of *Steganacarus magnus*, but it is much less abundant than the latter. *Cepheus* sp. apparently becomes a wood-borer as an adult, although it occurred in low densities in the branches examined. The nymphal form is a bizarre creature (Fig. 9.8), with a dorso-ventrally flattened shape which enables it to crawl beneath

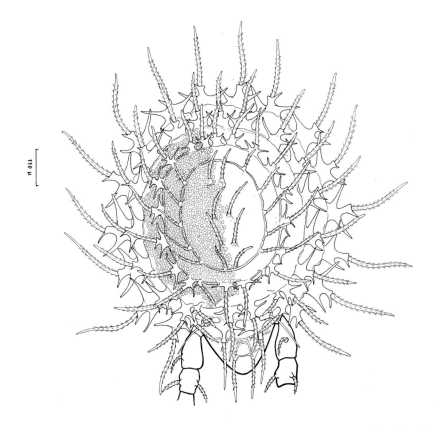

110 μ

Fig. 9.8 Nymphal stage of *Cepheus* sp., a dorso-ventrally flattened cryptostigmatid mite which lives in bark crevices.

the bark. The adult was found only in softer wood, but both nymphs and adults also occurred in the surrounding litter.

The wood frass produced by these burrowers supports a varied fauna of commensals and coprophages which may be considered, rather loosely, as "associates". Examples of this group include nematodes which were frequently found in the burrows of *Steganacarus magnus* immediately below the body of the mite, and faecal-feeding mites which included at least three species of the genus *Oppia*, *Galumna formicarius*, *Scheloribates laevigatus* and *Schwiebia talpa*. These coprophages ranged freely through the cavities of the decaying branches among the faecal material. *G. formicarius* was restricted to this habitat apparently, for it was never encountered in the surrounding litter. *S. talpa* showed a preference for the cavities beneath the bark, where it sometimes occurred in reproductive aggregations. *S.*

laevigatus was frequently associated with the *Hermannia* nymphs in the lenticels of yellow birch bark, although it also occurred deeper in the wood. In addition to these groups, Collembola occurred in low numbers beneath the bark, both of yellow birch and hemlock. In deciduous woodlands Collembola are often one of the most conspicuous faunal elements in decaying wood, and their low numbers in this case is undoubtedly a reflection of their low densities in the nearby forest floor. In some yellow birch branches there were what appeared to be inordinately large numbers of predatory mites belonging, in the main, to the genus *Gamasolaelaps*. These carnivores were not restricted in their distribution in the branches, although they occurred in highest numbers in channels in the bark where ample food was available, in the form of soft-bodied immature mites, Collembola, nematodes, enchytraeids and insect larvae.

If we now compare the general characteristics of the mite fauna of the two types of branches, several interesting differences appear, although it would be unwise to exaggerate the importance of these in view of the restricted nature of this study. It may be noted that although the overall density of mites per branch is higher in hemlock than in yellow birch, the fauna of the latter comprises a markedly larger number of species (12) than the former (6). Only three of the six species recorded from hemlock occur in any abundance. From this it may be deduced that the habitat provided by the yellow birch is a less rigorous one than that provided by the hemlock, and the fauna is less specialized, as a consequence. This conclusion is substantiated by the high proportion of burrowing forms in the hemlock fauna. Secondly, the abundance of predators appears to be related to the abundance of the "associate" fauna, and consequently is much greater in the yellow birch than in the hemlock. This is to be expected, since the bulk of the hemlock fauna is located in narrow borings, effectively closed to predators by the packed faecal material. On the other hand, the free-ranging commensals, coprophages and fungivores which are such a feature of the larger cavities in yellow birch, are more readily available to the predators. The close correspondence between the distribution patterns of associates and predators in the various zones in yellow birch adds weight to this interpretation. The abundance of both of these trophic groups is greatest in the bark cavities, rather lower in the heartwood and under the bark, and lowest on the outer surface of the bark. Finally, there is a clear distinction between the two kinds of branches in regard to the depth distribution of two of the three wood-boring species common to both, namely *Steganacarus magnus* and *Rhysotritia ardua*. Both of these species were more abundant in heartwood than in bark in yellow birch, whereas their peak densities in hemlock occurred in the bark, and they were totally absent from the heartwood. On the other hand, if one looks at the overall depth distribution of the fauna in the two types of branch, there is a common pattern characterized by a high density in the bark, and greatly reduced numbers on its outer surface. The surface fauna is probably subject to considerable variation, depending on the microclimate around the branches, the time of day, the amount of moss and lichen growth on the bark, the extent to which the branch is covered by leaf litter, and so on.

An Experimental Approach

The idea that the fauna of decaying branches consists of a number of ecological groupings which may react in different ways to environmental changes, has been taken a good deal farther by Fager (1968). This author studied the microarthropod fauna of decaying oak wood near Oxford by means of samples taken from natural fallen branches and synthetic branches, or "logs". The latter consisted of oak boxes packed with oak sawdust designed to provide four possible variants of packing and nutrient quality. These logs were then placed out on an oak woodland floor for varying periods of time from 4 to 10 months. Comparisons of the fauna which developed in these logs with that in natural branches showed that these artificial models were good substitutes. Moreover, since their initial characteristics were known, and the environmental factors to which they were exposed could be monitored accurately, the faunal counts obtained could be subjected to rigorous statistical procedures designed to eliminate all sources of variation except the one being examined. The details of the method used, which involved an analysis of recurrent groups considered briefly in Chapter 2, need not concern us further here, but some of the main conclusions of this work can be summarized as follows.

The total faunal content of the logs could be divided into (1) one major group of 23 species which represented the basic fauna of decaying oakwood all through the year and in all types of logs, and (2) nine additional groups, comprising from two to five species, which were influenced by some environmental features. Conspicuous among the basic fauna were nine species of mite and at least seven species of Collembola, together with the woodlice *Trichoniscus pusillus* and *Oniscus asellus*, Diptera belonging to the families Cecidomyiidae and Chironomidae, and the coleopteran *Denticollis linearis* and members of the Aleocharinae. Some of the smaller associations comprised species which varied in their abundance in logs from season to season. This phenomenon has already been considered earlier in this chapter. Some species associations could be recognized which preferred natural branches to synthetic logs, and this included various Collembola, the lumbricid *Bimastos eiseni*, the centipede *Lithobius crassipes* and dipteran Cecidomyiidae. A preference for moist logs rather than dry ones was suggested by the distribution of Diptera belonging to the families Cecidomyiidae, Ceratopogonidae and Psychodidae, certain Collembola *(Isotoma olivacea, Isotomurus palustris* and *Neelus minimus)* and mites *(Cilliba cassidea, Thyrisoma lanceolata* and *Geholaspis longispinosus).* A stronger association with enriched logs, compared with unenriched ones, was shown by the lumbricids *Dendrobaena rubida tenuis* and *Bimastos eiseni,* the collembolans *Dicyrtoma minuta, Isotoma westerlundi* and *Hypogastrura purpurescens,* and the mites *Damaeus onustus, Xenillus tegeocranus, Zercon triangularis* and *Eviphis ostrinus.* Fager also found that although there was a considerable amount of variation in the dominant species from log to log, there was a consistent pattern of faunal structure, such that the average synthetic log contained 900 individuals distributed among 41 species. Of these 41 species, two contributed 50% of the total number of individuals, three species contributed a further 25%, twenty-four species were represented by between two and seventy-four individuals, while twelve species occurred only as single individuals. Thus, we can visualize the animal community of decaying logs as consisting of a basic group of species populations, in which the numerically dominant species may vary from log to log, added to which are various other populations whose specific characteristics will depend upon the time of year, the nutrient quality of the log, and probably also its moisture content. This basic, descriptive information is vital to any detailed understanding of the ecological events occurring in decaying wood, and although we are still far from the point at which we can describe in detail the functioning of complex communities, studies such as the one just discussed bring us a stage nearer to this goal.

Cryptozoa

A distinction is sometimes made between the soil and decaying wood fauna and that living beneath stones, rotting logs and tree bark (Dendy, 1895). The latter has been termed "cryptozoa" and has been the subject of a detailed study by Cole (1946). Since this fauna, if it has an ecological identity of its own, forms a link between that of decaying wood already discussed, and that of habitats such as rocks and trees, considered later in this chapter, we might usefully review some of the main findings of Cole's work. The approach used was, to some extent, artificial, for it consisted of observations on the fauna occurring under oak boards placed directly on the ground in a woodland where black oak and hickory predominated. Altogether, more than 130 species of animal were recorded from beneath these boards, although many of these were probably accidental occurrences and could not be considered as truly characteristic of this niche. Of the various groups which tended to predominate under boards, compared with other neighbouring habitats, the most conspicuous were molluscs, such as the slug *Agriolimax reticulatus* and the snail *Retinella rhoadsi*, the woodlouse *Tracheoniscus rathkei*, spiders belonging to the families Lycosidae, Argiopidae, Thomisidae and Clubionidae, the centipede *Lithobius forficatus*, the flat-backed millipede *Scytonotus granulatus*, various Collembola, the roach *Parcoblatta pennsylvanica*, the cricket *Gryllus assimilis*, beetles belonging to the families Carabidae, Staphylinidae, Histeridae and Pselaphidae, and a variety of ants. In addition to this typical fauna, certain insects such as leaf hoppers, beetles of the families Nitidulidae, Coccinellidae and Chrysomelidae, and dipteran Mycetophilidae and Phoridae, occurred in abundance under boards only at certain seasons. In these cases, the cryptozoic niche provided the shelter required for hibernating stages or for reproductive aggregations of juveniles. Many of these seasonal immigrants did not feed extensively in these sheltered locations and, taken as a whole, this element of the cryptozoa was heterogeneous in regard to food habit. In contrast, the more characteristic cryptozoa element consisted, for the most part, of scavengers and omnivorous feeders, and Cole suggested that it was the lack of feeding specificity which distinguished the cryptozoa from the fauna of soil, humus and decaying wood.

In summary, this work suggested that the fauna inhabiting the spaces beneath stones and logs has a distinct identity which may, however, be obscured during the spring and autumn when various temporary residents invade this habitat, and when some, at least, of the permanent members migrate down into the deeper soil or into decaying wood to aestivate or hibernate. The presence of these two components, the permanent and the temporary, remind us of the situation occurring in decaying logs, and although the latter includes a wood-feeding element largely lacking in the cryptozoa, there is evidently a considerable amount of overlap between these two faunas. A similar overlap could undoubtedly be found to exist between the cryptozoa and the surrounding litter fauna. Perhaps we should recognize that the cryptozoic niche is one in a spectrum of habitats, sharing some faunal characteristics with its neighbours, while at the same time providing a particular combination of environmental features which favour the development of a basic, characteristic group of animal populations. In this case, the faunal characteristics are determined primarily by the responses of individual species populations to the moist, sheltered conditions provided. We can now go on to look at some of the faunal characteristics, often not far removed physically from the cryptozoic type, in which the physical factors of the environment are much less favourable, namely the exposed habitats of rocks and trees.

Rocks and trees

Strictly speaking, the fauna of habitats located above the surface of the soil does not fall within our terms of reference. However, these habitats often grade imperceptibly into those of the soil, and examples can be provided in which the general (but not necessarily the specific) composition of the fauna may be similar, in some situations above ground, to that of the soil. This short digression examines the faunal characteristics of the rather specialized habitats provided by vegetation growing on the exposed surfaces of rocks and on the trunks and branches of trees. Of course, it will be realized that an extremely diverse assemblage of animals is associated with the aerial parts of vegetation, and we have already seen that this fauna includes many specialized forms which belong to subsystems remote from that of the soil and litter. We are more concerned, here, with the fact that some of these habitats above the ground serve as temporary, or even permanent, extensions of the natural habitat range of certain members of the soil fauna.

The term "epigeal fauna" is often used to distinguish the animals living in or on aerial vegetation from the fauna of the soil and litter. There are difficulties in adhering rigidly to this distinction, as we have seen. However, part of the epigeal fauna consists of various mites and Collembola which have evidently evolved from rather unspecialized, soil-dwelling ancestors, and have become adapted to a much more restricted mode of life associated, for example, with growths of moss, lichens and fungi on rocks and trees. The adoption of this new mode of life may represent an escape from the soil, a greater independence, particularly with regard to moisture requirements, and an increasing "terrestrialness". However, the price of this emancipation has been, paradoxically, a greater specialization, a greater degree of restriction to one particular kind of diet, in many cases. Here we have good examples of the phenomenon described by Elton (1966), namely that as one proceeds vertically upwards away from the soil and into the aerial vegetation, specific associations between animals and plants become intensified. As these associations become stronger, contact with the soil diminishes. In the following survey, which is devoted mainly to the distribution patterns of microarthropods, we will see that varying degrees of specialization occur, even within a single taxonomic group such as the cryptostigmatid mites, and there is no reason to suppose that the patterns described may not be broadly applicable to other taxonomic groupings. In introducing this topic, we can look at some of the main features of the habitats concerned.

The Habitats

In a detailed study, to which we will refer repeatedly in the following pages, Travé (1963) defined three categories of rock and tree habitats with which cryptostigmatid mites are particularly associated. All of these involve growths of lichens, mosses and, to a lesser extent, higher plants, and they can be defined briefly as follows:

1. Habitats produced only by encrusting vegetation, and without the introduction of organic or mineral material derived from other sources. Such habitats are associated with crustose lichens of the genera *Verrucaria*, *Lecanora*, *Caloplaca*, *Buellia* and *Pertusaria*, and liverworts such as *Frulania* sp.
2. Habitats formed by the indigenous vegetation and introduced organic and mineral materials. The introduced materials, in this case, have less significance in determining the major ecological characteristics than does the indigenous vegetation which often includes the foliaceous lichens *Parmelia* and *Peltigera*, and mosses which are able to colonize thin organic and mineral deposits.
3. Habitats formed, as above, by indigenous vegetation and introduced organic and mineral materials, but different from category (2) above in that the introduced materials play a much greater part in determining the ecological characteristics of the habitat than does the indigenous vegetation, which consists of mosses, grasses and other herbaceous species. The habitat so formed may have an appreciable degree of soil development, and is often termed a "suspended soil". It frequently represents a situation intermediate between the categories mentioned above and the true soil.

These three categories may be applied both to vegetational growths on rocks and on the trunks of trees, as shown in Fig. 9.9, for the same species of vegetation are often found on both substrata.

The three habitat types differ from each other, sometimes considerably, in their microclimate and the kinds of food material they provide for microarthropods. The categories defined above may be regarded as representing a progression to a more varied range of habitats which approximates to the environment of the true soil fauna. In like manner does the species composition of the fauna increase in variety with an increase in the kind and number of available habitats. Generally speaking, habitats remote from the soil receive a much greater amount of solar radiation, and the variations in temperature, relative humidity and nutrient content are wider. These fluctuating conditions require that the species which become established be tolerant forms and moreover, since there is little opportunity for such species to migrate to other, more favourable, habitats, an isolated fauna develops, clearly distinct in its species composition from that of the soil. Travé (1963) noted that only 7·2% of the total cryptostigmatid fauna of the forest region of the eastern Pyrenees was common both to the soil and to habitats above the ground. Of the total cryptostigmatid fauna recorded, 68·4% was present in the soil of the forest floor, whereas only 24·4%

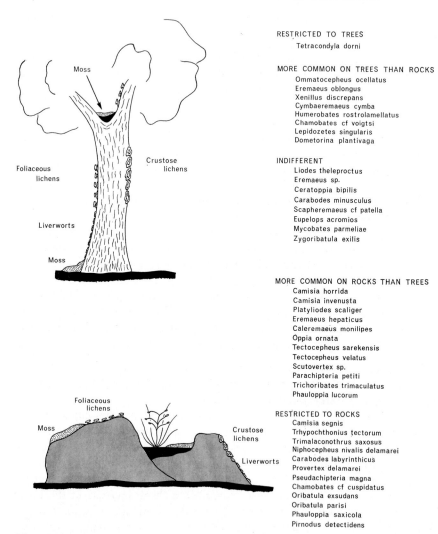

RESTRICTED TO TREES
Tetracondyla dorni

MORE COMMON ON TREES THAN ROCKS
Ommatocepheus ocellatus
Eremaeus oblongus
Xenillus discrepans
Cymbaeremaeus cymba
Humerobates rostrolamellatus
Chamobates cf voigtsi
Lepidozetes singularis
Dometorina plantivaga

INDIFFERENT
Liodes theleproctus
Eremaeus sp.
Ceratoppia bipilis
Carabodes minusculus
Scapheremaeus cf patella
Eupelops acromios
Mycobates parmeliae
Zygoribatula exilis

MORE COMMON ON ROCKS THAN TREES
Camisia horrida
Camisia invenusta
Platyliodes scaliger
Eremaeus hepaticus
Caleremaeus monilipes
Oppia ornata
Tectocepheus sarekensis
Tectocepheus velatus
Scutovertex sp.
Parachipteria petiti
Trichoribates trimaculatus
Phauloppia lucorum

RESTRICTED TO ROCKS
Camisia segnis
Trhypochthonius tectorum
Trimalaconothrus saxosus
Niphocepheus nivalis delamarei
Carabodes labyrinthicus
Provertex delamarei
Pseudachipteria magna
Chamobates cf cuspidatus
Oribatula exsudans
Oribatula parisi
Phauloppia saxicola
Pirnodus detectidens

Fig. 9.9 Some habitats above the ground colonized by microarthropods. (From Travé, 1963.)

of the species were rock-dwellers (saxicole) or tree-dwellers (arboricole).

The Fauna

The microarthropods of these habitats above the ground comprise a number of different groups, predominant among which are the Collembola (Ellis, 1974) and mites. Cassagnau (1961) reported that soil-dwelling species of Collembola are poorly represented, and that seven genera of saxicole and arboricole Collembola could be identified, namely *Xenylla*, *Anurophorus*, *Uzelia*, *Tetracanthella*, *Vertagopus*, *Willowsia* and *Entomobrya*. Among the mites, the Mesostigmata is less conspicuously represented than the Prostigmata and Cryptostigmata (Travé, 1963). Predatory Prostigmata, feeding mainly on Collembola, are very mobile forms, especially as nymphs and adults, and are frequently encountered in exposed places where their movement is not hindered by compact layers of organic material.

The list of Cryptostigmata that can be classified as saxicoles or arboricoles is a long one, despite the fact that the habitats above the ground are rarely, if ever, characterized by great faunal diversity. Attempts have been made to define a discrete association of such species, often referred to as the *Zygoribatula exilis* association, after one of its most frequent members. Other species included in this grouping are *Sphaerozetes piriformis*, *Minunthozetes pseudofusiger*, *Eremaeus oblongus*, *Phauloppia lucorum*, *Trhypochthonius tectorum*, *Scutovertex minutus*, *Camisia segnis* and *Cymbaeremaeus cymba* (Strenzke, 1952; Pschorn-Walcher and Gunhold, 1957; Knülle, 1957; Klima, 1959). This list is by no means complete, nor does it distinguish the different degrees of tolerance to various environmental factors shown by the species listed. For example, Strenzke (1952) sub-divided this association into two components, namely a hygrophile group consisting of *Z. exilis*, *S. piriformis*, *M. pseudofusiger* and *E. oblongus*, and a xerophile group including *P. lucorum*, *T. tectorum*, *S. minutus* and *C. cymba*. The effects of the microenvironment are clearly of paramount importance in determining the distribution of the various species of cryptostigmatid mites, and we now consider in more detail two of these effects, namely microclimate, and the effect of the food, which is governed by the character of the epiphytic vegetation.

The effect of microclimate. Colonization of the exposed habitats of moss and lichens requires that the animals concerned be able to withstand temperature and moisture extremes. Examples of such cryptostigmatid mites are *Provertex delamarei* and *Oribatula exsudans* (Travé, 1963). The xerophilous character of the saxicole and arboricole Cryptostigmata deserves to be emphasized for, as a general rule, these mites are noted for their dependence on high relative humidity. The survival of these tolerant species in exposed habitats is apparently related to their physiological state. For example, the arboricole *Humerobates rostrolamellatus* initially prefers a dry environment, under experimental conditions, but will reverse this preference to seek moisture when desiccated to a point at which one-third of the body weight is lost (Madge, 1964). Atalla and Hobart (1964) demonstrated experimentally that the larval, nymphal and adult stages of the cryptostigmatid *Eupelops acromios*, which is frequently found in the soil and in habitats above ground, showed a much higher resistance to low humidity (10–20%) than did species such as *Hermannia gibba* and *Platynothrus peltifer*, which are more restricted to the soil under natural conditions.

The microclimate of habitats above ground, particularly the temperature, is strongly influenced by the kind of substratum present. Ultimately, the temperatures prevailing in all of these habitats are closely related to the amount of solar radiation present. This amount will depend, in turn, on such factors as the amount of shade, the angle at which the sun's rays strike the surface of the habitat, the height of the sun above the horizon and the amount of cloud cover. In exposed, unshaded conditions where the influence of wind and rain may be a contributing factor, wider variations in temperature and relative humidity occur than in shaded conditions. This generalization is applicable both to the habitats located on the surface of tree trunks and branches, and to those on the surface of rocks. However, the former are more often shaded by foliage, the surface is less completely exposed to the effects of wind and water, and they offer an organic substratum on which epiphytic growths of moss and lichen can become established. The conditions for life are then less severe than in habitats developed on rocks, and it is not surprising to find a less isolated, less specialized fauna in the more sheltered sites. Travé (1963) could find only one species of cryptostigmatid mite, the xylophagous *Tetracondyla dorni*, which occurred only in the epiphytic habitats of tree trunks and branches, compared with 12 species which were restricted to rock habitats. Almost all of the species recorded in this work from arboreal habitats were also recorded from habitats associated with rock surfaces. Moreover, of the 28 species found in both habitat types, 12 (42%) were saxicole rather than arboricole, 8 (29%) were arboricole rather than saxicole, and a similar number were indifferent. These preferences, illustrated in Fig. 9.9, must not be interpreted incautiously for they may vary from one locality to another, particularly as microclimatic conditions vary. In England, the species *Carabodes labyrinthicus* is not restricted to rock habitats, and is often collected from foliaceous lichens growing epiphytically on tree trunks; similarly, *Camisia segnis* is frequent in arboreal habitats (Grandjean, 1936). However, these variations in distribution do not alter significantly the main conclusion formulated by Travé (1963), namely that the saxicole cryptostigmatid fauna is richer qualitatively and quantitatively than the arboricole fauna. This appears to be rather surprising, at first sight, in view of the fact that the arboreal habitats may be more amenable to microarthropods than rock surfaces, but it may reflect the high degree of adaptation of the saxicole fauna to its environment.

The effect of epiphytic vegetation. Provided that factors such as aspect and shade are equal, the microclimatic differences between epiphytic mosses and lichens are negligible. Nevertheless, clear faunistic differences often occur between these habitats. These differences may be a reflection of the fact that the two habitats differ in the kind of food material they provide for the saprophagous and phytophagous fauna. Cryptostigmatids such as *Pirnodus detectidens* and *Dometorina plantivaga* are noted for their lichen-feeding habit, and show a strong preference for, and sometimes a restriction to, the environments provided by lichen growths. Some degree of habitat separation is achieved in the case of these two species, since *P. detectidens* shows a strong preference for saxicole lichens, particularly *Pertusaria rupicola*, whereas *D. plantivaga* favours the arboricole lichens *Caloplaca ferruginea* and *Lecanora subfusca* (Travé, 1963). On the other hand, the distribution patterns of *D. plantivaga insularis* and *P. soyeri* in the Salvage Islands (Canaries) overlap to the extent that both species may occur in the same thallus of the lichen *Buellia subcanescens* (Travé, 1969). Other crypto-stigmatid mites associated with lichens include *Conoppia palmicinctum*, *Ommatocepheus ocellatus*, *Ameronothrus maculatus* and *Mycobates parmeliae* (Michael, 1883; Travé, 1963). In the case of the last-named species, the immature stages are restricted to lichens, whereas the adults feed also on mosses and liverworts, and have a distribution pattern ranging widely in habitats above ground. All of these species are adapted to life within the environment provided by lichen growths, for they are rarely, if ever, en-countered in the soil. Evidence of the kind of specialization involved here is provided by the burrowing habit of lichenophages, such as *Dometorina plantivaga* and *Ommatocepheus ocellatus*, the former being found, especially in the juvenile stage, in closed cavities in the thallus of crustose lichens where some protection against low relative humidities may be obtained. *O. ocellatus* is active only when the relative humidity reaches saturation; below this point, adults and immatures become completely immobile in the surface cavities of the lichen thallus which have been excavated by their feeding activities (Travé, 1963). Other species which require high relative humidities to trigger their normal activity are *Camisia segnis* and *Humero-bates rostrolamellatus*. Travé (1963) has described how the eggs of *C. segnis* are deposited within a protective coat in the thallus of the lichen, so that they are partly or completely covered.

However, many species of saxicole and arboricole Cryptostigmata are not restricted in their food preference to one particular vegetational substrate. Some of these, such as *Liodes theleproctus* and *Oribatula parisi*, maintain a preference for the lichen habitats, but others such as *Oppia ornata*, *Pseuda-chipteria magna*, *Parachipteria petiti* and *Chamobates* cf. *tricuspidatus* often choose moss in preference to lichens. To this list may also be added a number of species more closely associated with the soil, such as *Scutovertex* spp.,

Eupelops acromios, Ceratoppia bipilis, Tectocepheus spp., *Eremaeus* spp., *Zygoribatula* spp., *Phauloppia lucorum, Oribatula tibialis, Paraleius leotonycha* and *Metaleius strenzkei.* The presence of some, at least, of these species above the ground is probably due to their ecological plasticity, rather than their physiological specialization. This component also includes species which undertake daily vertical migrations to and from the soil, and do not qualify as true saxicoles or arboricoles.

Macrofungi

To complete this account of habitats above the ground we must briefly draw attention to the fauna associated with the large fruiting bodies of fungi often found growing on decaying wood. This habitat would generally be considered to belong to the first of the three types defined earlier in this chapter, and is often best represented by bracket fungi (Polyporaceae) and toadstools (Agaricaceae), which support a large and varied invertebrate fauna, particularly when the fruiting bodies start to decay. Prominent among this fauna are certain beetles, flies, slugs, mites and Collembola. This rich fauna can be regarded as consisting of various ecological types. Without going into too much detail for this topic has been dealt with by Elton (1966), a distinction can be made immediately between the species which breed and feed during the whole of the life cycle in this habitat, and those species which visit the fruiting bodies either to breed or feed, but not both. To the former category belong staphylinid beetles of the genus *Gyrophaena* which breed in bracket fungi and feed on the spores (Donisthorpe, 1935), and members of the beetle family Ciidae which represent an important destructive influence on the Polyporaceae (Paviour-Smith, 1968). These permanent inhabitants often form a close association with one species of fungus or with just a limited number of such hosts. Paviour-Smith (1960) defined two main groups of fungi, each with its own association of Ciidae species, such that a given species of beetle may breed in several species of fungi within its group, and one particular fungus may contain several different beetles breeding in it. These particular associations may be determined, to some extent at least, by different preferences of beetles for certain physical characteristics of the fruiting bodies. For example, the beetles *Octotemnus glabriculus* and *Cis boleti* are regular inhabitants of the rather stringy and tough fruiting bodies of *Polystictus versicolor*, whereas *Cis bilamellatus* and *C. nitidus* show a preference for more brittle Polyporaceae, the latter beetle choosing the woody *Ganoderma applanatum*, while the former selects the softer, spongy fruiting bodies of *Polyporus betulinus*.

Temporary inhabitants of these large fungi include flies belonging to the family Mycetophilidae (fungus gnats) which breed in this habitat, and a whole host of other insects, including carrion flies (*Lucilia*), bluebottles (*Calliphora*), dung flies and staphylinid beetles, together with slugs which are often very common on members of the Agaricaceae. Much still remains to be learned about the ecology of many of these groups, although it is likely that some of the flies, staphylinid beetles and slugs breed in soil, litter or decaying wood, and invade the fungi in search of food. As the fruiting body decays, some of these visitors may prolong their stay and may even breed in the rotting pulp. In any event, this community has strong links with the fauna of the ground, decaying wood and aerial vegetation, and in woodland has an important place in the overall ecological pattern.

Summary

This chapter deals with the faunal characteristics of a range of microhabitats which are physically distinct from, but have biological links with, the soil/litter subsystem. These are the microhabitats provided by decaying tree stumps, rotting logs, rocks and living trees, and macrofungi.

The extent to which the soil fauna participates in the decomposition of tree stumps depends on the type of decay and the moisture content of the wood. This participation is maximal in stumps undergoing moist decomposition, minimal in dry stumps and those undergoing the wet type of decomposition known as "white rot", in which fungi predominate over bacteria.

During moist decomposition of tree stumps, a faunal succession occurs, with soil animals becoming increasingly associated with later stages in the decay process. This is accompanied by a progressive increase in species diversity and an ecological shift from wood-feeding forms to saprophagous and fungivorous species.

An important microhabitat in woodlands and forests is provided by decaying logs lying on the surface of the leaf litter. The fauna of these comprises two components, namely a transient element which utilizes the logs as seasonal refuges, and a permanent fauna which is represented by animals which have extended their range of distribution from the soil and litter to these habitats above ground. There is a considerable amount of seasonal movement of animals between logs and the surrounding litter, and this activity may have two different functions. It may represent an attempt, on the part of the animals concerned, to maintain themselves in relatively constant microclimatic conditions. If the interior of a log is colder in winter and warmer in summer than the surrounding litter, then animals which move into logs during spring and back into the litter during autumn will avoid extreme temperature changes. On the other hand, movement into the logs may be associated with a particular event or stage in the life cycle. Breeding occurs here in the case of the millipede *Cylindroiulus punctatus* and some juvenile cryptostigmatid mites can satisfy specific nutritional requirements here. In these cases, the animals are selecting a microhabitat which has different characteristics from that in which they reside at other times of the year.

The permanent fauna of decaying logs, as illustrated by cryptostigmatid mites, consists of true wood-boring species, an associated fauna which may exhibit commensalism, and a predatory component. The microdistribution patterns of the predators are most closely related to those of the "associates", whereas the microdistribution of the wood-borers is a function of the kind of substratum. In yellow birch twigs, for example, wood-borers achieve peak densities in heartwood, whereas in hemlock twigs they are located mainly in the bark.

Mites and Collembola usually form an important part of the permanent fauna of decaying wood, although the numerically dominant species vary from one log to another. Factors which influence the species composition of this fauna include the nutrient quality of the wood, the time of year and probably the moisture content of the wood.

The spaces beneath stones and logs lying on the forest floor provide a discrete microhabitat with a distinctive fauna consisting of permanent and temporary elements. This fauna is termed the "cryptozoa" and the niche which it occupies is one of a spectrum extending from the soil and litter, at one extreme, to decaying wood, rocks and trees, at the other.

Growths of moss and lichens on rocks and trees are foci for the accumulation of decomposing organic material which are termed "suspended soils". The fauna of these microhabitats is generally less diverse and more specialized than that of the soil and litter. Among the mites and Collembola there are species which are highly adapted to withstand temperature and moisture extremes; examples can also be found among this fauna of extreme specialization and restriction to one kind of food such as certain species of lichens. However, the fauna also contains a group of species which ranges rather widely in habitats above the ground, together with an element which is clearly derived from the soil and litter fauna. This latter element also comes into prominence in the later stages of decay of the fruiting bodies of the larger fungi which are associated with decomposing wood.

REFERENCES

ATALLA, E. A. R. and HOBART, J. (1964). The survival of some soil mites at different humidities and their reaction to humidity gradients, *Ent. exp. appl.*, **7**, 215–28.

BANERJEE, B. (1967). Seasonal changes in the distribution of the millipede *Cylindroiulus punctatus* (Leach) in decaying logs and soil, *J. Anim. Ecol.*, **36**, 171–7.

CASSAGNAU, P. (1961). Écologie du sol dans les Pyrénées centrales, *Actualités Scientifiques et Industrielles*, 1283, Hermann, Paris.

COLE, L. C. (1946). A study of the Cryptozoa of an Illinois woodland, *Ecol. Monogr.*, **16**, 51–86.

DENDY, A. (1895). The cryptozoic fauna of Australasia, *Rep. Aust. Assoc. Adv. Sci.*, **6**, 99–119.

DONISTHORPE, H. (1935). The British fungicolous Coleoptera, *Ent. mon. Mag.*, **71**, 21–31.

ELLIS, W. N. (1974). Ecology of epigeic Collembola in the Netherlands, *Pedobiologia*, **14**, 232–37.

ELTON, C. S. (1966). "The Pattern of Animal Communities", Methuen, London.

FAGER, E. W. (1968). The community of invertebrates in decaying oak wood, *J. Anim. Ecol.*, **37**, 121–42.

GRANDJEAN, F. (1936). Les Oribates de Jean Frédéric Hermann et de son père, *Ann. Soc. ent. Fr.*, **105**, 27–110.

KLIMA, J. (1959). Die Zönosen der Oribatiden in der Umgebung von Innsbruck, *Natura tirol.*, 197–208.

KNÜLLE, W. (1957). Die Verteilung der Acari, Oribatei im Boden, *Z. Morph. Okol.*, **46**, 397–432.

KÜHNELT, W. (1961). "Soil Biology with Special Reference to the Animal Kingdom" (English edn., trans. Walker, N.), Faber, London.

LLOYD, M. (1963). Numerical observations on movements of animals between beech litter and fallen branches, *J. Anim. Ecol.*, **32**, 157–63.

MADGE, D. S. (1964). The humidity reactions of oribatid mites, *Acarologia*, **6**, 566–91.

MICHAEL, A. D. (1883). "British Oribatidae", Ray Society, London.

PARKIN, E. A. (1940). The digestive enzymes of some wood-boring beetle larvae, *J. exp. Biol.*, **17**, 364–77.

PAVIOUR-SMITH, K. (1960). The fruiting bodies of macrofungi as habitats for beetles of the family Ciidae (Coleoptera), *Oikos*, **11**, 43–71.

PAVIOUR-SMITH, K. (1968). A population study of *Cis bilamellatus* Wood (Coleoptera, Ciidae), *J. Anim. Ecol.*, **37**, 205–28.

PENNEY, M. M. (1967). Studies on the ecology of *Feronia oblongopunctata* (F.) (Coleoptera, Carabidae), *Trans. Soc. Brit. Ent.*, **17**, 129–39.

PSCHORN-WALCHER, H. and GUNHOLD, P. (1957). Zür Kenntnis der Tiergemeinschaft in Moos- und Flechtenrasen an Park- und Waldbäumen, *Z. Morph. Okol.*, **46**, 342–54.

SAVELY, H. E. (1939). Ecological relations of certain animals in dead pine and oak logs, *Ecol. Monogr.*, **9**, 321–85.

STRENZKE, K. (1952). Untersuchungen über die Tiergemeinschaften des Bodens: Die Oribatiden und ihre Synusien in den Böden Norddeutschlands, *Zoologica*, **37**, 1–167.

TRAVÉ, J. (1963). Écologie et biologie des Oribates (Acariens) saxicoles et arboricoles, *Vie et Milieu*, suppl. **14**, 1–267.

TRAVÉ, J. (1969). Sur le peuplement des Lichens crustacés des Iles Salvages par les Oribates (Acariens), *Rev. Ecol. Biol. Sol*, **6**, 239–48.

WALLACE, H. R. (1953). The ecology of the insect fauna of pine stumps, *J. Anim. Ecol.*, **22**, 154–71.

WALLWORK, J. A. (1957). "The Acarina of a Hemlock-Yellow Birch Forest Floor". Ph.D. Thesis, University of Michigan.

Chapter 10

The Soil Fauna of Deserts: Hot Deserts

The generally accepted definition of the term "desert" is that of an area in which the annual evapotranspiration exceeds rainfall. Although it is common to think of a desert as a hot, barren, sandy wasteland, this definition also embraces the so-called "cool" deserts, such as the Great Basin Desert of North America and the Gobi Desert of Asia, and may even be extended to include the cold Antarctic continent which is treated in the next chapter. However, most of the world's desert areas are contained within two broad bands lying above and below the equator, respectively, in the subtropical zones. Deserts usually develop in the interior of large land masses where they are isolated from the wetter coastal regions by high mountain ranges.

It is in these desert habitats, cold, cool and hot, that biological activity and productivity reach their lowest levels. Low rainfall and extremes of

Fig. 10.1 Low plateau, southern Mojave desert dominated by sagebrush (*Artemisia*) (foreground). (Photo. by author.)

Fig. 10.2 High plateau, southern Mojave desert dominated by the Joshua Tree (*Yucca brevifolia*) and juniper. (Photo. by author.)

Fig. 10.3 Desert near San Felipe, Baja California, Mexico, dominated by the coachwhip, or ocotillo (*Fouquieria splendens*). (Photo. by author.)

temperature limit the growing season of plants and the activities of animals to a short period of the year. Some desert animals, such as birds and most large mammals, avoid such unfavourable conditions by migrating to other localities where the environmental effects are less severe. However, many deserts are far from sterile, biologically, and their inhabitants comprise highly specialized plants and animals. It is true that in some regions, such as the Sahara and the Simpson Desert of central Australia, where rainfall is low and lacking altogether in some years, plant and animal life is hard to find. However, even here, the seeds of plants and eggs of crustaceans can lie dormant in the dry soil for years, awaiting the next rains. In other deserts, an annual rainfall of up to 15 cm allows the establishment of a variety of perennial plants, such as the sage (*Artemisia*), saltbush (*Atriplex*), creosote bush (*Larrea*), acacias, euphorbias, yuccas, agaves and a wide range

Fig. 10.4 Desert near Tucson, Arizona, dominated by the giant saguaro cactus (*Carnegia gigantea*). (Photo. by author.)

of cacti and succulents (Figs. 10. 1–4). This vegetation provides food and refuge for a diverse assemblage of animals, and association between plant and animal, often for mutual benefit, is a marked feature of desert life. Here, the struggle for survival is directed mainly against the physical and climatic influences of the environment; ecosystems are relatively simple in structure and limiting factors often clearly expressed.

The desert soil environment

The soil is of vital importance to many members of the desert fauna although, at first sight, the environment it provides would appear to be less than hospitable. Desert soils are usually sandy in texture, formed from the climatic weathering of quartzitic rocks. They are unstable in the high winds that often prevail in desert regions, partly because of their sandy nature and partly because the ground vegetation does not form a complete, protective cover. Very little organic material becomes incorporated into the soil profile which, as a result, maintains a mineral character (Fig. 10.5).

However, the microclimate below the surface of the desert soil is much less extreme than that at, or above, the surface. Measurements taken above and below the ground in hot desert regions show that temperature profiles

Fig. 10.5 Profile, cut to a depth of about 1 m, of a desert soil near Palm Springs, California. (Photo. by author.) Note the mineral character and lack of structure in the profile.

exhibit a marked decline below the soil surface, and this trend is particularly evident under the shade of vegetation (Wallwork, 1972). Moreover, the subterranean root systems of desert plants serve to concentrate the available moisture in what has been called the "leaf catchment area" (see Cloudsley-Thompson and Chadwick, 1964). At these sites are frequently located the relatively cool and moist burrow systems and nest chambers of desert vertebrates, such as kangaroo rats (*Dipodomys*), pocket mice and gopher tortoises. These animals forage for food on the soil surface during the cooler periods of the 24-hour cycle and withdraw to their soil refuges at hotter times of the day. The importance of the soil as a habitat for animal life, both vertebrate and invertebrate, is reflected in the fact that there is a strong selection for animal groups that are predominantly soil-dwelling in habit in hot deserts.

In some hot deserts, other microhabitats suitable for colonization by soil animals are provided by accumulations of organic material under large shrubs and trees and in rock crevices. Juniper bushes and piñon pine occur at higher elevations (> 1230 m) in the Mojave Desert of southern California, and provide a dense, if localized, shade effect. Within this shaded area, the fallen needles are matted together to form a stable litter layer which lies directly on the mineral soil. In other parts of this desert, where oases develop around springs and pools, rich growths of mosses, horsetails and grasses decompose to form a black, saturated, peat-like soil.

The range of microhabitats available to soil animals in hot deserts is much more restricted than that of the other habitat types discussed in previous chapters. The fauna that occurs here is mainly composed of xerophiles, species which are adapted in some way to survive in dry conditions. We can now consider the general characteristics of this fauna.

General characteristics of the desert soil fauna

The distinction between surface dwellers and subterranean forms largely disappears in hot deserts, since many of the commoner members of the fauna have their nests and resting areas in the soil but forage for food on the surface. However, a distinction can be made between animals which are capable of burrowing in the soil and those which cannot. A further distinction can be made between animals which are active during the day and those which are active at night. These two characteristics can be combined to provide a simple classification of the desert soil fauna:

Burrowers
    a. Surface active forms.
        (i) Day-active e.g. ants

   (ii) Nocturnal, e.g. small mammals, scorpions
 b. Subterranean forms, e.g. certain insects and reptiles
Non-burrowers
  a. Surface active forms, e.g. reptiles and beetles
  b. Subterranean forms, e.g. microarthropods.

Animals residing permanently in hot deserts may be adapted morphologically, behaviourally or phenologically to avoid drought and the extremes of temperature that prevail for most of the year. They may also be adapted physiologically to tolerate these conditions, and in many cases the survival of a species depends on a combination of these adaptations. The water

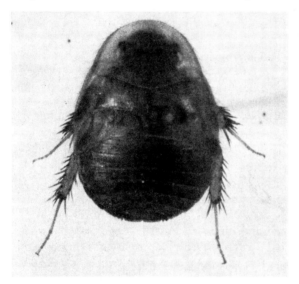

Fig. 10.6 The desert cockroach *Arenivaga* (nymph) from Baja California, Mexico. (Photo. by G. Ott.) With its flattened body and spiny legs, this animal is superbly adapted for burrowing in soft sand.

balance of desert animals such as the kangaroo rat and the desert cockroach *Arenivaga* (Fig. 10.6) is delicately poised and depends not only on the ability of these animals to restrict water loss from the body by morphological and physiological means, but also on their ability to avoid, by burrowing in the desert soil, the environmental conditions which would place them under stress.

   The environment of the hot desert exerts a strong selective pressure on the fauna, favouring burrowing species and those with efficient water budgets. In many deserts, such selection has brought into prominence three faunal groups, namely small mammals (especially rodents), reptiles and arthropods. The reptiles and the arthropods are pre-eminently suited to dry environments by virtue of their uricotelic method of excretion, and by their ability to curtail water loss from the body surface. The success of small mammals, on the other hand, depends on a combination of behaviour patterns and an ability to utilize the little water that is available in a very efficient manner. These various survival mechanisms can now be examined in more detail.

Mechanisms of survival

As we have already noted, there are two main categories of survival mechanisms available for the desert soil fauna, namely those of avoidance and those of tolerance.

Avoidance: The Burrowing Habit

The burrowing habit is by no means restricted to desert animals, but this ability is an obvious advantage in the desert environment where it provides a means of escape from the surface to the more suitable subterranean environment. Such burrowing animals include, among the vertebrates, the North African toad *Bufo mauritanicus*, the gopher tortoise *Testudo polyphemus*, many lizards including the so-called "horned toad" *Phrynosoma* and the iguana *Dipsosaurus*, and many small mammals such as ground squirrels, gerbils and kangaroo rats, together with some larger forms, for example the desert fox *Fennecus zerda*. The burrowing invertebrate fauna is equally varied and includes, among others, isopod crustaceans such as *Hemilepistus* and *Venezillo*, scorpions, solifugids and spiders. Insects are represented by burrowing ants, crickets, myrmeleonid "ant-lions", cicadas, termites, certain bees and wasps, and scarabaeid beetles, in addition to the desert cockroach already mentioned.

Avoidance: Activity Patterns

Avoidance of surface drought and temperature extremes is also achieved by behaviour patterns which restrict the activity of desert animals to periods of the 24-hour cycle when environmental conditions are least severe. Tenebrionid beetles are markedly nocturnal, as are many of the desert rodents which feed on seeds and other plant material. The nocturnalism of predatory forms, such as the "sidewinder" snake *Crotalus cerastes* (Fig. 10.7) and the scorpions, is probably more closely related to the activity rhythms of the prey than to changes in environmental conditions, for these animals are well adapted, physiologically, to survive in the harsh extremes of the desert.

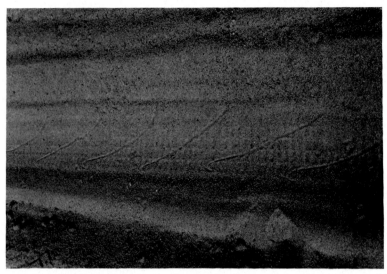

Fig. 10.7 Tracks of the sidewinder rattlesnake *Crotalus cerastes* made during a nightly foray in the desert near San Felipe, Baja California, Mexico. (Photo. by author.)

However, some desert animals such as sand lizards and harvester ants are active during the daytime, although this activity is more pronounced in the early and late parts of the day than at the mid-day period. Harvester ants observed in the desert of the northern part of the Mexican province of Baja California were active at the surface during the early part of the day, but retired underground just before noon when the temperature of the soil surface rose above 30°C. Other diurnal forms, such as beetles of the genus *Meloe* and the tenebrionid *Stenocara phalangium* have extremely long legs, and it has been suggested (Cloudsley-Thompson and Chadwick, 1964) that these insects are able to maintain their activity patterns since their bodies are raised high enough above the hot sand surface. Lizards can often spend a considerable part of the day above ground, sheltering from the sun under vegetation, or avoiding high soil surface temperatures by climbing into the aerial parts of shrubs. The North American desert iguana, *Dipsosaurus dorsalis*, frequently occurs in association with the creosote bush, *Larrea divaricata*. Another lizard, the small *Uta*, seeks protection among the fallen stems of the Joshua Tree, *Yucca brevifolia*, in the Californian Mojave Desert. When disturbed, sand lizards move very quickly across the surface of the soil, aided by their flattened feet, fringed with scales, which enable them to obtain secure purchase on the soft sand.

Some desert animals avoid unfavourable conditions by becoming quiescent. The presence of a diapausing pupal stage in the life cycles of Diptera,

Coleoptera and Lepidoptera provides an excellent mechanism for survival during unfavourable conditions. At least some of these insects show a remarkable facility for seeking moist oviposition sites in an apparently arid environment. In the Mojave Desert, for example, active Diptera larvae have been taken from the moist, pulpy wood at the base of old flowering stalks of the agave *Nolina* during the driest part of the year. On the other hand, nematodes and certain molluscs evidently tolerate hot, dry conditions by passing into a state of anabiosis. The production of resistant cysts or eggs is a feature of the protozoans, nematodes and crustaceans inhabiting temporary pools.

Tolerance

In addition to these avoiding mechanisms, the survival of many desert species also depends on the ability to tolerate extreme environmental conditions by physiological means. Such adaptations are obviously more important in large mammals, such as the camel, than in the smaller forms which

Table 10.1

Transpiration of arthropods from various habitats measured in dry air at temperatures between 20°C and 30°C. The values are given as micrograms/cm²/hr/mm Hg (from Edney, 1967, © 1967 by the American Association for the Advancement of Science).

Species	Habitat	Transpiration
Isopod crustaceans		
Porcellio scaber	Hygric	110
Venezillo arizonicus	Xeric	32
Hemilepistus reaumuri	Xeric	23
Insects		
Blatta orientalis	Mesic	48
Glossina palpalis Ads.	Mesic	12
G. morsitans Pupae	Xeric	0·3
Tenebrio molitor Larv.	Xeric	5
T. molitor Pupae	Xeric	1
Thermobia domestica Ads.	Xeric	15
Myriapods		
Lithobius sp.	Hygric	270
Glomeris marginata	Hygric	200
Arachnids		
Pandinus imperator	Mesic	82
Androctonus australis	Xeric	0·8
Galeodes arabs	Xeric	6·6
Ixodes ricinus	Mesic	60
Ornithodorus moubata	Xeric	4

can seek refuge in the soil. Nevertheless, many soil-dwellers are adapted physiologically to the low water content of the desert environment, and great emphasis is placed on water conservation. As we have already mentioned, reptiles and insects are particularly suited in this respect for they are mainly uricotelic, and uric acid needs little water for its elimination from the body. In the case of ureotelic small mammals, such as the kangaroo rat *Dipodomys*, the urine is extremely hypertonic and is concentrated with an efficiency that is five times greater than that in Man (Schmidt-Nielsen, 1964), with the result that these animals can rely almost entirely on metabolic water for their requirements.

There is no evidence that desert-dwelling insects can tolerate a lower body water content than their counterparts in more moist surroundings, but mechanisms for reducing water loss are undoubtedly well developed. In particular, the cuticle is relatively impermeable to the passage of water from the interior of the animal to the external environment; in other words the basic transpiration is low (Table 10.1). This physiological characteristic is often re-inforced by morphological and behavioural adaptations. For example, Warburg (1968) has pointed to the ability of certain isopods (woodlice) to conglobate, i.e. roll into a ball like pill millipedes (Fig. 10.8). This behaviour is shown by such desert isopods as *Armadillo officinalis*, *Venezillo arizonicus* and *Armadillo albomarginatus*. As Fig. 10.9 shows, this ability is not universal in desert woodlice, nor is it confined to the inhabitants of arid

Fig. 10.8 Conglobation in the pill millipede *Glomeris marginata* (Villers). (Photo. by A. Newman-Coburn.)

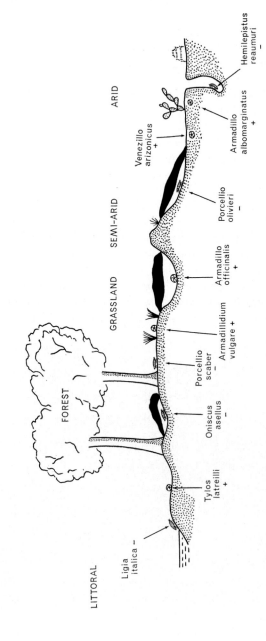

Fig. 10.9 The distribution of various isopods in terrestrial habitats (+ conglobating species; − non-conglobating species). (From Warburg, 1968.)

regions. The ability to conglobate does appear to be related, however, to the mode of life of the species concerned, for it is developed in those forms which live on or near the surface of exposed soil. Warburg (1965) has demonstrated that conglobating animals lose considerably less water than nonconglobating ones, owing to the fact that the curling of the body effectively closes up the ventral surface and curtails water loss from the respiratory organs located there.

The capacity for lowered transpiration must be well developed in desert insects that can survive for long periods without imbibing free water. An example of this is provided by the tenebrionid *Eleodes* sp., illustrated in Fig. 10.10, which can survive for at least two months in the laboratory on dry sand, without food or free water, and without showing any apparent ill effects. There is no evidence that such survival depends on the insect's

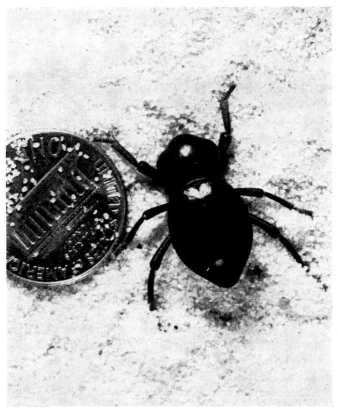

Fig. 10.10 The tenebrionid beetle, *Eleodes* sp., an inhabitant of desert and semi-desert soils in southern California. (Photo. by G. Ott.)

ability to increase the efficiency of oxidative metabolism. However, the possibility of active uptake of water from the surrounding atmosphere is suggested by Edney's (1967) studies on the desert sand roach *Arenivaga*, which can take up water from unsaturated air against a potential difference of several hundred atmospheres. Mechanisms of this kind may be important in those insects that are active in deserts all the year round. According to Leouffre's (1953) observations on the insects of the Sahara, the Tenebrionidae and Curculionidae are examples of this group.

The soil microfauna

So far we have been mainly concerned with the more conspicuous animals living in desert soils. This macrofauna has a very important place in the desert ecosystem and it is, perhaps, not surprising that the attention of ecologists and physiologists has been focused in this direction (Bodenheimer, 1953; Leouffre, 1953; Missone, 1959; Schmidt-Nielsen, 1964; Edney, 1967, 1968; MacMillen and Lee, 1967; Warburg, 1968). However, the microfauna also has a place in this system, particularly where organic matter accumulates in any quantity, although such sites have been the subject of very few investigations. Wood (1971) completed a wide survey of the mite, collembolan and crustacean fauna of a range of arid, semi-arid and sub-humid soils under a variety of vegetation types in Australia, and recorded density estimates for total microarthropods of the order of 2000–3000 per m² from desert sites. This author was able to describe a clear change in the faunal composition as the moist, sub-humid habitats gave way to increasingly arid semi-desert and desert. In the wettest parts of this sequence, mites outnumbered Collembola by about 2:1, whereas the numbers of these two groups were approximately equal in desert areas. Further, among the mites the ratio of Cryptostigmata to Prostigmata decreased with increasing aridity while, in this same sequence, the species *Folsomides deserticola* became predominant in the collembolan fauna.

Personal observations have also been made on the microarthropod fauna of juniper litter in the Joshua Tree National Monument, located in the southern part of the Californian Mojave Desert (Wallwork, 1972) (Figs. 10.1–2). This study extended over a period of 9 months in a region which has an average yearly temperature of 19·6°C, and an average yearly precipitation of 10·4 cm. The vegetation consists of low shrubs, such as *Cassia*, *Krameria*, the creosote bush *Larrea*, acacias, cacti such as *Opuntia* spp., *Echinocereus*, *Echinocactus* and *Mammillaria*, and agaves (*Agave*, *Nolina* and *Yucca* spp.). The Joshua Tree *Yucca brevifolia* is conspicuous in this area, and at higher elevations above 1230 m scattered juniper and piñon pine also occur. The soil is highly quartzitic and has a low organic content, and for the most part supports no detectable microfauna, although ants and carabid beetles, which can burrow down into the soil, are often abundant. However, under large shrubs and small trees, accumulations of litter occur. These serve not only to provide food for microarthropods, but also to ameliorate the temperature of the soil surface and to prevent moisture loss from the surface of the mineral soil. This effect is particularly striking when reinforced by the shade of the vegetation. The temperature of exposed mineral soil approaches a value which is normally intolerable to micro-arthropods, even during one of the coolest months of the year (during the summer, the temperature of the soil surface rises to well over 40°C). In the shade of juniper, the surface temperature may be more than 15°C lower, and here microarthropod populations are established and maintained throughout the year.

The juniper litter ecosystem is a relatively simple one and, at least in Joshua Tree, is dominated numerically by mites and Collembola and, to a lesser extent, Diptera larvae. In contrast to Wood's Australian study mentioned earlier, the Joshua Tree investigation revealed that mites, represented mainly by species belonging to the orders Cryptostigmata, Astigmata and Prostigmata, formed almost 90% of the microarthropod fauna, while the Collembola, represented mainly by two species belonging to the xerophilous genera *Folsomides* and *Xenylla*, comprised less than 10%. This preponder-ance of mites may be attributed to the high organic content of the juniper site. In more exposed areas where the organic content of the soil is low, mite populations, particularly those of the Cryptostigmata, are relatively small. In Joshua Tree, three species of Cryptostigmata, namely *Joshuella striata*, *Haplochthonius variabilis* and *Eremaeus magniporus*, contributed approxi-mately one-half of the total number of mites collected, the astigmatid mite *Glycyphagus* sp., which is probably phoretic as a juvenile on other arthro-pods, was responsible for one-third of the total, and the predatory prostig-matids *Spinibdella cronini* and *Speleorchestes* sp. about 10%. These various groups of mites and Collembola are noted for their ability to survive in dry conditions, and it is evident that we have, here, a highly specialized fauna, selected by, and highly adapted to, hot desert conditions.

From samples taken regularly over a period of 9 months in this juniper site, it was estimated that the total microarthropod density in litter and mineral soil was slightly in excess of 1500 individuals per m². This is slightly lower than Wood's Australian estimate (Table 10.2) and considerably less than that normally encountered in temperate soils. Further, although there is a tendency, in the desert environment, for a condensation or compression of the vertical faunal profile in the direction of the soil, the Joshua Tree study revealed that some vertical stratification of species populations occurred in the soil/litter sub-system. The bulk of the microarthropod fauna

Table 10.2

Comparison between the composition and abundance of the microarthropod fauna of desert soils in southern California and southern Australia.

Group	Average Density (nos./m²)		Number of Species	
	California	Australia	California	Australia
Total microarthropods	1632	2720	33	?
Cryptostigmata	720	150	10	8
Astigmata	435	0	1	0
Prostigmata	245	680	6?	11
Mesostigmata	20	220	1	4
Collembola	110	1370	3	6

(approximately two-thirds) was located in the 5–8 cm of mineral soil underlying the surface litter layer. This litter, which consisted of dry juniper needles, had a thickness of about 3 cm, and the rather wide fluctuations in temperature and moisture that occurred here may have been responsible for

Table 10.3

Densities and relative abundance of the main microarthropod groups in juniper litter and underlying mineral soil in the Joshua Tree National Monument (from Wallwork, 1972).

Group	Average Numbers per Sampling Unit ($\pm$ SE)		Relative Abundance (%)	
	Litter	Mineral	Litter	Mineral
Acari	8·8 $\pm$ 1·4	20·2 $\pm$ 2·8†	94·7	86·0
Collembola	0·1 $\pm$ 0·09	2·2 $\pm$ 0·8*	1·6	9·4
Pterygote Insecta	0·3 $\pm$ 0·1	1·0 $\pm$ 0·25†	3·7	4·3
Total microarthropods	9·2 $\pm$ 1·4	23·4 $\pm$ 3·0†	100·0	99·7

* Significant at $0·01 < P < 0·05$.   † Significant at $P < 0·01$.

the low faunal densities. The data provided in Table 10.3 illustrate this point. However, a detailed analysis of microdistribution patterns indicated that groups of species could be defined which were associated either with the litter or with the mineral soil.

A feature common to the microarthropod populations studied in the Mojave juniper site and those surveyed by Wood in Australian deserts is a period of peak density occurring during the winter months. In the Mojave, there is also a subsidiary peak in April, caused mainly by an increase in size of litter-dwelling populations. These peaks coincide with periods of recruitment, as a new generation of individuals is produced. The presence of two peaks, one in winter and the other in spring, is a reflection of the fact that certain species of mites reproduce mainly in winter, whereas others produce young later in the year (Fig. 10.11). In these respects, desert microarthropods behave in a similar way to their counterparts in cooler organic soils of the temperate zone.

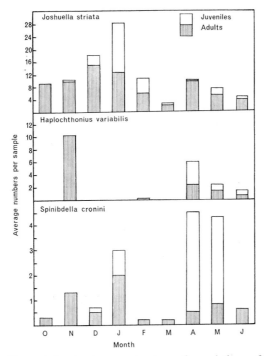

Fig. 10.11 Monthly variation in the age structure of populations of *Joshuella striata*, *Haplochthonius variabilis* (Cryptostigmata) and *Spinibdella cronini* (Prostigmata) in the desert soil of a Mojave juniper site. (From Wallwork, 1972.)

Concluding remarks

There is no doubt that the soil fauna of hot deserts is a very specialized one. Survival is achieved by a combination of morphological, physiological and behavioural adaptations, in which a premium is placed on the ability to restrict water loss from the body, and either to tolerate or avoid temperature extremes. Life cycles are also often geared to seasonal changes in climate in order to ensure that the appearance of a new generation coincides with the most favourable environmental conditions.

The very sharp separation of wet and dry seasons is an important feature of hot desert ecosystems. It determines the biological composition of the system largely through the degree of environmental stress that is imposed. It also influences, in a general way, the operational characteristics of the system since periods of intense biological activity alternate with periods of inactivity. In these respects, hot desert systems show some similarities with those of cold deserts and, since the latter will be considered in the next chapter, further discussion will be postponed until then.

REFERENCES

BODENHEIMER, F. S. (1953). Problems of animal ecology and physiology in deserts, *Desert Research. Spec. Publs Res Coun. Israel*, **2**, 205–29.

CLOUDSLEY-THOMPSON, J. L. and CHADWICK, M. J. (1964). "Life in Deserts", Foulis, London.

EDNEY, E. B. (1967). Water balance in desert arthropods, *Science, N.Y.*, **156**, 1059–66.

EDNEY, E. B. (1968). Transition from water to land in isopod crustaceans, *Am. Zool.*, **8**, 309–26.

LEOUFFRE, A. (1953). Phénologie des insectes du Sud-Oranais, *Desert Research. Spec. Publs Res. Coun. Israel.*, **2**, 325–31.

MACMILLEN, R. E. and LEE, A. K. (1967). Australian desert mice: independence of exogenous water, *Science, N.Y.*, **158**, 383–5.

MISSONE, X. (1959). Analyse zoogéographique des Mammifères de l'Iran, *Mem. Inst. r. Sci. nat. Belg.* (2me ser.) **59**, 1–157.

SCHMIDT-NIELSEN, K. (1964). "Desert Animals", Clarendon Press, Oxford.

WALLWORK, J. A. (1972). Distribution patterns and population dynamics of the micro-arthropods of a desert soil in southern California, *J. Anim. Ecol.*, **41**, 291–310.

WARBURG, M. R. (1965). Water relations and internal body temperature of isopods from mesic and xeric habitats, *Physiol. Zool.*, **38**, 99–109.

WARBURG, M. R. (1968). Behavioural adaptations of terrestrial isopods, *Am. Zool.*, **8**, 545–59.

WOOD, T. G. (1971). The distribution and abundance of *Folsomides deserticola* Wood (Collembola, Isotomidae) and other microarthropods in arid and semi-arid soils in southern Australia, with a note on nematode populations, *Pedobiologia*, **11**, 446–68.

Chapter 11

The Soil Fauna of Deserts: Cold Deserts

Outside the high alpine zones of mountains, the cold desert biome is found mainly in the Antarctic continent and its adjacent islands. Here, the ground is covered by a thick layer of ice the year round, and the mean monthly air temperature rarely rises above freezing and drops well below −20°C in winter.

In this cold polar desert, life is restricted, for the most part, to the coastal fringes of the continent where, during the short summer season, rocky ground may be exposed as snow and ice turn to meltwater. Even so, no truly terrestrial vertebrate (except for Man) has succeeded in becoming established here. The impoverished fauna consists of a few groups of invertebrates, mainly Protozoa, nematodes, collembolans and mites, which are associated with the vegetation of mosses, terrestrial algae and grasses. These animals are apparently cold-adapted, although little is known of their overwintering stages. As in the hot desert (Chapter 10), the terrestrial ecosystems of the Antarctic are relatively simple and, as such, their appeal to the ecologist is obvious. However, the fauna of Antarctica is of interest also to the zoogeographer, since considerable evidence is now accumulating to suggest that this continent may have formed a link with the other southern continents, and could have provided a dispersal route for plants and animals between, for example, South America and Australasia.

Climatic zones

The south polar region has been sub-divided into a number of zones, on the basis of their climatic characteristics (Holdgate, 1964). Listed in order, from south to north, these are:

(1) The continental zone; (2) The maritime, or oceanic, zone; (3) The sub-Antarctic zone; (4) The cold south temperate zone. The relationship between these various zones is shown in Fig. 11.1.

The continental zone comprises the bulk of the Antarctic continent proper, excluding parts of the Peninsula. In this zone, mean monthly air temperature never exceeds freezing point, and in winter drops to below

Fig. 11.1 Map of the south polar regions to show the distribution of the various zones.

−20°C. There is little precipitation and the vegetation consists mainly of lichens.

The oceanic, or maritime, zone includes the Peninsula region and the island groups of South Shetland, South Orkney, South Sandwich and Bouvet. Mean monthly air temperatures never exceed 1·5°C here, and for at least one month of the year the mean air temperature is above freezing. During winter, this temperature rarely drops below −10°C. Some precipitation occurs, mainly in the form of rain. Here, as in the continental zone, plant and animal activity is restricted to a very short summer season.

The sub-Antarctic zone, consisting of the islands of South Georgia, Prince Edward, Crozet, Kerguelen and Macquarie, is dominated by vascular plants which form a moorland type of vegetation. The mean monthly air temperature never exceeds 8°C and is above 5·4°C for less than seven months of the year. However, the mean is above freezing for at least half the year. These conditions are warm enough to allow an appreciable growth of plants and animals in an environment which is reminiscent of the Arctic tundra.

North of the sub-Antarctic is the cold south temperate zone where woody vegetation occurs, at least in lowland areas. Here, the mean monthly air temperature never falls below freezing, and is above 5·4°C for more than seven months. This temperature is above 8·5°C for at least two months of the year. This zone and that of the sub-Antarctic cannot be considered as true cold deserts, but they are included in this account for a number of reasons. Firstly, as one proceeds northwards from the South Pole, community diversity increases progressively, and it is logical to follow this increase into the warmer fringes of the south polar region. Secondly, although the four zones have a distribution which is broadly circum-polar, and each is distinct in a geographical sense, plant and animal distributions sometimes cross these boundaries. Geographical and biological zonations do not always coincide. Thirdly, the interpretation of zoogeographical distribution patterns in the continental zone cannot be achieved by considering this zone in isolation from the neighbouring zones to the north which may provide source areas for certain species of plants and animals.

Vegetation and microclimate

As we have already noted, a moorland type of vegetation predominates in the sub-Antarctic zone. Further south, particularly in the Peninsula region of the maritime zone, two main formations have been described, namely a cryptogam formation and a phanerogam formation.

The cryptogam formation can be divided into about half a dozen sub-formations of which the *lichen and moss cushion* sub-formation is the most varied and most extensive (Longton, 1967). This type is found on dry substrata, particularly cliff faces and scree slopes. Among the lichens are the brightly coloured (yellow and orange) *Caloplaca* and *Xanthoria* species; common mosses in this type are species of *Pohlia*, *Bryum*, *Drepanocladus* and *Tortula*. A *moss turf* sub-formation also occurs, consisting of densely packed clumps of *Polytrichum* and *Dicranum* species, the former predominating on dry slopes, the latter on wetter sites. On many wet sites, however, a *moss carpet* sub-formation is usually present, often covering a thin layer of peat. The common mosses of this type are *Drepanocladus uncinatus* and species of *Brachythecium* and *Acrocladium*. In wet sites where there is some running water during the growing season, the moss carpet is often broken by a *moss hummock* sub-formation. Tall moss species, such as *Bryum algens*, *Tortula* and tall *Drepanocladus* species, comprise most of this type.

The phanerogam formation is largely confined to drier sites, particularly on north-eastern and north-western facing slopes. The two main vascular plants present are *Deschampsia antarctica* and *Colobanthus quitensis*. The southern limit of this formation occurs in the Peninsula region.

In the continental zone, the diversity of the macroscopic vegetation is considerably reduced. According to Rudolph (1963), three components can be identified in the vegetation at Hallett Station, Victoria Land, namely the green alga *Prasiola crispa* which grows as a sheet over rock surfaces, a moss component (*Bryum argenteum*) and several crustose and foliose lichens. Apparently the distribution of this vegetation is determined mainly by the availability of water in summer and presence of nitrogenous debris associated with nesting sites. Even so, the total amount of cover was recorded as only 15·4%. The moss component predominates in wet sites, whereas the lichen component is typical of drier sites where the water level does not reach the surface (Janetschek, 1967).

The presence of a vegetation cover, such as that provided by a moss mat, has a profound effect on the microclimate to which the soil fauna is exposed.

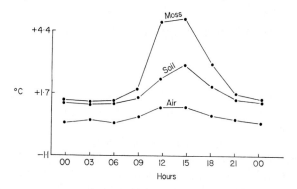

Fig. 11.2 The mean temperatures of air, soil (at 2·5 cm) and moss (at 2·5 cm) on Signy Island from 16 January to 31 March 1962. (From Holdgate, 1964.)

Direct radiation on to the ground and vegetation produces a warming effect which raises the temperature well above that of the air. This effect is illustrated by the data presented in Fig. 11.2. Here, mean temperatures of air, the mineral soil of a gravel patch, and of a mat of *Polytrichum-Dicranum* moss on the maritime Signy Island are monitored over a 24-hour period. It is obvious that the moss microhabitat provides a much more favourable summer microclimate than does the mineral soil, and that both of these are warmer than the air. As we will see later in this chapter, the moss microhabitat is favoured by many members of the invertebrate fauna, and high population densities occur here.

The microclimate at ground level in the continental zone is subject to a considerable amount of variability and is influenced by such factors as altitude, micro-relief, the presence of free water, snow cover and, of course, vegetation cover. Janetschek (1967) carried out a series of detailed temperature measurements of a number of sites in south Victoria Land, recording air temperatures 1 m above the ground, the temperature of the soil surface, of the cavities beneath stones, and at various depths in the soil. He was able to conclude that, at a given time of day, the air temperature was distinctly lower than that of the ground surface, and that this difference increased with altitude. Temperature inversions occurred occasionally, but these were mainly restricted to very exposed sites particularly where a fresh snow cover had developed. Maximum fluctuations in temperatures over a 24-hour period occurred at the surface of dry barren soils; in more moist soils, the amplitude of the temperature wave was reduced. From the point of view of the soil fauna, the most important features here are: (1) that the temperature at the ground surface may be well above freezing while the air temperature is as low as −10°C, and (2) moist soils not only provide suitable humidity conditions but also have less extreme temperature fluctuations than do dry soils. Thus, the maximum concentration of animal life can be expected to occur in the upper few centimetres of moist soils, and this has been found to be the case.

Antarctic soils and their fauna

Within the continental and maritime zones, the truly terrestrial fauna is essentially invertebrate in character. The vertebrates that are encountered on land, such as seals, penguins, petrels, shearwaters and albatross, belong to the marine fauna, for they obtain their food from this source. However, it must be borne in mind that sea birds play an important role in transporting large quantities of organic material from the Antarctic ocean to colonial nesting sites on the land. The fish and planktonic organisms which abound in the polar seas provide a rich food supply for gulls, petrels, penguins and terns, and the faecal material these birds deposit on land supports a variety of terrestrial invertebrate populations. In this hostile environment, soil formation occurs virtually under abiotic conditions and is restricted generally to ice-free areas (Ugolini, 1970). In the maritime zone, four groups of soils can be recognized, namely (1) mineral soils, embryonic in character, ranging from lithosols to fine glacial debris, (2) organic soils, which may be sub-divided further into thick deposits of peat underlying moss growths of *Polytrichum* and *Dicranum*, and thinner protoranker deposits which are less acid than the deep moss peats, (3) brown earths associated with growths of *Deschampsia* and *Colobanthus*, and (4) deposits of excreta, rich in nitrogen, calcium and phosphate, produced by marine birds and mammals (Allen and Heal, 1970).

The main groups of invertebrates represented in the Antarctic are the Protozoa, Rotifera, Nematoda, Tardigrada and Arthropoda. All of these, except the last, are essentially members of the soil-water fauna, but they have the ability to produce encysted stages during periods of drought and, in many cases, are widely distributed throughout the south polar region.

Among the protozoan fauna, members of the Testacea predominate in habitats where vegetation is present, and occur in densities which are comparable with those of temperate moorland and woodland (Heal, 1965).

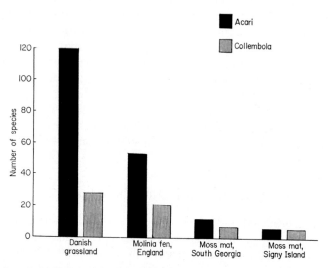

Fig. 11.3 Comparison of the number of species of Acari and Collembola in temperate and Antarctic localities. (From Tilbrook, 1967.)

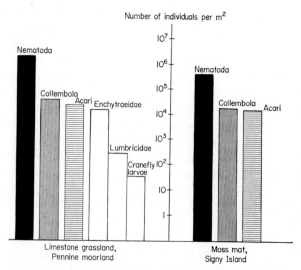

Fig. 11.4 Comparison between the densities of mesofaunal populations in temperate and Antarctic sites. (From Tilbrook, 1967.)

Some preliminary observations made on Signy Island in the maritime zone indicate that the composition of the protozoan fauna varies with the habitat. Testaceans, ciliates and small flagellates are the most conspicuous groups in moss peat, whereas the testaceans are replaced by naked amoebae in soil taken from a penguin colony. In moraine soil, the Protozoa are represented almost exclusively by flagellates and ciliates (see Tilbrook, 1970a).

Very little information is available about the ecology of rotifers and tardigrades in the south polar region. They have been recorded as members of the moss water fauna in the continental zone (Sudzuki, 1964) and Tilbrook (1970a) found them in the highest numbers in the wettest bryophytes, for example clumps of the moss *Pohlia nutans*, in the maritime zone. This particular moss appears to be a favoured habitat for the nematodes (Tilbrook, 1967), although these animals also build up population densities comparable with those of temperate soils in soils under *Deschampsia* and *Colobanthus* (see Fig. 11.4).

The arthropod fauna, which consists of Collembola and mites for the most part, is much poorer in terms of numbers of species than the microarthropod fauna of temperate soils, although population densities are of the same order of size (Figs. 11.3 and 4). The taxonomy of these microarthropods is better known than that of the other groups of Antarctic soil animals and, consequently, more details of their ecology are available. Much of what follows in this chapter is concerned with them.

Table 11.1

Numbers of families, genera and species, and percentage endemism (in parentheses) of free-living terrestrial microarthropods in the south polar region (from Wallwork, 1973).

Total numbers of	Cryptostigmata	Prostigmata	Collembola
Families	13 (0)	8 (0)	6 (0)
Genera	26 (34·6)	14 (?)	27 (18)
Species	41 (63·4)	41 (95·1)	47 (70·2)

Approximately 130 species of free-living terrestrial microarthropods have been recorded from the south polar region, with an almost equal representation of Collembola and mites belonging to the Orders Prostigmata and Cryptostigmata (Table 11.1). At the species level, this is a highly endemic fauna which has colonized microhabitats ranging from nests and faeces, through soils, terrestrial algae and lichens to mosses. In many of these sites, the Collembola and Prostigmata are numerically dominant, although cryptostigmatid mites may be locally abundant, particularly in shore microhabitats.

Community structure

During the last decade, a considerable amount of biological research has been carried out in Antarctica, and much of this has been published in a series of comprehensive works edited by Carrick *et al.* (1964), van Oye and van Mieghem (1965), Gressitt (1967) and Holdgate (1970). The following account is based partly on the work of contributors to these volumes. Here, we examine the terrestrial community structure of each of the major Antarctic zones, starting from the continental zone in the south and working northwards. Generally speaking, biotic diversity increases northwards and, at least in the sub-Antarctic, as we move from "pioneer" shore microhabitats, such as nests and faeces, into more terrestrial sites, such as soils, terrestrial algae and mosses.

Continental Zone

Within the continental zone, the habitats available to the terrestrial fauna are of two kinds: (1) Cavities under weathered rocks, stones and moraine deposits lacking any kind of vegetation cover; (2) Growths of moss and lichens.

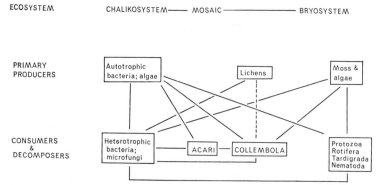

Fig. 11.5 The structure of terrestrial ecosystems in the continental zone of Antarctica. (From Janetschek, 1963.)

These two habitats have been designated the chalikosystem and bryosystem, respectively, by Janetschek (1963, 1967, 1970). Although detailed analyses of the community structure of these two systems have not been carried out, some preliminary data are available and these are summarized in Fig. 11.5. The chalikosystem, which probably represents one of the early stages in soil formation, extends farther inland (i.e. south) and also to a greater altitude (up to 2000 m) than the bryosystem which is more conspicuous in coastal areas. The former system is colonized by a soil-water fauna of bdelloid rotifers, tardigrades and rhizopod protozoans, together with the truly terrestrial collembolans *Antarcticinella monoculata* and *Neocryptopygus nivicola*, and the prostigmatid mites *Nanorchestes antarcticus* and *Stereotydeus mollis*. These forms probably subsist on bacteria, fungi and possibly algae, the only plant life recorded to date in this system. In the extremely barren chalikosystem soils, there is often only one arthropod species present, the mite *Nanorchestes antarcticus*, which has been recorded as far south as 85° 32′S.

The fauna of the bryosystem consists of a rather more varied assemblage, although its general character is similar to that of the chalikosystem fauna. Protozoans, nematodes, rotifers and tardigrades comprise the aquatic element. The microarthropods are represented by the collembolans *Isotoma klovstadi*, *Friesea grisea* and *Cryptopygus* sp., and about half a dozen species of prostigmatid mites. Some, at least, of these microarthropods are present in more exposed sites where organic debris accumulates, for example at the sites of penguin nests and carcasses. There is no sharp demarcation between the fauna of the chalikosystem and that of the bryosystem, particularly where the two systems occur side by side. Under such conditions, the collembolan *Gomphiocephalus hodgsoni* and the lichen-feeding cryptostigmatid mite *Maudheimia pretonia* occur. The latter has been recorded from the northern part of Victoria Land, and its close relative, *M. wilsoni*, from a comparable latitude on the other side of the continent in Queen Maud Land.

More precise definitions of these community systems must be based on a detailed knowledge of the tolerance of the various species to such environmental conditions as low temperatures and drought. From the little information that is available, it is apparent that microarthropod species differ in their cold-hardiness. Adults of the cryptostigmatid mite *Maudheimia wilsoni* can survive temperatures as low as −30°C under experimental conditions (Dalenius, 1965), and it has been suggested that normal locomotion and possibly breeding may occur in sub-zero temperature conditions. No information is available about the overwintering stages of this species, or their temperature tolerance, but it is possible that cold-resistance increases as winter approaches, possibly through the secretion of oil droplets. Experiments on the collembolan *Isotoma klovstadi* demonstrated a tolerance similar to that of *M. wilsoni* (Pryor, 1962). In contrast, the collembolan *Gomphiocephalus hodgsoni* is apparently less tolerant (Janetschek, 1963) and succumbs in temperature conditions between −20°C and −28°C, the optimal condition for normal activity being about 11°C. The upper lethal temperature for this Antarctic species (33°C) corresponds very closely with that of some temperate zone soil mites (Wallwork, 1960). However, it would be unwise to make too detailed comparisons on the basis of these preliminary observations, for experimental procedures have not yet been standardized. Also, little attention has been paid to the possibility that cold tolerance may vary within a given species from summer to winter, or that the results obtained may reflect not only responses to given temperature conditions, but also to relative humidity. There is some evidence for a relationship between species composition and the relative humidity of the soil; Wise *et al.* (1964) found that *Gomphiocephalus hodgsoni* inhabits rather drier sites than do prostigmatid mites.

Maritime Zone

In the more northerly maritime zone, the terrestrial community is more complex, and although the chalikosystem and bryosystem are still recognizable, they commonly form a mosaic with tufts of the mosses *Polytrichum, Drepanocladus, Pohlia, Brachythecium, Dicranum, Tortula* and *Bryum*, various lichens and liverworts, the green alga *Prasiola crispa* and the vascular plants *Deschampsia antarctica* and *Colobanthus quitensis* (Longton, 1967; Mackenzie Lamb, 1970; Gimingham and Lewis Smith, 1970).

General characteristics of the fauna. The arthropod fauna of these vegetational types consists of less than 10 species of Collembola, about 20 species of mites mostly belonging to the Prostigmata and Cryptostigmata, two chironomid Diptera, namely the wingless *Belgica antarctica* and the winged *Parochlus steineni*, the coleopteran *Cartodere apicalis* and, in the South Shetland Islands at least, occasional spiders, psocids and thrips (Tilbrook, 1967, 1970a; Strong, 1967; Covarrubias, 1965; Saiz *et al.*, 1968; Holdgate, 1967). Clearly the maritime zone supports a richer fauna than the continental zone, although some of the species recorded are only occasional finds, and may be accidental introductions. The dominant species in most of the habitats studied to date is the collembolan *Cryptopygus antarcticus* which, according to Tilbrook (1967), occurs in densities of over 200,000 per m² in the South Sandwich Islands. This particular collembolan is widely distributed in the maritime and sub-Antarctic zones. Tilbrook (1970b) studied some aspects of its biology and noted that apparently it could withstand low temperature conditions, for it was usually concentrated in the surface layer of moss mats. This author also suggested that *C. antarcticus* was probably a microphytic and detritus feeder, although its precise ecological status remains to be determined.

Also widespread in this zone are the prostigmatid mites *Nanorchestes antarcticus, Tydeus tilbrooki* and *Protereunetes minutus*. Samples of mites taken from clumps of *Polytrichum* and *Dicranum* on Argentine Island consisted mainly of *T. tilbrooki* and *P. minutus*, according to Tilbrook (1967) who recorded densities of the order of 50,000 per m². These density estimates compare very favourably with those for soil Collembola and mites of temperate regions, but perhaps such a comparison should be treated with some caution, for the Antarctic estimates are extrapolations from relatively small samples, and there is often an extremely wide variation in numbers from one sample to another. Similar reservations must be made about the other groups of invertebrates occurring in the maritime Antarctic, notably the protozoans, nematodes and tardigrades. These occur as members of the aquatic fauna in moss clumps throughout the world, and it is no surprise to find them abundant in Antarctica, particularly since many of them can encyst during unfavourable conditions. Tilbrook (1967) estimated nematode densities of about 1·2 million per m² on Signy Island in the South Orkney group, and tardigrades in the order of 800,000 per m² on the South Sandwich Islands. These minute forms are less conspicuous than the larger microarthropods, and since most of the quantitative information so far accumulated relates to the latter, it is with these that we are primarily concerned. These arthropods apparently feed mainly on fungi, algae, lichens, organic debris and, possibly, bacteria. Predators are relatively uncommon in this ecosystem, being represented, in many instances, by species of the prostigmatid mite genus *Rhagidia* and the mesostigmatid *Cyrtolaelaps racovitzai* which feed mainly on Collembola.

Distribution patterns. The horizontal distribution of the microarthropods is strongly aggregated, and appears to be influenced primarily by relative humidity and, only secondarily, by temperature conditions (Strong, 1967). There is very little evidence of any strong association between individual animal species and particular vegetation types. Commonly-occurring forms, such as the collembolan *Cryptopygus antarcticus* and the mites *Cyrtolaelaps racovitzai* and *Nanorchestes antarcticus*, are widespread among moss, lichen, algae, vascular plants and mineral soil (Tilbrook, 1967). There are, however, marked quantitative differences between the fauna of different vegetation

Table 11.2

Average densities and relative abundance of various microarthropods in different habitats in the Maritime Antarctic, based on samples of 1000 cc (from Covarrubias, 1965).

| | HABITAT | | | | | | | |
| | Moss | | Lichens | | Algae | | Grasses | |
Fauna	Density	Rel. abund.	Density	Rel. abund.	Density	Rel. abund.	Density	Rel. abund.
Acari	326·7	4·7	2197·9	87·5	36,287·6	98·3	2·7	0·3
Collembola	4899·1	95·2	331·2	12·5	681·9	1·9	1323·2	99·7
Diptera larvae	4·8	0·2	—	—	—	—	—	—
Psocoptera	—	—	0·2	0·01	—	—	—	—

types, and these are clearly shown by Covarrubias' (1965) study of the distribution of Collembola and mites on the South Shetland Islands and Peninsula region (Table 11.2). Collembola were most abundant in moss growths and grasses, where they comprised between 95% and 99·7% of the total arthropod fauna, whereas mites were more successful among terrestrial algae and lichens where they comprised 98% and 87%, respectively, of the total microarthropods. It is tempting to interpret these differences in horizontal distribution pattern in terms of differences in feeding habit. Indeed, the numerical abundance of mites in lichen and terrestrial algae may be a reflection of the fact that these plants can provide a source of food. However, although mosses apparently attract large populations of Collembola, it is doubtful if this vegetation provides an important source of nutriment.

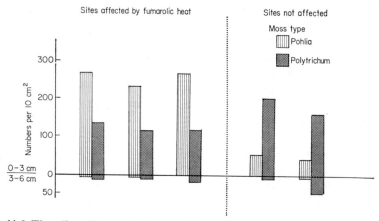

Fig. 11.6 The effect of fumarolic heat on the numbers of the collembolan *Cryptopygus antarcticus* on Candlemas Island, Maritime Antarctic. (From Tilbrook, 1967.)

Generally speaking, clumps of *Polytrichum* afford a more suitable habitat than the compact, waterlogged growths of *Pohlia*, although the latter is favoured on fumarolic sites (Fig. 11.6). This suggests that the physical appearance of the moss is less important in influencing collembolan distribution than are its microclimatic characteristics. On the other hand, growths of moss, in general, provide more favourable microhabitats than the underlying stratum, at least during the short summer season. Analyses of vertical distribution patterns through the profile of moss clumps have revealed that between 84% and 95% of the microarthropod fauna is located in the upper 3–5 cm of the vegetation (Covarrubias, 1965; Tilbrook, 1967; Schlatter *et al.*, 1968). This may well represent a selection for the more open texture of the moss compared with the compacted substratum beneath.

Shore habitats. Another major terrestrial ecosystem in the maritime Antarctic is provided by the habitats present in the marine littoral and lake shore systems. Such habitats often consist of loose stones sheltering organic debris, and the nesting sites of marine birds. The cryptostigmatid mites *Alaskozetes antarcticus*, *Halozetes belgicae* and *Magellozetes antarcticus* occur here and feed on organic material and lichens. The distribution of *Magellozetes antarcticus* is more strongly localized than that of the other two, for this species appears to be restricted to barren rock rubble. A similar restriction is shown by the collembolan *Archisotoma brucei* which occurs, for the most part, only among rocks and seaweed near the high-water mark on the shore. Accumulations of faecal material produced by marine birds and seals are often richly colonized by Collembola, whereas nesting material is more strongly favoured by mites. Microarthropod densities of more than 7,000 and 1,000, respectively, have been recorded by Covarrubias (1965) from nesting sites and faeces. In the former, more than 98% of this fauna consisted of mites, whereas in the latter an almost identical proportion was contributed by the Collembola.

Sub-Antarctic Zone

Northward, the fauna of the maritime Antarctic mingles with that of the sub-Antarctic zone to form a heterogeneous complex (Wallwork, 1966). Here, there is greater diversity and the food web is more complex (Fig. 11.7).

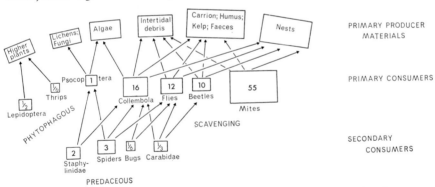

Fig. 11.7 Diagrammatic food web for an average sub-Antarctic island. Numbers in squares represent average numbers of species per isle. (After Gressitt, 1970.)

In the soils of South Georgia, for instance, Cryptostigmata belonging to the cosmopolitan genera *Oppia*, *Platynothrus*, *Edwardzetes*, *Liochthonius* and *Eobrachychthonius* are represented. They are joined by members of the Podacaridae, a family which occurs more frequently in Antarctica than anywhere else in the world, and is well represented by the genera *Podacarus*, *Alaskozetes*, *Halozetes* and *Antarcticola* in the marine littoral zone. Although

much more ecological work remains to be done on these forms, some preliminary observations have been made on their distribution patterns, and a
certain amount of habitat separation is evident. The principal habitats comprise clumps of moss, nesting materials of various sea birds, rock crevices
and grass tussocks, and the distribution of eight of the most abundant
species of cryptostigmatid mite in these habitats is indicated by the data
presented in Table 11.3. Two species groupings emerge from this analysis,

Table 11.3

Frequency of eight species of Cryptostigmata in various microhabitats on South Georgia*
(from Wallwork, 1972b).

Species	Total No. of Occurrences	Moss	Nests	Grass	Rock	Other	Total
Edwardzetes elongatus	87	*51·7*	*27·5*	1·1	14·9	4·6	99·8
Oppia crozetensis	52	*48·8*	*45·5*	1·9	3·8	—	100·0
Globoppia intermedia longiseta	40	*47·5*	*35·0*	5·0	12·5	—	100·0
Antarcticola georgiae	50	*56·0*	*30·0*	—	12·0	2·0	100·0
Podacarus auberti occidentalis	85	*23·5*	*43·5*	23·5	4·7	4·7	99·9
Porozetes polygonalis quadrilobatus	38	15·8	13·1	**39·5**	**31·6**	—	100·0
Magellozetes antarcticus	34	14·7	17·7	**29·4**	**38·2**	—	100·0
Sandenia georgiae	23	13·0	—	—	**87·0**	—	100·0

* Frequency values expressed as % of total occurrences in all microhabitats.

one associated mainly with moss clumps and nesting materials (italic
figures in the table), and the other with rock crevices and grass tussocks
(figures shown in heavy type). Not enough information is available, at this
stage, on the ecology of the species concerned to identify the reasons for
this habitat separation, but the fact that it may occur is important, since it
provides a mechanism for maintaining species diversity in an ecosystem
where resources are severely limited. A similar situation apparently occurs
on Macquarie Island, 3,000 miles away from South Georgia on the other
side of the sub-Antarctic zone. Here, the cryptostigmatid mite family
Podacaridae is richly represented (Wallwork, 1963; Watson, 1967). This
family has achieved a greater degree of evolutionary success in the south
polar region than anywhere else in the world, and on Macquarie Island
adaptive radiation is well illustrated by members of the genus *Halozetes*.
The group is basically marine littoral in its distribution, occupying niches
which, in temperate regions, are occupied by members of the related family
Ameronothridae. However, in Antarctica, the genus *Halozetes* has a range
of distribution which extends from the seashore into more terrestrial habitats. On Macquarie Island, for example, *Halozetes marinus* is exclusively

a denizen of the marine littoral, occurring mainly among algae on rocks in the upper *Porphyra* zone where its distribution overlaps with that of *H. intermedius*. This latter is not confined to the inter-tidal zone, however, for it also occurs in crevices and nest debris above the shore. A wider tolerance to environmental conditions is evidently shown by *H. crozetensis*, a species which has become established in a variety of habitats ranging from sea level to an altitude of about 250 m. These include marine algae, moss clumps, grassland leaf litter, rock crevices, algae and lichens in caves, and nest material. A similar pattern of distribution is shown by *H. belgicae*. Finally, *H. macquariensis* shows the greatest degree of terrestrialness, occurring in rock crevices, grass tussocks and muddy ground around freshwater pools, but being rare or absent from the inter-tidal zone.

Antarctic biogeography

The poverty of life in the Antarctic region is surprising. In comparable latitudes in the Arctic, a variety of grasses and low shrubs are able to grow successfully, and the soil fauna is correspondingly more varied. The factors which are probably responsible for this difference fall into two categories, i.e. they are ecological or geographical.

Average monthly summer temperatures are lower in the Antarctic than in the Arctic. This is a critical season for growth and reproduction, particularly among plants, and the paucity of the Antarctic flora may be related to the restricted growing season. This has repercussions on the soil fauna for the absence of a diversified flora means that the organic input into the soil, through leaf litter, is reduced. Further, soil erosion is an important factor in the Antarctic, where high winds are prevalent. As a result of these effects, Antarctic soils are thin, embryonic in character, low in organic content and relatively unstable. Habitat diversity is low and the soil fauna, as a consequence, is limited.

However, it must also be remembered that the Antarctic region is geographically isolated at the present time. It can be argued that the absence of a land connection between the maritime Antarctic and South America in the post-Pleistocene period could have prevented the southward spread of vascular plants, apart from *Deschampsia* and *Colobanthus*. Extensive glaciation during the Pleistocene would have eliminated many stocks of land animals, particularly invertebrates. Re-colonization by movement of cold-tolerant species into the region from South America and Australasia would have to occur over wide stretches of inhospitable ocean, and would be a slow process.

To what extent, then, has the isolated position of Antarctica influenced the character of its terrestrial invertebrate fauna? An attempt was made to answer this question by examining the present distribution patterns of the three most important groups of free-living terrestrial arthropods, namely the Collembola, and two groups of mites, the Prostigmata and Cryptostigmata (Wallwork, 1973). The conclusions emerging from this review can be summarized as follows:

1. Endemism, at the specific and generic levels, is high among the fauna of the continental zone. It is very unlikely that evolution could have proceeded at a rapid enough rate to allow such a distinctive fauna to develop during the post-Pleistocene period. It appears reasonable to suggest, therefore, that the species present in the continental zone are relics of an ancient fauna which survived the Pleistocene glaciation.

2. None of the families represented in the continental fauna is endemic. This suggests that the ancient fauna had an extensive distribution across a southern land mass incorporating the continental zone, the maritime zone, parts of the sub-Antarctic and the cold south temperate zone. This distribution pattern became fragmented when continental drift occurred during the Jurassic and Cretaceous. Microarthropod populations isolated by this event proceeded to evolve independently to produce the high levels of specific endemism which have now been recorded.

3. The Pleistocene glaciation probably had a catastrophic effect on the terrestrial invertebrate fauna of the south polar region. However, this effect would be less severe in the warmer coastal areas than in inland localities, and inter-tidal groups, such as the cryptostigmatid Podacaridae, could survive, subsequently to radiate into more terrestrial habitats.

4. There is a high level of specific endemism in the sub-Antarctic zone, indicating that this zone can be considered as a faunal province district from that of the cold temperate zone to the north and the other parts of the Antarctic to the south. This zone consists of a series of island groups some of which, for example South Georgia and Macquarie, are considerably older than the rest, and probably continental in origin. The generic composition of the microarthropod fauna of these two, now remote, localities is very similar, which suggests that at one time their faunal links were closer than they are today. The ancient distribution patterns mentioned earlier across the continental and maritime zones probably embraced these two sub-Antarctic island groups and may well have extended farther north into South America and Australasia.

5. However, in addition to this relic element in the fauna of South Georgia and Macquarie, there is a much more recent element consisting of species which evidently have been able to penetrate southwards. This element in the South Georgia fauna is South American in origin, whereas that of Macquarie has strong affinities with the New Zealand fauna. A few of these penetrant species have been able to move down into the maritime zone,

probably through the islands on the Scotia Ridge, but they do not occur in the continental zone.

6. The sub-Antarctic zone also includes a number of island groups of recent, volcanic, origin, such as the Crozets, Heard and possibly Kerguelen. The microarthropod fauna of these islands is poor, and appears to be derived mainly from that of the older Macquarie Island, by post-Pleistocene dispersal.

These findings point to the existence of a relic terrestrial invertebrate fauna in the south polar region which has survived the Pleistocene glaciation by virtue of its cold tolerance and its ability to seek refuges in warmer coastal habitats. This fauna is joined, in parts of the sub-Antarctic and maritime zones, by a penetrant element which has dispersed from the cold temperate zone during the post-Pleistocene period.

Desert soil faunas: a synthesis

It is obvious that there is scope for much more extensive studies of desert ecosystems. Even so, the picture presented by the information reviewed in this chapter and the previous one suggests that, although particular groups of the soil fauna may vary in their importance from one desert site to another, this fauna is essentially a highly specialized one.

In comparing the diversity and distribution patterns of the soil fauna of hot and cold desert soils, several similarities and differences are evident. The virtual absence of burrowing vertebrates in polar regions is due, in part, to the geographically isolated position of these regions, to their extremely cold climate, and to the fact that the sub-soil is permanently frozen. On the other hand, burrowing rodents are characteristic animals of hot deserts. Again, although in deserts both hot and cold there is a vertical compression of the faunal profile in the direction of the ground, some stratification of species populations exists, at least as far as the microarthropods are concerned. In cold deserts the bulk of this fauna is concentrated in the surface organic layer, which comprises growths of moss and lichens for the most part, whereas in hot deserts most of the microarthropod fauna is found in mineral soil underlying the surface litter. In both types of system, the fauna is species-poor, and in hot deserts, as we saw in the previous chapter, population densities may be as much as fifty times lower than in cool temperate soils. However, although the cold desert fauna is poor in terms of the number of species present, population sizes are comparable with those found in cool temperate soils.

Oases are a feature of both types of desert system and their soil fauna is of considerable interest. In the southern Mojave desert, the wet, peaty soils occurring in oases support a microarthropod fauna which is entirely different from that of the surrounding desert. Included here are at least two species of semi-aquatic cryptostigmatid mite which occur very commonly in wet soils throughout the cooler parts of the temperate region, namely *Hydrozetes terrestris* and *Malaconothrus gracilis* (Wallwork, 1972a). The presence of these species in isolated refuges suggests that they can disperse across hostile desert, and that the main factor influencing their distribution is the availability of suitable habitats.

In the cold Antarctic desert, similar refuges are provided by fumaroles. Holdgate (1964) has pointed out that a rich fauna of mites and collembolans occurs in the warm, moist and luxuriant mosses surrounding these volcanic vents, and this fauna has been investigated quantitatively by Tilbrook (1967) (see Fig. 11.6). His results show that Collembola and nematodes, in particular, have much higher population densities in fumarolic sites than in those not affected by heat. In the maritime Antarctic, the microarthropod fauna associated with fumaroles includes three species of cryptostigmatid mite which are not present in other, colder parts of the zone, although they do occur in the warmer sub-Antarctic zone to the north. Here again, then, is another example of distribution patterns being determined by the availability of suitable sites, rather than by the influence of geographical barriers.

One of the most attractive features of desert ecosystems, as far as the ecologist is concerned, is their relative simplicity. The faunal component consists of a small number of species which are specialized for life in extreme environments. The factors which influence the composition and distribution of the fauna and flora can often be identified without too much difficulty. Seasonal variations in temperature and moisture conditions are sharply demarcated, and reproductive rhythms can be correlated with these. Desert ecosystems are maintained by short energy impulses which are injected during a restricted growing season when plants increase in biomass, flower and produce organic substrates on which consumer animals can subsist until the next growing season. The identification of the principal components is technically feasible, and mathematical models can be devised to describe this system and to predict its behaviour under various conditions (Noy-Meir, 1973). For example, precipitation can be identified as the "input" which drives hot desert ecosystems, whereas temperature conditions may provide the driving force for biological activity in cold deserts. This type of analysis is still in the early stages of development, but it has important applications for agriculture. If deserts are to be converted into food-producing regions, extensive and expensive management will be required. A prior knowledge of the consequences of such actions could save time, money and effort.

REFERENCES

ALLEN, S. E. and HEAL, O. W. (1970). Soils of the Maritime Antarctic Zone. *In* "Antarctic Ecology", 693–96 (Holdgate, M. W., ed.), Academic Press, London and New York.

CARRICK, R., HOLDGATE, M. W. and PRÉVOST, J. (1964). "Biologie Antarctique: Antarctic Biology", Hermann, Paris.

COVARRUBIAS, R. (1965). Observaciones cuantitativas sobre los invertebrados terrestres antarcticos y preantarcticos, *Inst. Antarctico Chileno*, publ. no. 9 (1966), 7–63.

DALENIUS, P. (1965). The acarology of the antarctic regions. *In* "Biogeography and Ecology in Antarctica", 414–30 (van Oye, P. and van Mieghem, J., eds.), Junk, The Hague.

GIMINGHAM, C. H. and LEWIS SMITH, R. I. (1970). Bryophyte and lichen communities in the Maritime Antarctic. *In* "Antarctic Ecology", 752–85 (Holdgate, M. W., ed.), Academic Press, London and New York.

GRESSITT, J. L. (1967). "Entomology of Antarctica", Antarctic Research Series, Amer. Geophys. U. **10**, 395 pp.

GRESSITT, J. L. (1970). Subantarctic entomology and biogeography, *Pacif. Insects Monogr.*, **23**, 295–374.

HEAL, O. W. (1965). Observations on testate amoebae (Protozoa, Rhizopoda) from Signy Island, South Orkney Islands, *Br. Antarct. Surv. Bull.*, **6**, 43–7.

HOLDGATE, M. W. (1964). Terrestrial ecology in the Maritime Antarctic. *In* "Biologie Antarctique: Antarctic Biology", 181–94 (Carrick, R., Holdgate, M. W. and Prévost, J., eds.), Hermann, Paris.

HOLDGATE, M. W. (1967). The antarctic ecosystem, *Phil. Trans. R. Soc.* B 252, 363–83.

HOLDGATE, M. W. (1970). "Antarctic Ecology", vol. 2, Academic Press, London and New York.

JANETSCHEK, H. (1963). On the terrestrial fauna of the Ross Sea area (Preliminary report), *Pacif. Insects*, **5**, 305–11.

JANETSCHEK, H. (1967). Arthropod ecology of South Victoria Land. *In* "Entomology of Antarctica", 205–93 (Gressitt, J. L., ed.), Amer. Geophys. U., Washington.

JANETSCHEK, H. (1970). Environments and ecology of terrestrial arthropods in the high antarctic. *In* "Antarctic Ecology", 871–85 (Holdgate, M. W., ed.), Academic Press, London and New York.

LONGTON, R. E. (1967). Vegetation in the Maritime Antarctic, *Phil. Trans. R. Soc.*, B 252, 213–35.

MACKENZIE LAMB, I. (1970). Antarctic terrestrial plants and their ecology. *In* "Antarctic Ecology", 733–51 (Holdgate, M. W., ed.), Academic Press, London and New York.

NOY-MEIR, I. (1973). Desert ecosystems: environment and producers. *A. Rev. Ecol. Syst.*, **4**, 25–51.

VAN OYE, P. and VAN MIEGHEM, J. (1965). "Biogeography and Ecology in Antarctica", Junk, The Hague.

PRYOR, M. E. (1962). Some environmental features of Hallett Station, Antarctica, with special reference to soil arthropods, *Pacif. Insects*, **4**, 681–728.

RUDOLPH, E. D. (1963). Vegetation of Hallett Station area, Victoria Land, Antarctica, *Ecology*, **44**, 585–86.

SAIZ, F., HAJEK, E. R. and HERMOSILLA, W. (1968). Ecological research on Robert Island (South Shetland). Colonisation of introduced litter by subantarctic soil and moss arthropods, *Paper presented to SCAR symposium on Antarctic biology, Cambridge*, 11 pp.

SCHLATTER, R., HERMOSILLA, W. and DI CASTRI, F. (1968). Ecological research on Robert Island (South Shetland). Altitudinal distribution of terrestrial arthropods, *Inst. Antarctico Chileno*, publ. no. 15, 26 pp.

STRONG, J. (1967). Ecology of terrestrial arthropods at Palmer Station, Antarctic Peninsula. *In* "Entomology of Antarctica", 357–71 (Gressitt, J. L., ed.), Amer. Geophys. U., Washington.

SUDZUKI, M. (1964). On the microfauna of the Antarctic region I. Moss-water community at Langhovde, *JARE Scient. Rep.* E 19, 1–41.

TILBROOK, P. J. (1967). The terrestrial invertebrate fauna of the Maritime Antarctic, *Phil. Trans. R. Soc.* B 252, 261–78.

TILBROOK, P. J. (1970a). The terrestrial environment and invertebrate fauna of the Maritime Antarctic. *In* "Antarctic Ecology", 886–96 (Holdgate, M. W., ed), Academic Press, London and New York.

TILEBROOK, P. J. (1970b). The biology of *Cryptopygus antarcticus*. *In* "Antarctic Ecology", 908–18 (Holdgate, M. W., ed.), Academic Press, London and New York.

UGOLINI, F. C. (1970). Antarctic soils and their ecology. *In* "Antarctic Ecology", 673–92 (Holdgate, M. W., ed.), Academic Press, London and New York.

WALLWORK, J. A. (1960). Observations on the behaviour of some oribatid mites in experimentally-controlled temperature gradients, *Proc. zool. Soc. Lond.*, **135**, 619–29.

WALLWORK, J. A. (1963). The Oribatei (Acari) of Macquarie Island, *Pacif. Insects*, **5**, 721–69.

WALLWORK, J. A. (1966). Some Cryptostigmata (Acari) from South Georgia, *Br. Antarct. Surv. Bull.*, **9**, 1–20.

WALLWORK, J. A. (1972a). Mites and other microarthropods from the Joshua Tree National Monument, California, *J. Zool., Lond.*, 168, 91–105.

WALLWORK, J. A. (1972b). Distribution patterns of cryptostigmatid mites (Arachnida: Acari) in South Georgia, *Pacif. Insects*, 14, 615–25.

WALLWORK, J. A. (1973). Zoogeography of some terrestrial micro-Arthropoda in Antarctica, *Biol. Rev.*, **48**, 233–59.

WATSON, K. C. (1967). The terrestrial Arthropoda of Macquarie Island, *ANARE Scientific Reports B (Zool.)*, **99**, 90 pp.

WISE, K. A. J., FEARON, C. and WILKES, O. R. (1964). Entomological investigations in Antarctica, 1962–63 season, *Pacif. Insects*, **6**, 542–70.

Chapter 12

Distribution and Diversity: Synthesis

An attempt has been made in the preceding pages to review some of the more important aspects of the distribution and diversity of the soil fauna. This review raises two important questions, namely (1) do natural associations of species exist in the soil/litter subsystem as objective realities? and (2) if they do, why are they usually characterized by a high species diversity?

The first of these questions has great ecological significance for it explores the possibility that the composition of the soil fauna is related to a specific range of environmental variables and this, in turn, may provide a convenient method for comparing different habitats in functional, as well as structural, terms. There is no short answer to this question for the distribution of soil animals must be viewed against a whole complex of environmental effects. Studies on the distribution patterns of particular species along ecological gradients of one kind or another have shown repeatedly that these populations

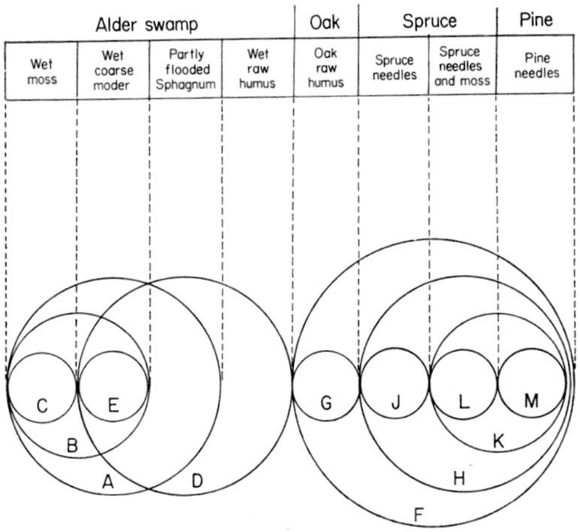

Fig. 12.1 The distribution of groups of cryptostigmatid mites in forest habitats in Germany. (From Knülle, 1957.) The species composition of the groups A–M is given in Appendix I.

overlap, and the patterns must be considered as a mosaic rather than a linear sequence (Fig. 12.1). In this event, it may be unrealistic to attempt to define a soil "community". The alternative approach is to consider that associations of species populations grade into each other in a "continuum" characterized, possibly, by "central" species which are more frequent than intermediates.

It is obvious that the greater the disparity between two habitats, the greater is the chance that the species composition will be entirely distinct in the two cases. More often than not, however, comparisons do not involve extremes, and faunal differences must be expressed in quantitative, rather than qualitative, terms. For example, Macfadyen (1952) considered the influence of plant type on the horizontal distribution of soil animals in a *Molinia* fen, and concluded that although no one species of mite or Collembola was restricted to any one plant type, individual species, and indeed whole groupings of species, showed marked differences in density between *Molinia*, *Deschampsia* and *Juncus* clumps. For the Collembola, the ratio of total population size between these three plant types was 3:3:1. For the cryptostigmatid mites, the ratio was 9:6:1.

However, significant differences in the density of individual species may not be indicative of the existence of distinct communities. Crossley and Bohnsack (1960) found significant variations in the abundance of four species of cryptostigmatid mites (*Tectocepheus velatus*, *Scheloribates* sp., *Peloribates* sp. and *Zygoribatula* sp.) from place to place in a pine forest floor. However, using a method of analysis based on joint occurrences, these workers could define only one recurrent group consisting of 9 common and abundant species. This suggests that the total picture was a simpler one, in that only a single association of species was being sampled, than the heterogeneity of individual species' distribution would have indicated at first sight.

Synusiae and isovalent groups

The identification of recurrent groups of species, often referred to as "Synusiae", with macroenvironmental factors has been pursued extensively by European ecologists for a number of years (Strenzke, 1952; Knülle, 1957; Franz, 1963). Obviously, these studies are of comparative value only within the limits imposed by the geographical distribution of the species concerned. In addition, the use of synusiae to indicate habitat characteristics is often misleading if the macroenvironment is considered as the effective habitat, for soil animals are influenced by the rather different complex of micro-environmental factors. It must also be remembered that species comprising a particular "association" are not necessarily restricted to it by definition.

Knülle (1957) illustrates this point in an analysis of the synusiae of crypto-stigmatid mites in sand heath, woodland, swamp woodland, moorland, grassland, salt marsh and the moss vegetation covering compact soil. This range of habitats was conveniently represented in the coastal plain of northern Germany. In many cases, the species belonging to a recurrent group were not restricted to the corresponding habitat type. For example, the mites *Parachipteria willmanni*, *Fuscozetes fuscipes*, *Oppia translamellata*, *Galumna elimata* and *Malaconothrus gracilis* are well represented in the soils of wet moorland, acid forest soils and those of swamp woodland and wet grassland. Obviously, the effective environment for these species does not correspond to any single major vegetational type. However, their distribu-tion may be influenced by the amount of moisture in the soil. Using this factor as a guide, Knülle defined groups of species on the basis of their frequency and abundance. The distribution of each group reflected the tolerance of its members to various degrees of soil moisture, arbitrarily defined as dry, "fresh", moist and wet. Often these groups, which are called "isovalents", cross vegetational boundaries in their horizontal distribution, and each may contain representatives from more than one synusia. This is illustrated in Fig. 12.2 which also shows how the isovalent groups link

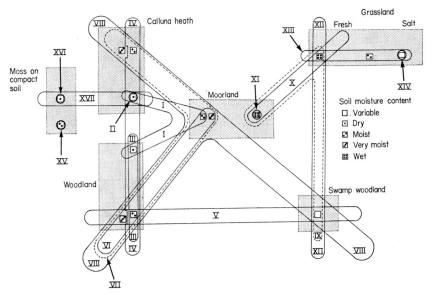

Fig. **12.2** The relationship between isovalent groups of cryptostigmatid mites (I–XVII) and synusiae in north German soils. (From Knülle, 1957.) Soil moisture charactersitics are indicated on the figure. The species composition of the isovalent groups is given in Appendix II.

various synusia together and establish the extent of their faunal similarities. The species composition of the various isovalent groups is worth recording, for many of the species included have a wide geographical distribution, at least in the northern hemisphere. The lists are given in Appendix II at the end of the chapter. Moritz (1963) has also used this approach to define ecological groups of cryptostigmatid mites in four woodland habitats in northern Germany.

Figure 12.2 shows that the distribution of any one isovalent group is not governed exclusively by the moisture factor. The species in group X, for example, are characteristic of wet soils, but their distribution is restricted to such soils in grassland and moorland and does not extend into swamp woodland. On the other hand, some isovalents are distributed over a range of soil moisture conditions, indicating tolerance on the part of the component species to variations in this environmental factor. Species of group I occur in dry soils of heath and woodland and in slightly moist soils of moorland. Again, the members of group XIII occur in a range of grassland conditions.

One difficulty in defining an isovalent group is that the various species of which it is composed may have differing tolerances to various environmental factors. Clearly, a species cannot be assigned to an isovalent group until its tolerance has been determined. If this tolerance is very broad, it may be impossible to assign the species. Such is the case with some of the very common cryptostigmatid mite species, such as *Tectocepheus velatus*, *Scheloribates latipes*, *Oribella lanceolata*, *Oppia nova* and *Oribatula tibialis*. One way of determining the tolerance of a species to various environmental factors is to plot its distribution in a range of habitats, and thereby obtain an estimate of its ecological "valence". Ecological valence expresses the extent to which environmental factors can vary within the tolerance zone of a species, and can be given a numerical value if the distribution in a range of habitats is known (Strenzke, 1952). The predictive value of this method is limited because it relies on information derived from observations on natural distribution patterns, but it is capable of being expanded and refined by using this information in conjunction with the more precise results obtained from physiological and behavioural experiments in the laboratory. For example, it is self-evident that if a particular species occurs consistently in a well-defined habitat, say a wet mixed moorland, the conditions to which it is exposed are tolerable. However, the absence of this species from other, different habitats does not necessarily imply that the conditions prevailing here are intolerable. The species in question may simply not have had the opportunity to disperse to these other habitats. The effective barrier in such cases is a dispersal, rather than an ecological, one. Laboratory experiments designed to test survival under different environmental conditions can shed much light on this question, although little attention has been paid to this so far.

In short, the concept of the "community" recognizes that a recurrent group of species can be defined for a given habitat locality, but does not imply any ecological restriction to that locality. The concept of the "isovalent", on the other hand, implies an exclusive group of species. It is more synthetic than that of the community, at least in theory, because it involves comparisons between various "communities" in which the distributions of the various species are correlated with the totality of the environmental effect. Exceptions to this occur locally where a small number of environmental factors are extreme as, for example, in highly saline marshes, in deserts and in impoverished habitats above the ground. Here the fauna is so specialized that little or no distinction can be made between the characteristic "community" and the isovalent group. Examples of this inclusiveness are given in Fig. 12.2. Returning to this figure, we can see that the concepts of the "community" and the "continuum" are not mutually exclusive. Both have an important place in the description of the horizontal distribution of soil animals.

We can now turn to the second question posed at the beginning of this chapter.

Mechanisms of habitat separation

We have remarked on the fact that the soil fauna is very diverse. Some idea of this diversity can be gained from the data presented in Table 12.1 which have been compiled from Fig. 12.2 and the species lists given in Appendix II. Although only one group of the soil fauna, the cryptostigmatid mites, is considered, the picture is a familiar one for we have seen it repeated with other groups in the previous chapters. Woodlands and grasslands support a more varied fauna than dry heathland or wet moorland, while the "extreme" habitats provided by growths of moss on compact soils are colonized by relatively few species. However, apart from these extremes, it is obvious that the soil fauna shows an astonishing diversity. Astonishing, because at first sight many of the soil animals appear to be unspecialized feeders. Many mites, Collembola, nematodes and enchytraeids feed on decomposing organic material and fungi. They appear to have common nutritional requirements which would bring them into direct competition with each other. Again, a variety of woodland predators, such as carabid beetles, mesostigmatid mites and pseudoscorpions feed mainly on Collembola. It is pertinent to ask, therefore, how such diversity can exist in the face of inter-specific competition for food? The principle of competitive exclusion (Gause, 1934) maintains that when two species populations have broadly similar requirements for a resource which is in limited supply, they cannot co-exist; one

11*

Table **12.1**

The distribution of isovalent groups of cryptostigmatid mites among various terrestrial habitats in northern Germany (compiled from data provided by Knülle, 1957).

Site	Isovalent Groups	Total Number of Species
"Fresh" woodland	III, IV, V, VI, VII, VIII	53
Wet grassland	IX, X, XII, XIII	39
Swamp woodland	V, VIII, IX, XII	38
Moist woodland	V, VI, VII, VIII	33
"Fresh" moorland	I, VI, VII, VIII	28
Dry woodland	I, III, IV	25
Moist moorland	VI, VII, VIII	23
Wet moorland	IX, X, XI	22
Dry sand heath	I, II, IV, XVII	15
Salt grassland	XIII, XIV	14
"Fresh" sand heath	IV, VII, VIII	13
"Fresh" grassland	XIII	9
Moist sand heath	VII, VIII	9
Moss on "fresh" soil	XV	4
Moss on dry soil	XVI, XVII	3

will succeed and the other will be driven into extinction. Anderson and Healey (1972) have put forward three hypotheses to explain co-existence and the high degree of diversity occurring among the soil fauna:

1. There is an excess of food available to soil animals. These populations are being regulated below the level of their resources either by predation, climate or some other environmental factor.

2. There exist undetected differences in resource utilization.

3. There are undetected microhabitat differences between species.

The first of these hypotheses could be applicable to sites in which there is a progressive accumulation of organic material in the soil. Such sites include acid moorland, forest peatland and some heathlands where peat deposits are being formed. Following the arguments of Hairston *et al.* (1960) it is not to be expected that this alternative would apply to systems such as grasslands and deciduous woodland where the turn-over of organic material is relatively rapid and little or no net accumulation occurs. However, Anderson and Healey (1972) have pointed out that populations of detritus-feeding animals may exist in non-equilibrium situations in which their densities and the level of nutrients fluctuate irregularly from year to year. These workers visualized the possibility that species which are competing actively for resources at one time and place may be sharing an excess at others. Support for this point of view is provided by Freeman (1967) who found no evidence that competition for food was an important factor in the ecology of larval Tipulidae inhabiting a range of plant types in the New Forest, southern England. Even so, it is unlikely that the hypothesis is applicable to predator/prey systems in the soil. Unpublished observations by G. F. Herrett on the functional response of the pseudoscorpions *Neobisium muscorum* and *Chthonius ischnocheles* to a range of psocid and collembolan prey densities indicate that these predators are compulsive feeders and can only be satiated after prolonged exposure to excess prey. The mesostigmatid mite *Pergamasus crassipes* behaves in the same way. A situation can be imagined in which a predatory mite or pseudoscorpion could chance on an aggregation of Collembola under natural conditions and, for a short period, the predator could be exposed to excess prey. However, the highly mobile Collembola undoubtedly would be able to disperse quickly before the predator became satiated.

As far as the second hypothesis is concerned, many of the major groups of soil animals do show some specialization in their feeding habits which will encourage efficient utilization of food resources. Soil nematodes, for example, can be divided into several feeding types, bacteriophages, phytophages, saprophages, predators and so forth. Again, some soil mites are predominantly fungivorous, others are saprophages, wood-feeders, pollen- or liquid-feeders, while a few species will feed on living plant material. However, the species diversity within each of these nutritional groups can be quite large, even within a small volume of soil, and competition for food would seem to be inevitable. Such competition between species with broadly similar diets would be minimized if different components of the diet were to be selectively assimilated into the body. Luxton's (1972) studies on the feeding biology of cryptostigmatid mites, discussed in Chapter 1, suggest that enzyme complements may vary from species to species in this, one of the most diverse groups of the soil fauna. A similar variability has been recorded among the Collembola (von Törne, 1967).

The third of Anderson and Healey's hypotheses is perhaps the most difficult to test for the features of the effective microhabitat can be quite varied, and quite variable for different species. As yet, we may not be able to comprehend the full extent of microhabitat diversity in the soil/litter system, but we can anticipate that such diversity will have spatial and temporal dimensions. As we will see, these two dimensions are sometimes inseparably linked. Three components of spatial diversity which merit attention are microclimate, structure and biochemical features.

Microclimate Diversity

Microclimatic variations occur both horizontally and vertically in the soil. Horizontal variations are most marked in open sites where the vegetation cover is discontinuous and a mosaic pattern of shaded and unshaded sites develops. Such sites are common in tussock grassland and heathland, and it was noted in Chapter 7 that temperature differentials of up to 10°C could be established under these conditions. Vertical temperature and moisture gradients also occur within the soil profile. Fluctuations in microclimatic conditions are most marked at the air/soil or air/litter interface and the amplitude of these fluctuations decreases with depth in the soil.

These horizontal and vertical gradients of microclimate have varying degrees of expression, some are steep while others are gradual. In general, horizontal gradients are rather more marked than vertical ones, and since the former are best expressed at the soil surface, it will be the surface-dwelling fauna which shows the most marked response. Evidence that this is so is provided by the distribution patterns of ants in grasslands (Chapter 4) and heathland (Chapter 7), and lycosid spiders in sand dunes (Chapter 7). Diversity is related at least in part, in these cases, to different preferences for microclimatic conditions, some species selecting shaded sites, others more exposed areas. These preferences provide a mechanism for the spatial separation of different species populations, even within a small area, such that a mosaic is formed with pockets of one species embedded within the horizontal range of another in a way that parallels the mosaic pattern of microclimate. The subterranean fauna is much less affected by these horizontal gradients, since surface variations in microclimate are considerably dampened down below the soil surface. However, as noted in Chapter 4, tropical earthworms of the family Eudrilidae show habitat separation which can be related to patterns of sun and shade.

Vertical gradients of microclimate in the soil undoubtedly contribute to microhabitat diversity, although the extent of this contribution is not easy to determine. Structural and biochemical gradients also make their contribution to this diversity, as we will see shortly. However, there is reliable evidence that differential responses to gradients of microclimate are shown by some of the commonest groups of soil animals. Among the Collembola and mites, for example, a relationship has been established between the resistance to desiccation and the location of a species in the soil/litter profile. Species of both of these groups of microarthropods which possess a tracheal respiratory system can colonize the upper layers of the leaf litter and even arboreal sites, whereas species which use the general body surface for respiration, and run a real risk of desiccation in exposed sites, are confined to the moist soil layers and to permanently damp vegetation (Table 12.2). Again, the temperature tolerances of cryptostigmatid mites have been shown to vary with the species (Wallwork, 1960; Madge, 1965) and to be related to vertical distribution patterns in nature.

Table 12.2

The relationship between habitat selection of various species of cryptostigmatid mites and certain physiological and ecological factors (adapted from Madge, 1964).

Species	Critical Temperature (°C)	Respiratory Surface	Favoured Habitat	Temperature and Moisture Conditions
Humerobates rostrolamellatus	60	Tracheae	Tree bark	Marked diurnal variations
Damaeus onustus	40–50	Tracheae	Woodland litter	Moderately variable
Steganacarus magnus	30–35	Reduced tracheal system	Litter and *Sphagnum*	Slight to moderate variability
Hypochthonius rufulus	None	Integument	*Sphagnum*, deep litter and humus	100% R.H.; small temperature variations

Structural Diversity

Like microclimate diversity, with which it is sometimes closely associated, structural diversity has both horizontal and vertical components. The association between these two types of diversity is most marked in open formations such as permanent pasture, heathland and moorland where a tussock form of vegetation predominates, and floristic diversity is at a maximum. Here, it is often difficult to determine whether the greater faunal diversity occurring within tussocks, in contrast to intervening areas, is due to the more equable microclimate or to the greater structural complexity of the microhabitat. Probably both factors are involved, and their influence may vary with the faunal group concerned. Duffey's (1962) studies on spider distribution in Wytham Woods, discussed in Chapter 4, certainly suggest that the diversity of the spider fauna is related to floristic diversity. Faunal comparisons between grazed and ungrazed grasslands (Chapter 4) indicate that other groups of the ground and soil fauna may respond in the same way although, here, the effect of microclimate is probably very important. Stradling (1968) found that the diversity of the ant fauna increased with the complexity of the vegetation on the sand dunes of Newborough Warren, Anglesey (see Chapter 7), although this is probably no simple relationship and undoubtedly involves responses to gradients of microclimate and also territorial behaviour.

Structural diversity in a horizontal plane is less evident in woodland and forest where ground vegetation may be lacking, than in grassland. But such diversity does occur and it is provided by microhabitats such as decaying logs, tree stumps, macrofungi, moss and lichen growths on rocks (see

Chapter 9), tree cones, deep and shallow leaf litter and suspended soils. Aoki (1967) carried out a survey of the cryptostigmatid mite fauna of many of these microhabitats in an oak/alder wood in Japan and found distinct differences in species diversity. Microhabitats closely associated with the soil harboured a richer fauna than the more remote ones. A similar study by Hammer (1972) pointed to some of the special adaptations shown by these faunas, for example the pollinivorous habit of cryptostigmatids associated with alder cones.

Structural diversity in the vertical plane, i.e. within the soil profile, is usually more marked in woodland and forest than in grassland. Woodland profiles are often well stratified with distinct layering of the organic and mineral profiles. Here, each layer has its own particular structural characteristics, and this situation contrasts with that in many grassland and arable soils where the activities of earthworms and Man cause the mineral and organic components of the soil to be mixed intimately together (Chapters 3 and 5). In compacted soils, the diameter of the soil cavities which form the living spaces for non-burrowing animals decreases progressively with depth (Haarløv, 1955). This provides a mechanism for habitat separation, allowing small species of nematodes, mites and Collembola, for example, to penetrate to depths in the profile which are inaccessible to larger species. The relationship between depth distribution and the body size of cryptostigmatid mites is discussed in Chapter 8 and documented by the data presented in Table 8.4 (p. 230).

Biochemical Diversity

Chemical differences between different types of leaf litter, and between different stages in the decomposition of a particular type of leaf litter, provide a range of variations in the chemical environment of soil animals. The effect of these variations will be most marked in saprophagous species, i.e. the animals which feed on leaf litter or its decomposition products. Again, this component of microhabitat diversity operates both horizontally and vertically.

As far as the soil fauna is concerned, the palatability and digestibility of leaf litter are functions of the carbon:nitrogen ratio and the polyphenol content. In the pure forest stands of temperate regions a fairly homogeneous litter type is produced, and horizontal gradients of C/N ratios and polyphenol content can only be detected when comparisons extend across major vegetational types. Such comparisons reveal that the species diversity of the microarthropods is high, and that of the macrofauna is low, in acidic sites where C/N ratios and polyphenol content are high, whereas the macrofauna of lumbricid earthworms, millipedes, woodlice and gastropod molluscs achieve greater prominence in nitrogen-rich sites. This ecological shift can be explained by the fact that many microarthropods, nematodes and enchytraeids feed on soil fungi which flourish in acidic soils, whereas most of the saprophagous macrofauna feed directly on leaf material and tend to select tissues which are rich in nitrogen and low in polyphenols. As could be expected, mixed forest stands will provide a more heterogeneous litter layer, and species diversity is usually greater here than in the litter of pure stands. Certainly, tropical forests present a great amount of heterogeneity as far as tree species are concerned, and it is worth recalling that the species diversity of the soil and litter fauna appears to be greater than in the pure stands of temperate regions, even though population densities are markedly lower than in the latter (Chapter 8).

Horizontal biochemical gradients develop in temperate sites where vegetational succession is occurring, for example eroding blanket bog (Chapter 6) and coastal sand dunes and salt marshes (Chapter 7). Changes in the species composition of the cryptostigmatid mite fauna in an eroding blanket bog sequence have been attributed to the changing nature of the food material available to these saprophages and fungivores (Block, 1966). This biochemical diversity provides a range of substrates which can be exploited by a variety of species. A similar situation occurs on sand dunes where variations in the organic content of the soil may contribute to habitat separation among the ant fauna, with *Tetramorium caespitum* selecting sites with greater amounts of organic material than those preferred by *Lasius* species (Chapter 7). However, this biochemical factor is only one in a whole complex of factors which encourage habitat separation and the establishment of a diverse ant fauna in sand dunes. There is a parallel here with the estuarine salt marsh (Chapter 7) where the ability to tolerate submersion in seawater is reflected in the zonation pattern of cryptostigmatid mites (Luxton, 1967). However, there are still some unexplained features of this zonation pattern, and it may be that biochemical variations, expressed through the seasonal availability of food, may also restrict certain species to particular sites on the marsh and, thereby, encourage spatial separation and an increase in diversity.

As noted already, different types of organic litter have different biochemical characteristics. The leaves of conifers have higher C/N ratios and greater amounts of polyphenols than do the leaves of deciduous trees and herbaceous ground vegetation. However, biochemical gradients are established within a particular litter type as fresh leaf material becomes progressively degraded into humus. The stages in this decomposition process are spatially separated in many soils, for as the leaf fragments become smaller and less recognizable they come to occupy lower horizons in the profile, and are covered by fresh litter. Each of these stages has its own particular biochemical characteristics. There is, for example, a progressive reduction in the poly-phenol content as organic material undergoes decomposition. Polyphenols are unpalatable to soil animals, and it is not until these compounds have been removed from the leaf litter by the activities of fungi and bacteria that this litter becomes acceptable as food for the fauna. This leaf litter micro-sere provides an example of a vertical biochemical gradient in which the diversity of substrates increases as intact leaves become broken down into fine fragments of raw humus and then mixed with faecal material to form a "fermentation" layer lying beneath the litter. The further decomposition of this layer releases inorganic nutrients, which are taken up by plants, and promotes humus formation. These two events reduce the biochemical diversity of the deeper humus layers. *In situ* observations can be made on the distribution patterns of small soil animals in this microsere, using the techniques of soil sectioning developed by Haarløv and Weis-Fogh (1955) and Anderson and Healey (1970). Such studies have shown that various species of cryptostigmatid mites maintain characteristic positions in the soil profile throughout the year. For example, Anderson (1971) found that *Carabodes labyrinthicus* was an inhabitant mainly of the leaf litter horizon in a sweet chestnut (*Castanea*) site in southern England, whereas *Nanher-mannia elegantula* was much more commonly associated with the fermen-tation layers. The two 'box' mites *Steganacarus magnus* and *Rhysotritia ardua* showed a distinct vertical separation, with the former concentrated in the lower litter and upper fermentation layers, and the latter in the lower fermentation zone and the upper part of the humus. This pattern of vertical distribution was very similar to that in a nearby beechwood soil profile which, however, was less compacted than that under the *Castanea*. This led Anderson (1971) to conclude that the vertical separation of species popula-tions was governed more by differential responses to the biochemical gradient than to the physical characteristics of the profile.

The Time Factor

The mechanisms for habitat separation reviewed so far in this chapter have dealt mainly with the spatial separation of potential competitors. The distances involved in such a separation can be quite small, even microscopic. On the other hand, it is also evident that competition for food can be mini-mized in truly sympatric species if there is a selective utilization of different chemical components of the diet. Such competition can be avoided by other, equally subtle means, particularly among predators. In woodlands, the pre-datory carabid beetle fauna consists of two components, namely day-active species, such as members of the genus *Notiophilus*, and a nocturnal element represented by members of the genus *Nebria*. As noted in Chapter 3, this separation of activity patterns during the 24-hour cycle minimizes inter-specific competition for food, and permits a greater diversity of predators than would be possible in open grasslands, for example, where only the day-active component is present.

The time factor operates not only diurnally, but also seasonally to produce habitat separation among potential competitors. A particularly good illus-tration of this seasonal effect is shown by the ant fauna of heathland sites. The timing of the main dispersal event, the nuptial flight, varies from one ant species to another, so that even if two species have similar microhabitat preferences, they could settle in different sites if their periods of flight activity did not overlap (see Chapter 7). Seasonal variations in the timing of activity patterns are also shown by carabid beetles of the genus *Calathus* on sand dunes (see Chapter 7) and by crane-flies (Tipulidae) in fen carr sites (Freeman, 1967). Such variations provide yet another mechanism for pro-moting species diversity by reducing direct competition for food and other resources.

Concluding remarks

The mechanisms for habitat separation discussed in the previous section are many and varied, as the summary presented in Fig. 12.3 illustrates. The species diversity of the soil fauna in general can be viewed against the totality of these effects, although it is obvious that they operate on a more restricted and selective scale for individual populations. Much of the evidence available suggests that the distributions of closely related, or ecologically similar, species populations form a mosaic pattern which is stable enough to allow a high species diversity to be maintained. This point is well illustrated by Freeman's (1967, 1968) studies on Tipulidae in the New Forest, southern England. In a fen carr site in this area, the larvae of ten tipulid species occurred, and their persistence together was attributed to the possibility that each species possessed a microhabitat in which it was superior to all other species. This idea of an "optimum" habitat is implicit in the concept of the isovalent group discussed earlier. It was noted, for example, that the "association" of species occurring in a particular site comprised represen-tatives from several different isovalent groups, in many cases. This being so, we would expect the range of microhabitats available to be fully exploited as species with slightly different preferences separate into slightly different microhabitats. This separation is not necessarily complete. Indeed, it was noted in Chapter 6 that larval *Tipula subnodicornis* have a distribution in moorland sites which extends from peaty soils under *Calluna*, *Sphagnum*, *Eriophorum* and *Juncus* to *Sphagnum* bogs, although densities in the latter site are low and evidently it does not represent the optimum habitat for these crane-fly larvae. The *Sphagnum* bog habitat probably serves as a

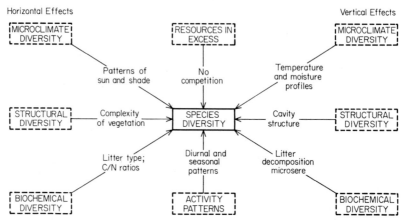

Fig. **12.3** A summary of the main factors affecting the species diversity of the soil fauna.

"refuge" in this case. Such refuges can make an important contribution to microhabitat diversity, and to species diversity, by allowing less successful competitors to co-exist with the more successful ones. Again, these refuges are not merely physical parts of the environment which are accessible to one species but not to another. The pressure of competition may cause of species to switch its diet at certain times of the year and thereby gain the benefit of a food refuge. Lewis (1965) noted that a woodland population of the centipede *Lithobius variegatus* changed its diet from animal prey to plant material during the winter when prey was scarce. Such a tactic will enable a species to survive in sub-optimal conditions.

The presence of refuges, of whatever kind, does not eliminate completely the possibility of competition between two co-existing species. But it may pave the way for the development of a mild form of stable competition which, as Williamson (1972) suggests, may be more common in natural situations than is presently realized. The mechanisms of habitat separation that we have been considering can do much to minimize competition, but even when this cannot be eliminated completely, there are grounds for supposing that the species diversity of the soil fauna will not be affected adversely.

The soil and its associated habitats are immensely complex, from the ecological point of view. We are beginning to understand the extent of the microhabitat diversity that occurs here, but there is still much more to be learned about the reaction of species populations to this diversity. Each component of microhabitat diversity can be regarded as a dimension along which the distribution of a species is approximately normal. If we consider the reaction of a population to all the effective components of its environment, we arrive at the concept of the fundamental niche, a multidimensional space, or hypervolume (Hutchinson, 1957). This volume represents the response of the population to all of the effective environmental factors, and it is within the limits of these responses that the population can survive, more or less indefinitely. The practical application of this concept to the soil fauna presents certain problems for, as Williamson (1972) argues, there is an infinite number of dimensions, and arbitrary decisions must be made on the choice of these. However, as Anderson and Healey (1972) maintain, the mosaic of microhabitats present even within a single soil horizon can be regarded as a large number of compartments, each with similar resources, but between which the movement of faunal populations is strongly restricted. Each of these compartments is, to some extent, an abstraction for the restrictions on faunal movement from one to another are imposed not only by structural and microclimatic factors, but also by feeding specificity and the timing of periods of activity. Add to this the possibility that a mild form of stable competition can exist within each compartment, and perhaps we can account for the high degree of diversity of the soil fauna.

REFERENCES

ANDERSON, J. M. (1971). Observations on the vertical distribution of Oribatei (Acarina) in two woodland soils, *C.R. 4eme Coll. Int. Zool. Sol*, Paris, 257–72.

ANDERSON, J. M. and HEALEY, I. N. (1970). Improvements in the gelatine-embedding technique for woodland soil and litter samples, *Pedobiologia*, **10**, 108–20.

ANDERSON, J. M. and HEALEY, I. N. (1972). Seasonal and inter-specific variation in major components of the gut contents of some woodland Collembola, *J. Anim. Ecol.*, **41**, 359–68.

AOKI, J-I. (1967). Microhabitats of oribatid mites on a forest floor, *Bull. Nat. Sci. Mus. Tokyo*, **10**, 133–38.

BLOCK, W. C. (1966). The distribution of soil Acarina on eroding blanket bog, *Pedobiologia*, **6**, 27–34.

CROSSLEY, D. A. and BOHNSACK, K. K. (1960). Long-term ecological study in the Oak Ridge area. III. The oribatid mite fauna in pine litter, *Ecology*, **41**, 785–90.

DUFFEY, E. (1962). A population study of spiders in limestone grassland, *J. Anim. Ecol.*, **31**, 571–99.

FRANZ, H. (1963). Biozönotische und synökologische Untersuchungen über die Bodenfauna und ihre Beziehungen zur Mikro- und Makrofauna. *In* "Soil Organisms", 345–67 (Doeksen, J. and van der Drift, J., eds.), North Holland, Amsterdam.

FREEMAN, B. E. (1967). Studies on the ecology of larval Tipulinae (Diptera, Tipulidae), *J. Anim. Ecol.*, **36**, 123–46.

FREEMAN, B. E. (1968). Studies on the ecology of adult Tipulidae (Diptera) in southern England, *J. Anim. Ecol.*, **37**, 339–62.

GAUSE, G. F. (1934). "The Struggle for Existence", reprinted 1964, Hafner, New York.

HAARLØV, N. (1955). Vertical distribution of mites and Collembola in relation to soil structure. *In* "Soil Zoology", 167–79 (Kevan, D. K. McE., ed.), Butterworths, London.

HAARLØV, N. and WEIS-FOGH, T. (1955). A microscopical technique for studying the undisturbed texture of soils. *In* "Soil Zoology", 429–32 (Kevan, D. K. McE., ed.), Butterworths, London.

HAIRSTON, N. G., SMITH, F. E. and SLOBODKIN, L. B. (1960). Community structure, population control, and competition, *Amer. Nat.*, **94**, 421–25.

HAMMER, M. (1972). Microhabitats of oribatid mites on a Danish woodland floor, *Pedobiologia*, **12**, 412–23.

HUTCHINSON, G. E. (1957). Concluding remarks. *Cold Spring Harbour Symposium Quant. Biol.* **22**, 415–27.

KNÜLLE, W. (1957). Die Verteilung der Acari: Oribatei im Boden, *Z. Morph. u. Ökol. Tiere*, **46**, 397–432.

LEWIS, J. G. E. (1965). The food and reproductive cycles of the centipedes *Lithobius variegatus* and *L. forficatus* in a Yorkshire woodland, *Proc. zool. Soc. Lond.*, **144**, 269–83.

LUXTON, M. (1967). The ecology of saltmarsh Acarina, *J. Anim. Ecol.*, **36**, 257–77.

LUXTON, M. (1972). Studies on the oribatid mites of a Danish beech wood soil I. Nutritional biology, *Pedobiologia*, **12**, 434–63.

MACFADYEN, A. (1952). The small arthropods of a *Molinia* fen at Cothill, *J. Anim. Ecol.*, **21**, 87–117.

MADGE, D. S. (1964). The water relations of *Belba geniculosa* Oudms. and other species of oribatid mites, *Acarologia*, **6**, 199–223.

MADGE, D. S. (1965). The effects of lethal temperatures on oribatid mites, *Acarologia*, **7**, 121–30.

MORITZ, M. (1963). Über Oribatidengemeinschaften (Acari: Oribatei) norddeutscher Laubwaldboden, unter besonderer Berücksichtigung die Verteilung regelnden Milieubedingungen, *Pedobiologia*, **3**, 142–243.

STRADLING, D. J. (1968). "Some Aspects of the Ecology of Ants at Newborough Warren National Nature Reserve, Anglesey", Ph.D. Thesis, University of Wales.

STRENZKE, K. (1952). Untersuchungen über die Tiergemeinschaften des Bodens: Die Oribatiden und ihre Synusien in den Böden Norddeutschlands, *Zoologica*, **37**, 1–167.

VON TÖRNE, E. (1967). Beispiele für indirekte Einflüsse von Bodentieren auf die Rotte von Zellulose, *Pedobiologia*, **7**, 220–27.

WALLWORK, J. A. (1960). Observations on the behaviour of some oribatid mites in experimentally-controlled temperature gradients, *Proc. zool. Soc. Lond.*, **135**, 619–29.

WILLIAMSON, M. (1972). "The Analysis of Biological Populations", Edward Arnold, London.

Appendix I

Species composition of the groups A-M in Fig. 12.1.

GROUP A
Malaconothrus gracilis, Steganacarus striculus, Phthiracarus testudineus, Fuscozetes fuscipes, Platynothrus peltifer, Rhysotritia ardua.

GROUP B
Nanhermannia comitalis, Tropacarus sp.*, Tectocepheus concurvatus, Xenillus tegeocranus.*

GROUP C
Oppia translamellata, Nanhermannia nana, Suctobelba forsslundi, Hypochthonius rufulus, Minunthozetes semirufus, Nothrus palustris, Gustavia fusifer, Oribella paolii, Liebstadia similis, Parachipteria willmanni, Euzetes globulus.

GROUP D
Tectocepheus velatus sarekensis, Trimalaconothrus novus.

GROUP E
Trimalaconothrus glaber, Oribatula tibialis, Trhypochthoniellus excavatus.

GROUP F
Suctobelba falcata, Nothrus silvestris, Suctobelba subcornigera, Pseudotritia duplex, Nanhermannia elegantula.

GROUP G
Suctobelba tuberculata, Suctobelba acutidens, Tectocepheus cuspidentatus, Nanhermannia areolata, Ceratozetes thienemanni.

GROUP H
Tectocepheus velatus, Eniochthonius grandjeani, Chamobates incisus.

GROUP J
Galumna lanceata, Eulohmannia ribagai, Oppia minus, Oppia maritima, Oribatella quadricornuta.

GROUP K

Parachipteria punctata, Chamobates cuspidatus, Pseudotritia minima, Adoristes ovatus, Pelops duplex, Euphthiracarus cribrarius, Cepheus cepheiformis.

GROUP L

Ceratozetes minimus, Melanozetes mollicomus.

GROUP M

Hermannia gibba, Scheloribates confundatus, Oppia quadricarinata, Suctobelba subtrigona, Damaeobelba minutissima, Suctobelba similis.

Appendix II

Species composition of the isovalent groups I–XVII shown in Fig. 12.2.

GROUP I
Carabodes minusculus, Trichoribates trimaculatus, Trhypochthonius cladonicola, Carabodes marginatus, Camisia biurus.

GROUP II
Tectocepheus tenuis, Humerobates rostrolamellatus, Camisia segnis, Eupelops bilobus, Eupelops acromios.

GROUP III
Parachipteria punctata, Euphthiracarus cribrarius, Eulohmannia ribagai, Porobelba spinosa, Brachychthonius marginatus, Damaeobelba minutissima, Hermannia gibba, Oribatella quadricornuta, Ceratozetes minimus, Suctobelba acutidens, Adoristes ovatus, Galumna lanceata, Liacarus coracinus, Ceratozetes thienemanni, Nanhermannia areolata, Suctobelba subtrigona.

GROUP IV
Chamobates schützi, Camisia spinifer, Oppia minus, Pseudotritia duplex.

GROUP V
Xenillus tegeocranus, Suctobelba perforata, Suctobelba trigona, Euzetes globulus, Nanhermannia elegantula, Oppia quadricarinata, Anachipteria deficiens, Diapterobates humeralis, Tropacarus sp., *Ceratoppia sexpilosa.*

GROUP VI
Suctobelba falcata, Suctobelba longirostris, Suctobelba similis, Suctobelba tuberculata, Oppia maritima, Oppia unicarinata, Phthiracarus piger, Eniochthonius grandjeani, Chamobates cuspidatus, Eupelops duplex, Cepheus cepheiformis, Pseudotritia minima, Tectocepheus cuspidentatus, Brachychthonius zelawaiensis.

GROUP VII
Nothrus silvestris, Galumna nervosa, Rhysotritia ardua, Brachychthonius berlesei, Chamobates incisus.

GROUP VIII
Carabodes labyrinthicus, Steganacarus striculus, Melanozetes mollicomus, Scheloribates confundatus.

GROUP IX
Parachipteria willmanni, Fuscozetes fuscipes, Oppia translamellata, Galumna elimata, Malaconothrus gracilis, Hypochthonius rufulus, Nanhermannia nana, Nothrus palustris, Platynothrus peltifer, Minunthozetes semirufus.

GROUP X
Trhypochthonius excavatus, Suctobelba palustris, Punctoribates sellnicki, Eupelops strenzkei, Nothrus pratensis, Brachychthonius furcillatus.

GROUP XI
Limnozetes sphagni foveolatus, Trimalaconothrus foveolatus, Trimalaconothrus sculptus, Trimalaconothrus novus, Allogalumna tenuiclavus, Peloptulus montanus.

GROUP XII
Astegistes pilosus, Phthiracarus testudineus, Oribella paolii, Nanhermannia comitalis, Heminothrus thori, Suctobelba forsslundi, Suctobelba singularis, Trimalaconothrus glaber, Tectocepheus concurvatus, Achipteria quadridentata, Eupelops hygrophilus, Brachychthonius ensifer, Brachychthonius scalaris, Gustavia fusifer.

GROUP XIII
Liebstadia similis, Eupelops occultus, Trichoribates novus, Ceratozetes mediocris, Scheloribates laevigatus, Peloptulus phaenotus, Oppia clavipectinata, Brachychthonius sellnicki, Scutovertex sculptus.

GROUP XIV
Oribatella arctica litoralis, Hermannia pulchella, Punctoribates quadrivertex, Ameronothrus schneideri, Trichoribates incisellus.

GROUP XV
Zygoribatula exilis, Sphaerozetes piriformis, Eremaeus oblongus, Minunthozetes pseudofusiger.

GROUP XVI
Phauloppia lucorum, Trhypochthonius tectorum.

GROUP XVII
Scutovertex minutus.

Author Index

Subject Index